THE FRONTIERS COLLECTION

THE FRONTIERS COLLECTION

Series Editors:
A.C. Elitzur M.P. Silverman J. Tuszynski R. Vaas H.D. Zeh

The books in this collection are devoted to challenging and open problems at the forefront of modern science, including related philosophical debates. In contrast to typical research monographs, however, they strive to present their topics in a manner accessible also to scientifically literate non-specialists wishing to gain insight into the deeper implications and fascinating questions involved. Taken as a whole, the series reflects the need for a fundamental and interdisciplinary approach to modern science. Furthermore, it is intended to encourage active scientists in all areas to ponder over important and perhaps controversial issues beyond their own speciality. Extending from quantum physics and relativity to entropy, consciousness and complex systems – the Frontiers Collection will inspire readers to push back the frontiers of their own knowledge.

Other Recent Titles

Weak Links
Stabilizers of Complex Systems from Proteins to Social Networks
By P. Csermely

Mind, Matter and the Implicate Order
By P.T.I. Pylkkänen

Quantum Mechanics at the Crossroads
New Perspectives from History, Philosophy and Physics
Edited by J. Evans, A.S. Thorndike

Particle Metaphysics
A Critical Account of Subatomic Reality
By B. Falkenburg

The Physical Basis of the Direction of Time
By H.D. Zeh

Asymmetry: The Foundation of Information
By S.J. Muller

Mindful Universe
Quantum Mechanics and the Participating Observer
By H. Stapp

Decoherence and the Quantum-To-Classical Transition
By M. Schlosshauer

The Nonlinear Universe
Chaos, Emergence, Life
By A. Scott

For a complete list of titles in The Frontiers Collection, see back of book

Mark P. Silverman

QUANTUM SUPERPOSITION

Counterintuitive Consequences of Coherence, Entanglement, and Interference

With 97 Figures and 13 Tables

 Springer

Mark P. Silverman
Trinity College
Department of Physics
Hartford, CT 06106
USA

Series Editors:

Avshalom C. Elitzur
Bar-Ilan University,
Unit of Interdisciplinary Studies,
52900 Ramat-Gan, Israel
email: avshalom.elitzur@weizmann.ac.il

Mark P. Silverman
Department of Physics, Trinity College,
Hartford, CT 06106, USA
email: mark.silverman@trincoll.edu

Jack Tuszynski
University of Alberta,
Department of Physics, Edmonton, AB,
T6G 2J1, Canada
email: jtus@phys.ualberta.ca

Rüdiger Vaas
University of Gießen,
Center for Philosophy and Foundations of Science
35394 Gießen, Germany
email: Ruediger.Vaas@t-online.de

H. Dieter Zeh
University of Heidelberg,
Institute of Theoretical Physics,
Philosophenweg 19,
69120 Heidelberg, Germany
email: zeh@uni-heidelberg.de

Cover figure: Image courtesy of the Scientific Computing and Imaging Institute,
University of Utah (www.sci.utah.edu).

ISBN 978-3-642-09097-4   e-ISBN 978-3-540-71884-0
DOI 10.1007/978-3-540-71884-0
Frontiers Collection ISSN 1612-3018

Cover design: KünkelLopka, Werbeagentur GmbH, Heidelberg

Printed on acid-free paper

9 8 7 6 5 4 3 2 1

springer.com

To Sue, Chris, and Jennifer

Preface

In the course of a long career as a physicist, I have investigated quantum mechanics with electron interferometry and microscopy, radiofrequency and microwave spectroscopy, coherent laser spectroscopy, nuclear magnetic and electron paramagnetic resonance, atomic beams, radioactive nuclei, and of course pencil, paper, and computers. Throughout this time I have had other scientific interests as well, but none has held such permanent fascination for me as trying to understand the structure and interactions of the quantum world. Perhaps one reason is that *everything* at some sufficiently deep level of fundamentality is a quantum system.

The range of topics discussed in this book reflect, in part, my diverse quantum interests, but all involve, in one way or another, the unifying element of quantum superposition. Although by its very etymology the word 'quantum' (Latin for 'how much') intrinsically stresses the idea of discreteness in nature, what is actually most distinctive about quantum mechanics, in my opinion, is the unique character and far-reaching consequences of the concept of *superposition* – a superposition not of forces or fluxes or any other directly measurable, tangible physical quality, but of abstract amplitudes conveying a measure of probability.

Quantum superpositions can be overtly spatial, as in the familiar example of superposed waves issuing from two slits leading to a periodic distribution of independent particles on a distant screen; overtly temporal as in the superposition of excited-state amplitudes resulting in ensemble quantum properties that oscillate in time, or subtly internal as in the spin-based symmetry restrictions on multiparticle wave functions from which different quantum statistics arise. Quantum superposition can occur in systems ranging in size and complexity from single elementary particles, to widely separated atoms in 'entangled' states, to mesoscopic electronic circuits, to degenerate stars of unimaginable densities and pressures. The number of interesting systems whose properties are attributable to superposition of quantum amplitudes is virtually limitless. A monograph, however, is not, but the technical essays included here provide a sampling of the diversity of quantum interference phenomena arising from conceptually different kinds of quantum superposition.

It was originally my intention to have completed this book in time for the tenth anniversary of its predecessor, *More Than One Mystery: Explorations in Quantum Interference*, which, I believe, was at the time (1995) the first book to be published that was devoted specifically to the subject of quantum interference. Alas, other research commitments and a heavy teaching schedule thwarted my plan, but I believe the updated and considerably expanded final product has actually benefited from the delay. Indeed, some of the quantum effects and experiments I proposed well over ten years ago were realized recently, some even as I was working on the corresponding chapters.

The title of the present book – which is not simply *MTOM 2*, or *Son of MTOM*, or some such reflection on its progenitor – deserves an explanation. Like many a physicist, I found the vivid imagery of Richard Feynman's colorful expressions captivating, although I did not always agree with the content. In particular, his contention (in the famous *Lectures*), which was subsequently repeated often by others, that the phenomenon of two-slit electron interference contains 'the *only* mystery' of quantum mechanics, bothered me. As a scientist developing novel quantum interference experiments involving correlated pairs of particles in entangled states, or single particles propagating around magnetic flux tubes in a space with 'holes', or the effects of potentials on the fluctuations of particles restricted by the spin–statistics relation, or other unusual examples of quantum superpositions conceptually distinct from Young's two-slit experiment, I thought Feynman's assertion to be misleadingly narrow. It was, after all, an offhand remark to stimulate a class of struggling CalTech undergraduates. And so I wrote a book drawn from my own researches illustrating why – if one adopts Feynman's idiom – there really is '*more* than one mystery' to quantum mechanics.

In the years since publication of *MTOM*, I have come to regret that title, for it appeared to suggest – particularly to someone unfamiliar with its origin – a point of view the very opposite to that which I held then and hold now. I do not regard quantum mechanics as a 'mystery'. Rather, it is a highly developed and well-understood physical theory, perhaps the most carefully scrutinized and best understood of all physical theories. The glib assertions by many scientists and science popularizers that 'nobody understands quantum mechanics' – another Feynman idiom – is balderdash. Competent physicists (as opposed to poorly informed science writers or science philosophers), who use quantum mechanics on a daily basis to elucidate successfully countless physical phenomena, clearly must understand the instrument with which they are working. When they pretend otherwise, I am reminded of the exchange between analyst Jack Ryan and Director of Central Intelligence William Cabot in the film adaptation of Tom Clancy's thriller, *The Sum of All Fears*:

JR: ...I don't think that adds up.

WC: It adds up. You just don't like what it adds up to.

What quantum mechanics 'adds up to' is that it is an irreducibly statistical theory, albeit unlike any necessitated simply by 'incomplete knowledge', with nonlocal features inexplicable from the perspective of classical physics. But something that is strange is not necessarily incomprehensible, although it may not be visualizable. Mathematicians, for example, may understand very well the principles of a 10-dimensional geometry even if no 10-dimensional figure can be drawn.

In this book I discuss quantum superposition – the "heart of the matter" to quote another Feynmanism – in its manifold variations. Technical essays in MTOM have been expanded throughout with new material. I have enlarged the chapter on the electron two-slit experiment to include discussion of controversial issues like Schrödinger's cat, Wheeler's delayed-choice thought experiment, and macroscopic manifestation of quantum interference – issues that I have found to be sensationally distorted, especially in publications for the general reader. The original chapter on correlated particles now constitutes two chapters with new sections on distinctions between quantum ensembles, correlated emission from excited atoms, coherence properties of thermal electrons, and more thorough consideration of the coherence properties of field-emitted electrons. Several sections were added to the chapter on the physics of chiral systems. Appendices have been added that clarify points raised in the text or provide supplementary technical discussions.

I have also added a chapter that addresses one of the most challenging problems in physics: the collapse of a sufficiently massive relativistic degenerate star to a black hole, depicted in the scientific and popular literature as an invisible gaping singularity in space-time where the laws of physics break down. There is no doubt that black holes must be among the most exotic denizens of the cosmos, but the wild flights of fancy that have become part of their mythology (infinite densities, 'worm holes' through space-time to distant parts of the universe, portals to other universes, etc.) must rival in absurdity the worst of what I have seen written about Schrödinger's cat or the many-worlds interpretation of quantum mechanics. It may be prosaic to say, but a star – even the most bizarre star – if it is a real physical object, and not merely a solution to some set of differential equations, cannot be an infinitely dense hole in space-time. We do not yet know all the laws of physics, nor fully understand the consequences of those laws we think we know, but we know enough, I believe, to conceive of mechanisms that would prevent the formation of such a comic-book caricature of a star. In the last chapter I make some suggestions whereby quantum superposition, operating in systems of extreme densities and pressures, could stabilize degenerate matter at a macroscopic size comparable to that of a neutron star or quark star.[1]

---

[1] The nature of dark matter in the universe is another outstanding problem of astrophysics and cosmology for which quantum superposition may provide the solution (by formation of a condensate of extremely low mass bosons). I have

I hope the reader will find the essays in this book as edifying and thought-provoking to read, as the author found them to write.

Trinity College,   *Mark P. Silverman*
January 2007

Books by Mark P. Silverman

- *And Yet It Moves: Strange Systems and Subtle Questions in Physics* (Cambridge University Press, 1993)
- *More Than One Mystery: Explorations in Quantum Interference* (Springer, New York, 1995)
- *Waves and Grains: Reflections on Light and Learning* (Princeton University Press, 1998)
- *Probing the Atom: Interactions of Coupled States, Fast Beams, and Loose Electrons* (Princeton University Press, 2000)
- *A Universe of Atoms, an Atom in the Universe* (Springer, New York, 2002)

---

discussed this solution in the book *A Universe of Atoms, An Atom in the Universe* (Springer, New York, 2002).

Contents

1 **The Enigma of Quantum Interference** 1
  1.1 The Most Beautiful Experiment 1
  1.2 Two-Slit Interference of Single Electron Wave Packets 3
  1.3 Confined Fields and Electron Interference 11
  1.4 'No-Slit' Interference of Single Photons:
     Superposition, Probability, and Understanding 22
  1.5 Macroscale Objects in Quantum Superpositions 27
  1.6 Quantum Mechanics and Relativity:
     The 'Wrong-Choice' Experiment 38

2 **Correlations and Entanglements I:**
  Fluctuations of Light and Particles 45
  2.1 Ghostly Correlations of Entangled States 45
  2.2 A Dance of Correlated Fluctuations.
     The 'Hanbury Brown Twiss' 54
  2.3 Measurable Distinctions Between Quantum Ensembles 60
  2.4 Correlated Emission from Coherently Excited Atoms 65
  2.5 The Quantum Optical Perspective 70
  2.6 Coherence of Thermal Electrons 77
  2.7 Comparison of Thermal Electrons and Thermal Radiation 86
  2.8 Brighter Than a Million Suns:
     Electron Beams from Atom-Size Sources 88
  2.9 Correlations and Coincidences: Experimental Possibilities 100
  2A Consequences of Spectral Width on Photon Correlations 106
  2B Chemical Potential at $T = 0$ K 107
  2C Probability Density of a Sum of Random Variables 108
  2D Correlated Fluctuations of Electrons at Two Detectors 109

3 **Correlations and Entanglements II:**
  Interferometry of Correlated Particles 111
  3.1 Interferometry of Correlated Particles 111

3.2   The Aharonov–Bohm (AB) Effect with Entangled Electrons . . 112
3.3   Hanbury Brown–Twiss Correlations of Entangled Electrons . . . 118
3.4   Correlated Particles in a Mach–Zender Interferometer 122

4   Quantum Boosts and Quantum Beats 135
4.1   Superposing Pathways in Time 135
4.2   Laser-Generated Quantum Beats 139
4.3   Nonlinear Effects in a Three-Level Atom 145
4.4   Quantum Beats in External Fields 155
4.5   Correlated Beats from Entangled States 159

5   Sympathetic Vibrations:
     The Atom in Resonant Fields 165
5.1   Beams, Bottles, and Resonance 165
5.2   The Two-Level Atom Looked at Two Ways 174
5.3   Oscillating Field Theory 182
5.4   Resonance and Interference:
     Tell-Tale Mark of a Quantum Jump 190
5.5   Quantum Interference in Separated Oscillating Fields 199
5.6   Ion Interferometry and Tests of Gauge Invariance 206
5A   Oscillatory Field Solution
     to the Two-State Schrödinger Equation 214
5B   Generalized Rotating Field Theory and Optically-Induced
     Ground State Coherence in a 3-State Atom 215

6   Symmetries and Insights:
     The Circulating Electron in Electromagnetic Fields 219
6.1   Broken Symmetry of the Charged Planar Rotator 219
6.2   The Planar Rotator in an Electric Field 222
6.3   The Planar Rotator in a Magnetic Field 233
6.4   The Planar Rotator in a Vector Potential Field 239
6.5   Fermions, Bosons, and Things In-Between 246
6.6   Quantum Interference in a Metal Ring 250
6A   Magnetic Hamiltonian of the Two-Dimensional Rotator 254

7   Chiral Asymmetry: The Quantum Physics of Handedness . . 257
7.1   Optical Activity of Mirror-Image Molecules 257
7.2   Quantum Interference and Parity Conservation 262
7.3   Optical Activity of Rotating Matter 272
7.4   'Electron Activity' in a Chiral Medium 281
     7.4.1   Longitudinal Polarization 285
     7.4.2   Transverse Polarization 287
7.5   Chiral Light Reflection 290
7.6   Chirality in a Medium with Broken Symmetry 299

8   **Condensates in the Cosmos:**
     Quantum Stabilization of Degenerate Stars 307
     8.1   Stellar End States 307
     8.2   Quantum Properties of a Self-Gravitating Condensate 311
     8.3   Quantum Properties of a Self-Gravitating System
            of Degenerate Fermions 314
     8.4   Fermion Condensation in a Degenerate Star 320
     8.5   Fermicon Stars vs Black Holes 333
     8.6   Can Ultra-Strong Magnetic Fields Prevent Collapse? 335
     8.7   Gravitationally-Induced Particle Resorption into the Vacuum . 340
     8A   Gravitational Binding Energy of a Uniform Sphere of Matter .. 346
     8B   Stability in a Self-Gravitating System with Negative Pressure . 347
     8C   Quark Deconfinement in a Neutron Star 349
     8D   Energy Balance in the Creation of the Universe 353
     8E   Particle Resorption in a Schwarzschild Geometry 355

References 361

Index 375

1

The Enigma of Quantum Interference

1.1 The Most Beautiful Experiment

In the late 1980s, I had the pleasure and distinction of being the first occidental professor invited to the then newly created Hitachi Advanced Research Laboratory (ARL) in Kokubunji, Japan.[1] The Hitachi Company was (and still is) a well known manufacturer of state-of-the-art electron microscopes, among many other high-tech instruments, but the ARL, which could be likened to a uniquely Japanese blend of the Princeton Institute for Advanced Studies and the original Bell Research Labs, was created to do basic, rather than profit-driven, research. My responsibility as Chief Researcher in quantum physics at the ARL was to think of innovative, but feasible, fundamental experiments for the electron microscopy group to do.

Being a teacher, as well as a research scientist, I knew how difficult it was for students to grasp the profound implications of quantum interference, especially as they had never observed this phenomenon themselves but only read about it in textbooks or saw it described on the chalkboard, usually in the form of the Young's two-slit experiment with electrons. As my first contribution, therefore, I proposed an experiment, employing the attenuated beam of a field-emission electron microscope, to illustrate and capture on videotape the growth of electron-interference fringes by detection of one electron at a time. The proposal was duly brought before the top management for consideration, and I learned later with some disappointment that my idea was declined, as it was "not the purpose of the lab to make films for school classrooms".

The idea, however, did not die. The experiment was eventually performed, the results were recorded on film, and, in gratitude for this and other ideas during my tenure, I was presented with one of the few existing copies of this extraordinary five-minute, silent, monochrome videotape (*Single-Electron Build-up of an Interference Pattern*) as a gift upon my leaving Japan. It is quite possible (although I do not know with certainty) that I may now possess the *only* existing copy, for Hitachi, once having seen the light that

---

[1] The Hitachi ARL has since moved to Hatoyama, Japan.

'classroom films' could attract potential scientists and engineers from Japanese high schools and universities, was quick to replace the original with a narrated full-length color production. Also, with development of the internet, the film has been made available for viewing at the Hitachi ARL website.

In 2002, science historian Robert Crease, writing a column for *Physics World*, asked his readers to "submit candidates for the most beautiful experiment in physics". Of the ten leading results[2] published in the September edition [1], the first on the list was Young's double-slit experiment applied to the interference of single electrons. I was gratified to see that other physicists shared my belief that there is something definably special about this most basic demonstration of a quantum interference phenomenon.

What is so special? What is it that makes the electron double-slit *experiment* 'beautiful', in analogy to a work of art or composition of music, forms of artistic expression to which the concept of beauty is more usually applied? To be sure, beauty is a subjective quality, but the beauty of a scientific experiment lies less in the eye than in the *mind's eye* of the observer. Asked what makes an experiment beautiful, scientists are wont to talk about the 'simplicity' or 'elegance' or 'economy' of a technique which manifests a difficult or subtle effect with readily available equipment. Such rhetoric, however, would hardly seem to describe the experiment I proposed.

A high-voltage field-emission electron microscope, wherein electrons are accelerated to energies greater than one hundred thousand volts and which, with its huge oil-filled drums to isolate the high-voltage transformers, takes up a good-sized room, is perhaps more a thing of awe than of beauty. Materially, the experiment depended on complex and sophisticated-not simple-technology. Conceptually, the outcome of the experiment did not simplify physicists' understanding of quantum weirdness, but delineated it starkly in one of its most extreme forms, thereby adding tinder, not water, to philosophical arguments that have burned for over 75 years. Furthermore, a state-of-the-art instrument of the kind employed in the experiment was available at the time (if not also presently) only in a handful of labs throughout the world. By the time definitive electron double-slit experiments were performed, quantum mechanics had already been tested in a myriad of ways and never found wanting, so the outcomes were neither surprising nor ever in doubt. Why, then, is the electron double-slit experiment one of the most beautiful experiments in physics? In my opinion, the most compelling reason is this: *the unambiguous directness of the observable results, despite being expected, have the capacity to shock the intellect into realizing, like no mathematical proof is capable of*

---

[2] The full list of 10 beautiful experiments cited in *Physics World* is: (1) Young's double-slit experiment with single electrons, (2) Galileo's experiment on falling bodies, (3) Millikan's oil-drop experiment, (4) Newton's decomposition of sunlight with a prism, (5) Young's light-interference experiment, (6) Cavendish's torsion-bar experiment, (7) Eratosthenes' measurement of the Earth's circumference, (8) Galileo's experiments with rolling balls down inclined planes, (9) Rutherford's discovery of the nucleus, (10) Foucault's pendulum.

achieving, that the formalism of quantum mechanics reveals a strange and disturbing reality. Experiments depending on complex apparatus, as is the case in the study of elementary particles, usually lead to complex outcomes requiring intricate analysis to extract the signature of some sought-for phenomenon. By contrast, the two-slit interference pattern builds up before one's eyes, one electron at a time, and every attempt to determine the specific path of an individual electron in that pattern is destined to end in futility.

The aesthetic quality of this experiment was aptly depicted-at least in part-by Crease in the closing remarks of his report:

> It is natural to call beautiful those [experiments] that captivate and transform our thinking, that make the result stand out clearly [...] in a materially embodied way, and that reveal that we are actively engaging with something beyond us.

1.2 Two-Slit Interference of Single Electron Wave Packets

In the more than eighty years that have passed since Louis de Broglie first proposed the wave-like behavior of particles, the idea of particle interference has become more-or-less familiar to most physicists. The term 'wave–particle duality', which once may have sounded like an oxymoron, for a long while now has been an integral part of physicists' working vocabulary. And yet, despite the fact that technological advances have greatly facilitated experimental demonstrations of the wave-like attributes of matter, these processes can be no more visualized today than when they first reached the consciousness of the physics community. The interference of massive particles remains an intriguing phenomenon which "has in it the heart of quantum mechanics", in the words Feynman used to introduce quantum concepts to his students. "In reality," he wrote, "it contains the only mystery."

I have reservations about Feynman's last remark, as well as a number of quotations by other eminent physicists, which are frequently repeated, often out of context or without regard to subsequent progress, to give an impression that "nobody understands quantum mechanics" (another Feynmanism [2]), but I will leave these issues for later. Let us now take stock of just how strange the self-interference of single-particle wave packets is, as judged by our ordinary experience, by examining the experiment I proposed to the Hitachi electron microscopy group.

In this demonstration [3], a field-emission electron microscope was employed to produce an electron version of Young's two-slit experiment with light as shown in Fig. 1.1. Electrons, drawn from a sharp tungsten filament by an applied electrostatic potential of about 3–5 kV, were subsequently accelerated through a potential difference of 50 kV to a speed of approximately 0.41 the speed of light ($c = 3 \times 10^8$ m/s). Subsequently split by an electrostatic

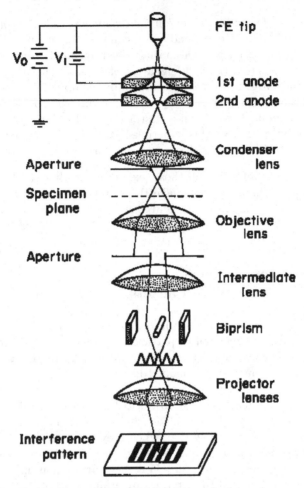

V_0 V_1

FE tip

1st anode
2nd anode

Aperture

Condenser
lens

Specimen
plane

Objective
lens

Aperture

Intermediate
lens

Biprism

Projector
lenses

Interference
pattern

Fig. 1.1. Electron interference with a field-emission electron microscope. Wave-front spitting occurs at the biprism. (Courtesy of A. Tonomura, Hitachi ARL)

device known as a Möllenstedt biprism (effectively the two slits of the apparatus), the two components of the electron beam recombined in the observation plane of the microscope where the build-up of a pattern of interference fringes was recorded on film and on a TV monitor.

The appearance of a fringe pattern is not in itself extraordinary. After all, if electrons were actually waves, then the experimental configuration would represent a type of wavefront-splitting interferometer, and there is nothing unusual about the linear superposition of waves to generate an interference pattern. What is startling, however, is the observed *emergence* of the fringe pattern in a microscope of approximately 1.5 m in length under conditions where the mean interval between successive electrons is over 100 km! Clearly any given electron was detected long before the succeeding electron emerged from the field-emission tip. Under these circumstances it is unlikely that there could have been any cooperative interaction between the electrons of the beam.

Electron detection events appear on the TV monitor one by one at random locations as illustrated in Fig. 1.2. The first few hundred scattered spots hardly hint of any organization. However, by the time some hundred thousand electrons have been recorded, stark alternating stripes of white and black stand out sharply as if made by the two-slit interference of laser light. Indeed, except for the large difference in wavelengths – about 500 nm for visible light and 0.005 nm (a tenth the diameter of a hydrogen atom) for the 50 keV electrons[3] – the uninformed observer could not tell whether the fringe pattern

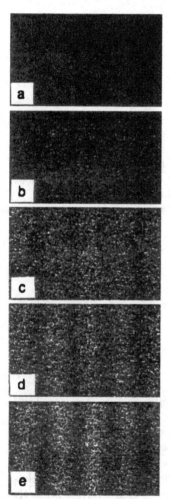

Fig. 1.2. Evolution of the electron interference pattern in time. Electrons arrive at the rate of approximately 1000 per second. The number recorded in each frame is: (**a**) 10, (**b**) 100, (**c**) 3000, (**d**) 20 000, and (**e**) 70 000. (Courtesy of A. Tonomura, Hitachi ARL)

---

[3] The electron volt (eV) is a unit of energy equal to that gained by a particle with the charge of the electron (e) transported through a potential difference of 1 volt. 1 eV $= 1.6 \times 10^{-19}$ J.

was created by light or by particles. If the location of each electron arrival is random, and there is no communication between electrons, how then can the overall spatial distribution of detected electrons manifest a coherently organized pattern? *That* is the enigma of quantum mechanics to which Feynman referred.

Since the two-slit electron interference experiment is conceptually the simplest, if not archetypical, example of quantum interference, it is worth examining it quantitatively in more detail, if only to introduce geometric and dynamical quantities that will be encountered again later. The speed β (relative to that of light) of an electron in the beam is deducible from the relativistic expression for energy conservation

$$E = \frac{mc^2}{\sqrt{1 - \beta^2}} = mc^2 + eV , \tag{1.1}$$

and takes the form

$$\beta = \frac{\sqrt{\left(1 + \dfrac{eV}{mc^2}\right)^2 - 1}}{1 + \dfrac{eV}{mc^2}} , \tag{1.2}$$

where m is the electron mass, e the magnitude of the electron charge, and V the accelerating potential. From relation (1.2), the de Broglie wavelength λ of the electron, a measure of the magnitude of the linear momentum p,

$$p = \frac{mv}{\sqrt{1 - \beta^2}} = \frac{h}{\lambda} , \tag{1.3}$$

can be expressed in the form

$$\lambda = \frac{\lambda_C}{\sqrt{\left(1 + \dfrac{eV}{mc^2}\right)^2 - 1}} , \tag{1.4}$$

where λ_C is the electron Compton wavelength

$$\lambda_C = \frac{h}{mc} = 2.43 \times 10^{-12} \text{ m} , \tag{1.5}$$

and $h = 6.26 \times 10^{-34}$ joule-second (Js) is Planck's constant. (Later in the book, we will also use the so-called 'reduced' Compton wavelength, $\lambda_C = \hbar/mc = 3.87 \times 10^{-13}$ m, where $\hbar \equiv h/2\pi$ is the reduced Planck's constant.)

Strictly speaking, the electron wavelength refers to monoenergetic electrons in much the same way as an optical wavelength characterizes perfectly monochromatic light. These are idealizations that are only imperfectly realized in nature. Relying again on optical imagery (which is useful, but can be misleading if taken too literally), one can describe the electron beam more appropriately in terms of wave packets as shown in Fig. 1.3. If the energy uncertainty (or energy dispersion) of the beam is ΔE, then by the Heisenberg uncertainty principle there is a characteristic time interval, the beam coherence time t_c

$$t_c = \frac{\hbar}{\Delta E} \, , \tag{1.6}$$

over which an electron wave packet emerges from the source. Propagating at a mean speed v, the wave packet has a characteristic length, or longitudinal coherence length

$$l_c = v t_c \, . \tag{1.7}$$

For a 50 keV beam with dispersion of 0.1 eV, the coherence time and length are respectively $t_c = 6.6 \times 10^{-15}$ s and $l_c = 7.9 \times 10^{-7}$ m. In order for wave packets to overlap and interfere, their optical path length difference between source and detector must not be much in excess of l_c. The coherence length can greatly exceed the de Broglie wavelength, and does so in the present experiment by five orders of magnitude.

In addition to a longitudinal extension, the wave packets also have a lateral extension as characterized by the transverse coherence length l_t

$$l_t \sim \frac{\lambda}{2\alpha} \, , \tag{1.8}$$

arising from the finite size of the source. Here 2α is the angular diameter of the source as seen from the diffracting object; equivalently, 2α is the beam divergence angle as seen from the viewing plane. To understand the relevant geometrical relations and the origin of expression (1.8), examine Fig. 1.4. Electrons emitted from the center of the source give rise to a diffraction pattern centered about the symmetry axis, i.e., at the origin O of the viewing

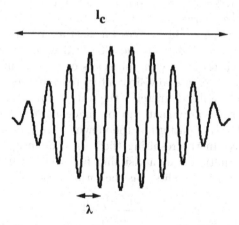

Fig. 1.3. Schematic diagram of an electron wave packet. The de Broglie wavelength λ is determined by the particle linear momentum. The longitudinal coherence length l_c is determined by the dispersion in particle energy

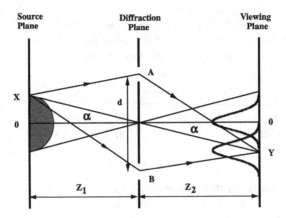

Fig. 1.4. The lateral or transverse coherence length l_t is determined by the angular dispersion of the beam. It is a measure of the maximum extent of wave front separation – here equivalent to slit separation d – at which a diffracting object is coherently illuminated by an extended source

screen. Electrons emitted from an above axis point X in the source plane produce a diffraction pattern centered at the below-axis point Y on the viewing screen. As in light optics, points X and Y are connected by a straight line through the center of the diffracting object (in this case the midpoint between two slits a distance d apart); the corresponding distances from the symmetry axis stand in the same ratio as the separations Z_1 between the source and diffraction plane and Z_2 between the diffraction plane and viewing screen.

One can readily show that the relative phase of the electron wave functions describing electrons arriving at point Y from the origin (O) and periphery (X) of the source is approximately $2\pi dX/\lambda Z_1$. If this phase difference is π radians, the crests of the waves at Y from source point X overlap the troughs of the waves at Y from the source point O – and the contrast or visibility of the interference fringes is zero. This defines (at least approximately) a maximum slit separation $d = \lambda/(2X/Z_1)$ – to be identified with the lateral coherence width – below which the diffracting object is coherently illuminated by the extended source and gives rise to visible interference fringes. The ratio $2X/Z_1$ (for $X \ll Z_1$) is the angular diameter of the source appearing in relation (1.8).

The above heuristic argument can be made more rigorous by actually integrating the diffraction pattern at the viewing screen over all contributing points of the source. In the resulting theoretical intensity distribution, the oscillatory term is multiplied by a visibility function V which, in the case of a diffracting screen with two slits, has the form [4]

$$V = \frac{|\sin(\pi d/l_t)|}{\pi d/l_t}, \tag{1.9}$$

where l_t is again the lateral coherence width of relation (1.8). It is seen that V vanishes for $d = l_t$. In any event, the point of importance is that, as a general condition for diffraction, the transverse coherence length must not be greatly inferior to the size of the diffracting object or aperture. In the Hitachi experiment under discussion, the divergence angle was on the order of 4×10^{-8} radians, and therefore $l_t \sim 140$ μm, a value larger than the approximately 0.5 μm radius of the anodic filament of the biprism, yet much smaller than the 10 mm separation of the two adjacent grounded electrodes.

Upon traversing the biprism, the components of the incident electron beam passing to one side or the other of the filament are deflected towards the observation plane and overlap there at an angle twice the deflection angle α, thereby giving rise to interference fringes with fringe spacing $\lambda/2\alpha$. The resulting fringe spacing of approximately 700 nm was subsequently magnified 2000 times by a projector lens.

The details of the detection process, which involved the use of sophisticated image processing apparatus for recording the number and locations of all electrons arriving within the field of view, will be left to the original literature. Suffice it to say that close to 1000 electrons arrived at the detector each second, and that a pattern of sharp fringes could be created in about one half hour. However, at an electron speed of $\sim 0.41c$ there is a mean spatial separation of approximately 1.23×10^5 m between two sequential electrons detected 1 ms apart. At any given moment during the experiment, therefore, there is likely to be only one particle traversing the apparatus. This feature must perplex anyone seeking to understand experiments of this kind at a deeper level than simply being able to predict the outcome (which, of course, is what quantum mechanics enables one to do).

So subtle and contrary to ordinary experience are the implications of electron self-interference, that one is usually not fully aware of the alternation in use of language required to describe the experiment. For example, to explain the action of the biprism one speaks of the deflection of the components of the beam to one side or the other of the filament. But where is there really a *beam*, for effectively only one particle at a time moves through the biprism? To be sure, knowing the spatial variation of the electrostatic field of the filament, one can calculate the deflection angle α of an incident electron. However, the supposition that an electron has with certainty actually taken one of two classically conceivable pathways through the apparatus theoretically leads to no interference effect, and, indeed, the experimental capacity to produce an interference pattern would be destroyed by the intrusive observation to test that supposition. Conversely, one can speak of the diffraction of waves around the filament. But where is there really a *wave*, since the electron is always produced and detected as an elementary corpuscular entity with discrete electric charge (4.8×10^{-10} esu or 1.6×10^{-19} C), mass (9.11×10^{-31} kg), and spin angular momentum ($\hbar/2$)?

It is frequently said or implied that the wave–particle duality of matter embodies the notion that a particle – the electron, for example – propagates like a wave, but registers at a detector like a particle. Here one must again exercise care in expression, so that what is already intrinsically difficult to understand is not made more so by semantic confusion. The manifestations of wave-like behavior are statistical in nature and *always* emerge from the collective outcome of many electron events. In the present experiment nothing wave-like is discernible in the arrival of single electrons at the observation plane. It is only after the arrival of perhaps tens of thousands of electrons that a pattern *interpretable* as wave-like interference emerges.

Likewise, there is a conceptually significant distinction between the dynamical variables with which particle and wave-like characteristics of matter are quantified, e.g., the variables connected by the Einstein and de Broglie relations. Characteristics like energy and linear momentum can well pertain to individual particles. By contrast, the corresponding quantities of frequency and wavelength – and ultimately, of course, the wave function – although commonly spoken of in the context of single-particle wave packets, actually characterize a hypothetical ensemble of particles all similarly prepared. Although one can in principle measure the mass, charge, or energy of a single electron (held, for example, in an electromagnetic trap), one can not measure its de Broglie wavelength except by a diffraction or interference experiment employing many such electrons.

And last, we must be aware that frequently used, but inaccurate, expressions like 'particle interference' or 'interference of particles' are but casual abbreviations of the more precise idea of interference of wave packets or wave functions or probability amplitudes – that is, of *mathematical functions* by means of which the probabilities of events (and related statistical quantities like transition rates and cross-sections) of physical particles can be calculated.

Thus, the original quantum mechanical enigma distills into this: Is the fabric of nature so constructed that the laws of motion are at best statistical, ultimately pertaining only to *systems* of particles or to a single particle observed *repeatedly*? Is it physically meaningless even to speak of certain attributes of a particle as objectively real if they can never be simultaneously observed and measured? Although there is little disagreement among physicists concerning the formulation and mathematical procedures of quantum mechanics, the interpretation of the theory and assessments of its fundamentality have evoked over the years a broad range of opinion for which the reader may consult the literature [5].

That electrons behave singly as particles and collectively as waves is indeed strange, but, Feynman's remark notwithstanding, this is not the only quantum mystery – if, by 'mystery', we mean an incapacity to give a deterministic explanation of a quantum phenomenon. Charged particles can do other things that are equally strange – indeed, in some ways stranger. They can interact with electric and magnetic fields through which, classically, they do not pass. They can arrive at detectors in classically inexplicable cluster patterns although emitted apparently randomly from their source. And, once part of a localized system of particles, they exhibit long-range correlations that strongly affect their subsequent self-interference well after the original system has apparently ceased to exist and the constituent particles have become widely dispersed. It is to be stressed, of course, that the rhetorical term 'mystery' does not refer in any way to an inability of quantum theory to account for the phenomena under discussion, but only to the insufficiency of our ordinary experience (i.e., classical physics) to permit us to imagine some tangible mechanism by which the processes might occur.

1.3 Confined Fields and Electron Interference

It has long been a fundamental proposition of modern physics (the origin of which dates back well before the creation of special relativity to at least the time of Michael Faraday) that material systems interact with one another, not instantaneously at a distance, but causally through the medium of a field. The first, and still the most familiar, implementation of this philosophical perspective of nature was in the area of electromagnetism. The mathematical embodiment of the field theory of electromagnetism is the set of Maxwell equations

$$\nabla \cdot \boldsymbol{E} = 4\pi\rho , \tag{1.10}$$

$$\nabla \cdot \boldsymbol{B} = 0 , \tag{1.11}$$

$$\nabla \times \boldsymbol{E} = -\frac{1}{c}\frac{\partial \boldsymbol{B}}{\partial t} , \tag{1.12}$$

$$\nabla \times \boldsymbol{B} = \frac{1}{c}\frac{\partial \boldsymbol{E}}{\partial t} + \frac{4\pi}{c}\boldsymbol{J} , \tag{1.13}$$

and the Lorentz force law

$$\boldsymbol{F} = \rho\boldsymbol{E} + \frac{1}{c}\boldsymbol{J} \times \boldsymbol{B} , \tag{1.14}$$

which, together, are considered to represent completely the classical interactions of electric ($\boldsymbol{E}$) and magnetic ($\boldsymbol{B}$) fields with each other and with charge (ρ) and current ($\boldsymbol{J}$) densities. The very expression of the laws of electrodynamics as differential equations seems to signify that all interactions take place locally, the charged particles being influenced only by electric and magnetic fields in their immediate vicinity. Every well-formulated problem in classical electrodynamics essentially reduces to determining the fields produced by a system of charges (stationary or moving) and reciprocally the forces exerted on these charges by the fields. To facilitate the solution of such a problem, vector and scalar potential fields, $\boldsymbol{A}$ and ϕ, related to the electric and magnetic fields by the derivative expressions

$$\boldsymbol{E} = -\nabla\phi - \frac{1}{c}\frac{\partial \boldsymbol{A}}{\partial t} , \tag{1.15}$$

$$\boldsymbol{B} = \nabla \times \boldsymbol{A} , \tag{1.16}$$

are ordinarily introduced. The electromagnetic potentials are not unique, but can be modified by a so-called gauge transformation

$$\boldsymbol{A} \longrightarrow \boldsymbol{A}' = \boldsymbol{A} + \boldsymbol{\nabla}\Lambda \,, \tag{1.17}$$

$$\phi \longrightarrow \phi' = \phi - \frac{1}{c}\frac{\partial \Lambda}{\partial t} \,, \tag{1.18}$$

with gauge function Λ which leaves the electromagnetic fields, and hence the Maxwell equations and Lorentz force law, invariant. In order to leave invariant, as well, the quantum mechanical equation of motion of a particle with charge q, the gauge transformation also modifies the wave function

$$\Psi \longrightarrow \Psi' = \Psi \mathrm{e}^{\mathrm{i}q\Lambda/\hbar c} \,. \tag{1.19}$$

The transformation function Λ is largely arbitrary although usually required to be a single-valued function in order that the contour integral of gauge-transformed vector potentials leads to the same value of magnetic flux Φ through Stokes' law

$$\oint_{C} \boldsymbol{A}\cdot\mathrm{d}\boldsymbol{S} = \iint_{S} \boldsymbol{B}\cdot\mathrm{d}\boldsymbol{S} = \Phi \,, \tag{1.20}$$

where C is the contour bordering the open and orientable surface S penetrated by the magnetic field lines. The fields $\boldsymbol{E}$ and $\boldsymbol{B}$ must, themselves, be unique for a given configuration of charges and currents, because they are directly related to electromagnetic forces. As a consequence, the classical perspective has been to regard $\boldsymbol{E}$ and $\boldsymbol{B}$ as the primary or fundamental fields and $\boldsymbol{A}$ and ϕ as auxiliary or secondary fields needed for calculational convenience only.

It is of historical interest to note, however, that Maxwell, who introduced these fields in his famous treatise [6], accorded a more physical significance to the vector potential.[4] Having initially termed $\boldsymbol{A}$ the "vector-potential of magnetic induction", Maxwell subsequently designated it the "electromagnetic momentum at a point" and interpreted $\boldsymbol{A}$ as representing the "direction and magnitude of the time-integral of the electromotive intensity which a particle placed at [a point] would experience if the primary current [in one of two interacting circuits] were suddenly stopped". In other words, Maxwell regarded $\boldsymbol{A}$ as a measurable quantity related to momentum, a conception that may be found, albeit sharpened by the use of modern terminology, in the contemporary physics literature [8]. Nevertheless, the requirement that physical observables be representable by gauge invariant expressions underlies a long-standing belief that the electromagnetic potentials, though intimately related to measurable quantities, are not themselves directly observable.[5]

---

[4] I discuss Maxwell's designations of electromagnetic fields and potentials more thoroughly in [7].

[5] In general, the vector potential $\boldsymbol{A}$ can be Fourier analyzed and decomposed into transverse and longitudinal components. The transverse component is invariant under a gauge transformation and corresponds to a measurable quantity; the longitudinal component is not invariant under a gauge transformation.

The theoretical demonstration in 1959 by Y. Aharonov and D. Bohm [9] (to be designated from this point on as AB) that the diffraction of charged particles can be influenced by electromagnetic potentials under conditions where the electromagnetic fields are *null* opened a new chapter in the study of quantum interference phenomena and gave rise to controversial issues, both theoretical and experimental, that to some extent are still debated. One of the principal questions raised by AB concerned whether or not the vector and scalar potentials were more fundamental than the electric and magnetic fields. AB argued that they were. The quantum implications exposed by the AB paper had actually been revealed some ten years earlier by Ehrenberg and Siday [10], who were investigating the refractive index of electrons in an electron microscope. This paper, however, was apparently not widely read, and it is through AB that physicists came to recognize the extraordinary consequences of electromagnetic potentials in quantum mechanics.

The basic mathematical relation underlying the AB effect seems to have been known at least as far back as the early 1930s by P.A.M. Dirac in a celebrated study of quantized singularities (magnetic monopoles) [11]. Expressed more generally to embrace both vector and scalar potentials, it is this. If Ψ_0 is the solution to the quantum equations of motion (e.g., the Dirac or Schrödinger equation) in the absence of electromagnetic interactions, then the corresponding wave function Ψ of a charged particle in the presence of a time-independent vector potential field and a spatially uniform scalar potential field takes the form

$$\Psi(\boldsymbol{r},t) = \Psi_0(\boldsymbol{r},t)e^{iS(\boldsymbol{r},t)} , \tag{1.21}$$

where the phase $S(\boldsymbol{r},t)$ is given by

$$S(\boldsymbol{r},t) = \frac{q}{\hbar c}\left(\int_{\boldsymbol{r}_0}^{\boldsymbol{r}} \boldsymbol{A}\cdot\mathrm{d}\boldsymbol{s} - \int_{t_0}^{t} c\phi\,\mathrm{d}t\right) . \tag{1.22}$$

The integration in the phase is over an arbitary space-time path between some point of origin $(t_0, \boldsymbol{r}_0)$ and destination $(t, \boldsymbol{r})$.

The above result is demonstrable by direct substitution of the wave function (1.21) into the wave equation

$$H\Psi = i\hbar\frac{\partial\Psi}{\partial t} , \tag{1.23}$$

where the Hamiltonian H is constructed from the field-free Hamiltonian H_0 by replacing the canonical linear momentum $\boldsymbol{p}$ with $\boldsymbol{p} - q\boldsymbol{A}/c$ and adding $q\phi$ to the potential energy. The resulting wave equation is then of the form

$$H'\Psi_0 = i\hbar\frac{\partial\Psi_0}{\partial t} , \tag{1.24}$$

where the transformed Hamiltonian

$$H' = U^\dagger H U - i\hbar U^\dagger \frac{\partial U}{\partial t} \tag{1.25}$$

reduces to the field-free Hamiltonian H_0 when $\partial \boldsymbol{A}/\partial t$ and $\boldsymbol{\nabla}\phi$ both vanish.

It is worth noting that the operator $\boldsymbol{p} - q\boldsymbol{A}/c$, which corresponds to the kinetic linear momentum $\boldsymbol{P}$ (equal to $m\boldsymbol{v}$ for a nonrelativistic classical particle), gives rise to a gauge-independent expectation value, whereas the expectation value of $\boldsymbol{p}$, the dynamical variable entering into the quantum commutation relation

$$[\boldsymbol{r}, \boldsymbol{p}] = i\hbar\mathbf{1} , \tag{1.26}$$

and serving as the generator of spatial translations, is gauge dependent. [In (1.26), $\mathbf{1}$ is the unit dyad or second-order tensor.] Although both operators are Hermitian, $\boldsymbol{P}$ is considered a dynamical observable, but $\boldsymbol{p}$ is not. That the kinetic and canonical linear momenta are not equivalent in the presence of electromagnetic potentials is not unique to quantum mechanics, but is known as well, although is less consequential, in classical mechanics [12].

One type of current configuration that ideally gives rise to a null magnetic field in a spatial region where the vector potential does not vanish is that of an infinitely long axial coil or solenoid. Within the solenoid the magnetic field $\boldsymbol{B}$ is parallel to the symmetry axis with a strength and orientation respectively determined by the magnitude and sense of the current circulation through the windings. The magnetic flux Φ within an infinitely long coil of radius R is simply the product of the field strength and the cross section area πR^2. Outside the solenoid the magnetic field is ideally null. The vector potential, however, forms cylindrical equipotential surfaces in both regions of space (with a sense of circulation opposite that of the electron current). In the Coulomb gauge, i.e., the gauge for which $\boldsymbol{\nabla} \cdot \boldsymbol{A} = 0$, the tangential (and only) component of $\boldsymbol{A}$ is a function of the radial coordinate r given by

$$\boldsymbol{A}(r) = \begin{cases} \Phi/2\pi r & (r \geq R) , \\ \Phi r/2\pi R^2 & (r \leq R) . \end{cases} \tag{1.27}$$

For any real solenoid, of course, there is a return field in the exterior region, but the magnitude of this field diminishes with increasing ratio of length to radius [13]. Besides the ideally infinite solenoid, a toroidal current configuration formed by joining the two ends of a finite solenoid accomplishes the same task of producing a confined magnetic field (although the expressions for the resulting vector potential field are not as simple).

Feynman has pointed out the dramatic consequence of the seemingly innocuous relations (1.21) and (1.22) in the context of a two-slit particle interference experiment [14] with an ideal solenoid placed behind the diffraction screen and between the slits (the long axis parallel to the slits). The experiment is conceptually simpler than the configuration originally analyzed by

AB which involved scattering of charged particles directly incident upon the solenoid. It is the Feynman version of the AB effect (actually first described in the paper by Ehrenberg and Siday) that has generally found its way into physics textbooks [15].

Reduced to its essentials, as schematically shown in Fig. 1.5, there are two types of classically indistinguishable pathways by which an incident particle can propagate from its source S to the detector D: by going clockwise above the solenoid (path I) or counterclockwise below the solenoid (path II). It is assumed that the particles never penetrate the solenoid, and that in the absence of a current through the solenoid there is complete symmetry above and below the forward direction of the beam. Since there is a vector potential (but no scalar potential) in the space accessible to the electrons, the probability amplitude for each pathway, when the solenoid interior contains an axial magnetic field, takes the form of (1.21) with ϕ equal 0 in the phase (1.22). Hence the total probability amplitude for a particle to be received at D can be written as

$$\Psi(D) = \frac{e^{i\Delta}}{\sqrt{2}} \left[1 + e^{i(\delta_0 + \delta)} \right] , \qquad (1.28)$$

where Δ is an inconsequential global phase, δ_0 is the relative phase (dependent upon optical path length difference) when the magnetic field is zero, and

$$\delta = \frac{q}{\hbar c} \left(\int_{\text{path I}} \boldsymbol{A} \cdot d\boldsymbol{s} - \int_{\text{path II}} \boldsymbol{A} \cdot d\boldsymbol{s} \right) = \frac{q}{\hbar c} \oint_C \boldsymbol{A} \cdot d\boldsymbol{s} = \frac{2\pi\Phi}{\Phi_0} \qquad (1.29)$$

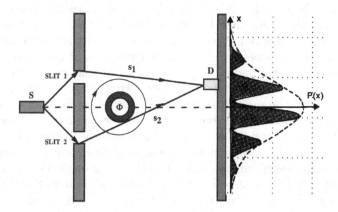

Fig. 1.5. Schematic diagram of the two-slit Aharonov–Bohm (AB) effect. A coherently split electron wavefront, issuing from source S, passes to one side or the other of a region of confined magnetic flux Φ and is recombined at the detector D. The resulting interference pattern, within an unshifted diffraction envelope, is influenced by the magnetic field through which the electrons do not pass

is the relative phase contributed by the confined magnetic field. Φ/Φ_0 is the ratio of magnetic flux through the solenoid to the unit of magnetic flux, or fluxon, which for an electron is

$$\Phi_0 = \frac{hc}{e} = 4.14 \times 10^{-7} \text{ gauss cm}^2 \ . \tag{1.30}$$

The contour C marking the integral in (1.29) is a closed path between S and D circumscribing the solenoid in a clockwise sense. If the magnetic field within the coil is directed into the plane of the paper, then the external vector potential field circulates in the same sense as the contour C, and the sign of the magnetic flux is positive. It follows from relations (1.28) and (1.29) that the probability of receiving a particle at D is

$$P(\text{D}) = |\Psi(\text{D})|^2 = \frac{1}{2}\left[1 + \cos\left(\delta_0(\text{D}) + 2\pi\frac{\Phi}{\Phi_0}\right)\right] \ . \tag{1.31}$$

The physical content of the above relation is that a magnetic field, from which region the charged particles are totally excluded, can influence the spatial distribution of the particles. In fact, the influence can be strong. For D located in the forward direction ($\delta_0 = 0$), there is constructive interference and thus 100% probability of receiving a particle in the absence of the magnetic field ($\Phi = 0$). However, with a solenoidal magnetic field of such strength that $2\pi\Phi/\Phi_0$ is an integral multiple of π radians, the probability $P(\text{D})$ is zero; the isolated magnetic field has converted a bright fringe (maximum) into a dark fringe (minimum) in the resulting interference pattern. What is one to make of such a phenomenon whereby particles can be apparently displaced from their 'intended' direction without the agency of an external force?

To some, the answer has been that the AB effect does not physically exist, that it is merely a mathematical self-delusion [16]. Although the details of the underlying reasoning must be left to the original literature, the core of the argument, which is not encountered much in discussions of fundamental quantum physics anymore, concerns the arbitrariness of gauge transformations. In short, one can find a gauge, the proponents claimed, for which the vector potential vanishes entirely from the equation of motion which thereby describes a system in an environment free of any electromagnetic influence. The argument is fallacious, however, for the proposed transformation removes not only the vector potential in the space accessible to the particles, but alters the magnetic field in the interior of the solenoid as well. As pointed out earlier, no gauge transformation is admissible that changes the physical configuration of the electromagnetic fields. Moreover, aside from the theoretical inconsistency of the argument, there is substantial experimental evidence in support of the existence of the AB effect.

To others, who accept the experimental confirmations of the predicted fringe shifts, the answer has been that the AB effect is essentially a consequence of, or is equivalent to, the classical Lorentz force. Feynman, for example, in his *Lectures*, described the action of the solenoid as essentially equivalent to that of a magnetic strip placed behind the two slits. A magnetic field, however, would displace the *forward* beam direction, i.e., the *center* of the single-slit diffraction pattern. This is not what should occur in the AB effect. Analyses of model configurations [17] have shown that, in the absence of a local magnetic field (i.e., a magnetic field in the space though which the charged particles propagate), the pattern of interference fringes is displaced within the enveloping diffraction pattern which is, itself, not affected by the magnetic field in the excluded region, as illustrated in Fig. 1.5. Were this not the case, the AB effect would conflict with the Bohr correspondence principle, which requires quantum mechanics to give results compatible with classical mechanics in a domain for which both theories are valid. In this regard it is interesting to note that there is usually more than one way to extract the classical limit of a quantum calculation – e.g., one can let Planck's constant h approach zero, or let some quantum number approach infinity – and the different ways are not always equivalent [18]. The AB effect, as described so far, occurs with unbound particles; the classical limit is suitably taken by letting $h \to 0$, in which case the spatial periodicity of the fringe pattern becomes infinitesimally small and the central diffraction spot undeflected.

To the majority of physicists concerned with this fundamental issue, there remains the final option of accepting the AB effect for what it appears to be: a force-free interaction with either a local vector potential field or a nonlocal magnetic field. The issue of a local versus nonlocal interaction is actually deceptive, for the two points of view are equivalent in representing the AB effect as an intrinsically nonlocal physical phenomenon. Although it is the case that charged particles interact directly with the vector potential field at their instantaneous positions, this local interaction in itself is not sufficient to produce the AB effect. The allowed paths of the particles must circumscribe a region of space within which the magnetic field is confined and from which the particles are excluded. The AB effect, therefore, reflects the global geometry (or topology) of the space accessible to the particles. In the simple two-slit configuration represented in Fig. 1.5, the ambient space has the topology of a doughnut.

What, in view of the AB effect, is a reasonable posture to take regarding the fundamentality of electromagnetic fields and potentials? One widely accepted interpretation (although perhaps not so widely known throughout the physics community as a whole) has been articulated by Wu and Yang [19], according to whom a complete description of electromagnetism is provided by the nonintegrable (i.e., path-dependent) phase factor $\exp\left[\mathrm{i}S(\boldsymbol{r}, t)\right]$ in (1.21). It is the phase *factor*, and not the phase (1.22) alone, that is physically meaningful, because the phase (which manifests the arbitrariness of the potentials) contains more information than is determinable by measurement. Conversely, the fields $\boldsymbol{E}$ and $\boldsymbol{B}$ contain less information than is measurable and therefore provides an incomplete description of electromagnetism when quantum processes are taken into account. Note that these are *classical* fields; the quantum processes refer here only to matter.

The experimental confirmation of the AB effect, upon which the post-Maxwellian interpretation of electrodynamics ultimately rests, has never been a simple matter, in part because of the extreme difficulty of producing a magnetic field with no leakage into the spatial region of the particles. In the earliest AB experiments confined magnetic fields were produced by either tiny ferromagnetic filaments (or whiskers) or microscopic solenoids. Although the results were in accord with theoretical predictions derived from the nonintegrable phase factor, the fact that the ideal conditions of the AB effect had not been met allowed critics to point to the classical Lorentz force as the causative agent for any fringe shifts. While a complete survey of AB experiments must be left to the literature,[6] it is relevant to mention here two experiments, one of the earliest and one of the most definitive, of particular conceptual interest.

In the 1962 experiment of Werner Bayh [21], the AB effect was recorded on photographic film by a 40 keV electron beam split and recombined by a system of three electrostatic biprisms. Between the first and second biprism, at a location where the extent of separation of the electron beam is greatest, was inserted a tiny tungsten coil about 5 mm in length and less than 20 μm in diameter to serve as the AB solenoid. Detailed calculations of the spatial variation of the coil magnetic field (based on an equation of Buchholz [22]) indicated that for a coil of 20 μm diameter and pitch of 6 μm the radial component of the magnetic field in the mid-plane of the coil and at a distance of 10 μm from the coil windings was weaker than the interior axial field by a factor of approximately 2×10^{-5}. Since the components of the split electron beam could be separated by 50–60 μm without exceeding the coherence condition which had to be met for interference to occur, Bayh concluded that the nonideal effects of the coil should be negligibly small.

The AB phase shift was demonstrated by fastening the film, upon which the interference pattern was to be recorded, to a small electric motor and advancing the film at a rate proportional to the rate of increase of current through the windings of the coil. The film was shielded except for a 0.5 mm wide slit oriented perpendicular to the interference fringes so that each narrow horizontal section through the interference pattern corresponded to a well-defined value of magnetic flux through the solenoid. The resulting interference pattern, shown in Fig. 1.6, showed the continuous lateral displacement of the fringes (for a total distance of roughly four times the fringe spacing) *within* the enveloping pattern (produced by diffraction around the biprism filament) which remained unchanged despite the variation in vector potential and magnetic fields.

What is especially interesting about this experiment is that, as a result of the time-varying magnetic flux, the fringe shifting may also be accounted for by an apparently purely classical argument, one based on Faraday's law of induction

$$\mathcal{E}(t) = -\frac{1}{c}\frac{\partial \Phi(t)}{\partial t} , \tag{1.32}$$

---

[6] Detailed reviews of AB experiments are provided in [16] and in [20].

where $\mathcal{E}(t)$ is the electromotive force

$$\mathcal{E}(t) = \oint_C \boldsymbol{E}\cdot\mathrm{d}\boldsymbol{s} \tag{1.33}$$

induced around the solenoid in the exterior region through which the electrons pass. The induced electric field does work on the electrons and thereby engenders a relative phase

$$\theta = \frac{1}{\hbar}\int e\mathcal{E}(t)\mathrm{d}t = \frac{e}{\hbar c}\int \mathrm{d}\Phi = 2\pi\frac{\Phi}{\Phi_0}\, , \tag{1.34}$$

which is exactly what one would have expected on the basis of the force-free AB effect.

The difficulty with this classical interpretation, however, is apparent once one realizes again that it is not a succession of classical wave fronts, but rather discrete and uncorrelated electrons that passed through the interferometer. With an energy of 40 keV, and hence a velocity of $\approx 0.38c$, an electron propagated from source to film – a distance on the order of 1 m – in about 0.01 μs. Although Bayh did not specify the rate at which the current increased through the solenoid windings, it was undoubtedly over a much longer time interval. Consequently, the interference pattern was created over a relatively long period of time by the arrival of a large number of independent electrons, each one sampling an effectively instantaneous value of the local vector potential field. Nevertheless, the fact that the space accessible to the electrons was not entirely devoid of a force field was a potential source of criticism.

In an effort to avoid such criticism, researchers at the Hitachi Advanced Research Laboratory produced the AB effect under conditions more closely

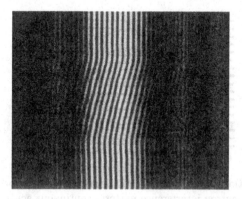

Fig. 1.6. Demonstration of the electron wave phase shift in the presence of a vector potential field (AB effect) in the Bayh experiment. The magnetic field is held constant in the upper and lower third of the figure; in the middle third the variation in interference fringes follows a linear variation in magnetic field strength. Adapted from [21]

duplicating the requirements of an ideal force-free environment than had been attained previously [23]. As far as I am aware, the Hitachi experiment remains the definitive test of the AB effect at the time this chapter is being written. The most important new feature of the Hitachi experiment, which, like earlier experiments, employed a Möllenstedt biprism to split the beam of an electron microscope, was the use of a microscopic ($\approx$ 4 μm diameter) toroidal ferromagnet, in place of a 'whisker' or solenoid, as the source of a confined magnetic field. Ideally, the magnetic field lines circulate within the toroid about the C_∞ symmetry axis (i.e., the axis perpendicular to the plane of the toroid). However, to guard against possible leakage of the magnetic field, the toroid was covered completely with a superconductive layer of niobium. When brought below the critical temperature $T_c = 9.2$ K, the niobium underwent a transition to the superconducting state and expelled magnetic flux from its interior, and therefore *into* the permalloy toroid, by means of the Meissner effect.[7] An additional layer of (nonsuperconductive) copper further helped reduce penetration of the 150 keV electrons into the toroid.

In the absence of a magnetic toroid, the split electron beam gives rise to the standard pattern of parallel fringes in the observation plane. With the magnet in place (above the biprism), theoretical analysis predicts an AB phase shift between components of an electron wave packet propagating through the central hole of the toroid compared with passage around the outer periphery. When the experiment was performed with an unshielded toroidal magnet, one could see in the resulting interferogram the continuous displacement of a light or dark fringe from the exterior region, across the body of the annulus, into the region of the hole as shown in Fig. 1.7. Of course, without the shielding layers electrons can penetrate the magnet, and critics could again attribute phase shifts to classical effects of the Lorentz force.

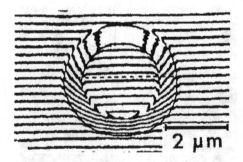

Fig. 1.7. Observation of the AB effect with an unshielded toroidal ferromagnet. The fringes in the electron interference pattern, continuous across the outline of the toroid within which the magnetic field is confined, are uniform, parallel, and shifted within the zero-field region of the hole. (Courtesy of A. Tonomura, Hitachi ARL)

---

[7] For a discussion of the Meissner effect see, for example, [24].

The use of toroids with a superconducting outer layer, however, had an unanticipated and potentially adverse side effect, namely, that the magnetic flux trapped within the annulus became quantized in units of one half a fluxon. This quantization condition

$$\Phi = \frac{nhc}{2e} = \frac{n\Phi_0}{2} \, ,$$ (1.35)

where n is an integer, pertains to *flux*, not fields, and should not be confused with the quantization of the electromagnetic field – i.e., the description of electromagnetic waves in terms of photons – which plays no role in the present system. Rather, the flux quantization condition is a consequence of the fact that the charge carrier within the superconductor is not a single electron, but a Cooper pair of electrons, and that the wave function of this pair is macroscopically coherent around the annulus. As a result of flux quantization within the toroid, the magnetic phase shift in (1.31) becomes $2\pi\Phi/\Phi_0 = n\pi$ and is either 0 (mod 2π) or π (mod 2π) according to whether n is an even or odd integer. In the first case the AB effect leads to no observable outcome. In the second case, however, there is a complete fringe reversal from maximum to minimum between the exterior region and the central hole. Experimentally, the fabricated toroids used in the Hitachi experiment produced a range of discrete flux values, both odd and even. The observation of the predicted 180° phase reversal shown in Fig. 1.8 provides the strongest evidence I know of that the spatial distribution of charged particles can be altered by a magnetic field which the particles never encounter directly.

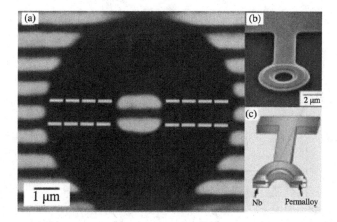

Fig. 1.8. AB effect experiment performed with a toroidal ferromagnet shielded by an outer superconducting layer. (**a**) The fringe reversal between the external rgion and the region of the hole represents an AB phase shift of 180°. (**b**) Photomicrograph of the toroid. (**c**) Schematic of cross-section showing external niobium layer and internal permalloy magnet. (Courtesy of A. Tonomura, Hitachi ARL)

1.4 'No-Slit' Interference of Single Photons: Superposition, Probability, and Understanding

Quantum mechanics is a statistical theory of particles and interactions. As such, it provides a probabilistic description of the outcome of observing many (in principle, infinitely many) similarly prepared particles or systems of particles, or, equivalently, many observations of a single system, as, for example, a trapped ion. Within the mathematical formalism of quantum mechanics, the seemingly illogical physical effects of self-interfering single-particle wave packets arise from the radically different rules for treating probabilities, as compared with the ordinary rules one is taught in a course on probability and statistics. Why the physical particles of nature must obey such rules (as countless experiments indicate that they do) may elicit debate, but the rules, themselves, are self-consistent and, when applied correctly, yield unambiguous, empirically testable outcomes. Within the framework of the quantum formalism, the results, therefore, are quite logical.

As an example explicitly illustrative of the difference between quantum and (for want of a better term) classical probability rules, let us consider the sequential passage of light, one photon at a time, through a polarizer (P_1), birefringent slab (B), and a second polarizer (P_2), as shown in Fig. 1.9. The example is of particular interest because, in contrast to the electron interferometry experiments discussed in the previous sections, which entailed complex and exotic apparatus and had never been performed, or even thought of, before the twentieth century (the electron was discovered only in the late 1890s), the propagation of light through polarizing materials is a well-understood process of classical physical optics (although not when performed one photon at a time).

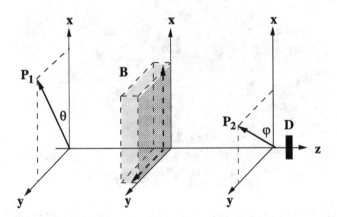

Fig. 1.9. Single photons pass through a polarizer, P_1 birefringent slab B (with principal axes along the coordinate axes), and polarizer P_2 to reach detector D. The fraction of incident photons reaching D, as computed by standard probability rules, differs markedly from the quantum mechanical prediction

In the context of classical optics, a monochromatic light wave (comprising a huge number of photons[8]) passes through polarizer P_1, which orients the electric vector of the transmitted wave at an angle θ to the vertical axis (x axis in a Cartesian frame where the light propagates along the z axis, and the y axis is out of the page). I will take the birefringent slab to be oriented with its principal axes along x and y, whereupon it is well-known that the proportion of the transmitted light intensity polarized along the x axis is $\cos^2 \theta$, and correspondingly $\sin^2 \theta$ for light polarized along the y axis. The total transmittance sums to 1, as expected, since the slab is presumed to be transparent.[9] From the classical optical rule known as Malus' Law, one can also state with certainty that the fractions of x-polarized light and y-polarized light passing polarizer P_2, with a transmission axis oriented at an angle ϕ to the vertical, are $\cos^2 \phi$ and $\sin^2 \phi$, respectively.

If the preceding experiment were performed one photon at a time, we can ask the question: What is the probability that a photon will pass polarizer P_2 if we know with certainty that all photons incident on the birefringent slab had first passed through polarizer P_1? Transcribing the transmittance data into classical probability statements, we can write:

- $P(\theta) = 1 =$ probability that a photon incident on B is polarized along the transmission axis of P_1.
- $P(x|\theta) = \cos^2 \theta =$ probability that a photon emerging from B is x-polarized, given that it was θ-polarized upon entering B.
- $P(y|\theta) = \sin^2 \theta =$ probability that a photon emerging from B is y-polarized, given that it was θ-polarized upon entering B.
- $P(\phi|x) = \cos^2 \phi =$ probability that an incident x-polarized photon passes P_2.
- $P(\phi|y) = \sin^2 \phi =$ probability that an incident y-polarized photon passes P_2.

The conditional probabilities sum appropriately to unity, as they exhaust all independent outcomes for the given condition or all conditions for a given outcome.

From Bayes' theorem,[10] we then deduce:

---

[8] The number of photons per second in a $P = 1$ mW beam of red HeNe laser light ($\lambda = 633$ nm) is given by: $dN/dt = P\lambda/hc \approx 3 \times 10^{13}$.

[9] To keep the example focused on the main point at issue, I am neglecting here peripheral considerations such as the reflection of light at the incident face and the spatial separation of the ordinary and extraordinary rays. I discuss the optics of birefringent media more thoroughly in my earlier book [7].

[10] Bayes' theorem, which follows directly from the standard rules governing probabilities, may be stated as follows. If $B_1, \ldots, B_n$ is a full set of mutually exclusive events, then the probability of an event B_k, given the occurrence of an event A, is

$$P(B_k|A) = \frac{P(A|B_k)P(B_k)}{P(A)} = \frac{P(A|B_k)P(B_k)}{\sum_{i=1}^{n} P(A|B_i)P(B_i)}.$$

- Probability that an x-polarized photon is incident on P$_2$ is

$$P(x) = P(x|\theta)P(\theta) = \cos^2\theta .$$

- Probability that a y-polarized photon is incident on P$_2$ is

$$P(y) = P(y|\theta)P(\theta) = \sin^2\theta .$$

Applying Bayes' theorem again to determine the probability, $P(\phi|\theta)$, that a photon which has passed polarizer P$_1$ will pass polarizer P$_2$, we arrive at the prediction based on classical probability

$$P_{cl}(\phi|\theta) = P(\phi|x)P(x) + P(\phi|y)P(y) = \cos^2\theta\cos^2\phi + \sin^2\theta\sin^2\phi . \quad (1.36)$$

Suppose that the transmission axis of P$_2$ is parallel to that of P$_1$, and both are at 45° to the vertical, i.e., $\phi = \theta = \pi/4$. It then follows from (1.36) that $P_{cl}(\phi = \theta = \pi/4) = \cos^4\theta + \sin^4\theta = 1/2$, which means that 50% of the photons which have passed P$_1$ are missing, even though the two polarizers are parallel. The predicted outcome is clearly incorrect.

To deduce the correct probability from the quantum mechanical formalism, we must introduce basis vectors that specify the polarization states of the photons. We are interested only in the polarization of the light, and can disregard in this section the spatial and temporal variations of a more complete quantum description. Thus, $|x\rangle$ and $|y\rangle$ form a complete set of basis states polarized along x and y, and $|e_1\rangle$ and $|e_2\rangle$ and $|f_1\rangle$ and $|f_2\rangle$ are likewise two sets of basis states, where the axes e_1 and e_2 are oriented at angles θ and $\theta + \pi/2$ to the vertical, and the axes f_1 and f_2 are oriented at angles ϕ and $\phi + \pi/2$ to the vertical. From the completeness relation and quantum rules for rotating state vectors, the sets of basis states can be expressed in terms of one another as follows:

$$\begin{aligned}
|x\rangle &= |e_1\rangle\langle e_1|x\rangle + |e_2\rangle\langle e_2|x\rangle = |e_1\rangle\cos\theta - |e_2\rangle\sin\theta , \\
|y\rangle &= |e_1\rangle\langle e_1|y\rangle + |e_2\rangle\langle e_2|y\rangle = |e_1\rangle\sin\theta + |e_2\rangle\cos\theta ,
\end{aligned} \quad (1.37)$$

$$\begin{aligned}
|f_1\rangle &= |x\rangle\langle x|f_1\rangle + |y\rangle\langle y|f_1\rangle = |x\rangle\cos\phi + |y\rangle\sin\phi , \\
|f_2\rangle &= |x\rangle\langle x|f_2\rangle + |y\rangle\langle y|f_2\rangle = -|x\rangle\sin\phi + |y\rangle\cos\phi .
\end{aligned} \quad (1.38)$$

Upon substitution of relations (1.37) into (1.38), we obtain a relation between polarization states corresponding to the two polarizers

$$\begin{aligned}
|f_1\rangle &= |e_1\rangle(\cos\theta\cos\phi + \sin\theta\sin\phi) + |e_2\rangle(\cos\theta\sin\phi - \sin\theta\cos\phi) \\
&= |e_1\rangle\cos(\theta - \phi) - |e_2\rangle\sin(\theta - \phi) .
\end{aligned} \quad (1.39)$$

The probability that a photon polarized by P$_1$ will be transmitted by P$_2$ is then obtained, according to the rules of quantum mechanics, by projecting one state on the other,

$$P_{qm}(\phi|\theta) = |\langle e_1|f_1\rangle|^2 = \cos^2(\theta - \phi) , \quad (1.40)$$

which leads to $P_{qm}(\phi|\theta) = 100\%$ for a configuration of parallel polarizers, $\phi = \theta$, as has long been known from classical physical optics. The relation (1.40) depends only on the angle between the two transmission axes, and not on the orientation of the polarizers relative to the principal axes of the birefringent slab. Equivalently stated, the final result does not depend on the polarization of the photons emerging from the slab.

It is worth noting explicitly that the experimental outcome, correctly summarized by (1.40), rather than by (1.36), is no less indicative of a quantum interference effect than is the electron two-slit experiment, even though the interference is not manifested as a fringe pattern on a viewing screen. The interference is evident in the sums of terms in the first line of (1.39), which, by means of a trigonometric identity, leads to the more compact expression of (1.40). To refer to this experiment as a 'no-slit interference of photons' is not strictly accurate, for the polarization basis states $|x\rangle$ and $|y\rangle$ are analogs of the two slits. However, because these states represent orthogonal polarizations, the x and y polarized components of a photon state vector cannot interfere until projected onto a common axis.

In contrast to the electron interference experiment, where the outcome – the buildup of an electron fringe pattern, one electron at a time – has surprising implications, what is surprising about this photon interference experiment is that the expected outcome should not be surprising to anyone familiar with classical optics, but becomes surprising *and incorrect* when analyzed by the standard rules governing probability.

Much has been made of the 'incomprehensibility' of quantum mechanics, by both scientists and science writers. To the detriment of communicating a meaningful understanding of epistemological issues raised by quantum mechanics, the history of the subject is replete with eminently memorable, but often nonsensical, aphorisms and images that are all too frequently quoted out of context or without regard to further conceptual developments. We have Bohr's comment that anyone who can think about quantum physics without getting giddy doesn't understand the first thing about it. We have the image of Einstein standing by his office window, explaining to a visitor that the insane asylum across the way houses those madmen who have *not* thought about quantum mechanics. We have Feynman telling his students that no one understands quantum mechanics. We have John Bell's remark that nobody knows what quantum mechanics says about any particular situation. And in one of the most egregiously inappropriate images of all, which pervasively decorates book covers and conference posters, we have Schrödinger's cat in a linear superposition of dead and alive macroscopic states. (Schrödinger, at least, had the sense to refer to his example as 'ridiculous'.) With physicists communicating sentiments like these to one another or to the general public, it is no wonder that popular exegeses of quantum mechanics may contain much of what Gell-Mann called 'quantum flapdoodle'.

The superposition and interference of probability amplitudes, which is, I believe, the most defining characteristic of quantum mechanics as it is presently formulated, does indeed lead to outcomes that stretch the mind's capacity to interpret. Nevertheless, it is an abuse of language and inaccurate representation of the state of physics to suggest at the present time that no one understands quantum mechanics.

For one thing, if by 'understanding' one means seeing the essential reason behind a successful theoretical description of nature, would it have been any less true before the development of quantum theory to say that no one understands classical mechanics? How would one explain the fact that massive (classical) objects appear to obey Newton's laws of motion? That '$F = ma$' is not regarded as a 'mystery' on a par with quantum interference by those who do not know why it is valid, is due only to the circumstance that familiarity gives the illusion of understanding. However, it was not until the development of quantum theory – and, in particular, the path integral formulation by Feynman, which demonstrated most convincingly (in my opinion) the relationship between quantum phases, classical action, and the variational principle leading to Newton's laws – that classical mechanics ceased to be a mystery.

For another thing, if the formalism of quantum mechanics is not applied appropriately, then the paradoxes that arise likely stem from error and confusion more than from nature's inscrutable ways. This is especially the case in the extrapolation of quantum interference effects of elementary particles or coherent systems of particles to the gross objects of ordinary experience. What is often forgotten or ignored in taking such artistic license is that not everything can be ascribed a wave function. Cats, rabbits, chickens, and other ludicrous examples of fauna and flora, which have been employed in popular accounts of quantum mysteries, do not have wave functions[11] – and, correspondingly, conclusions drawn regarding the superposition and interference of these functions may be misleading, to say the least. In appropriate quantum descriptions of macroscopic-sized objects, the coherence parameters that govern the extent to which the interference of probability amplitudes takes place are calculable, and the failure to observe a manifestation of such a macroscopic superposition is readily explicable. That a human being does not diffract upon walking through a doorway – as an electron beam does when traversing a slit of width comparable to its transverse coherence length – is no mystery. Moreover, as quantum mechanics is a statistical theory, *one* human being passing through a doorway would not, in any event, display wavelike behavior.

And last, much misunderstanding of quantum mechanics resides in the perpetuation of the term 'wave function', which carries the regrettable connotation of a mysterious nonobservable, yet physical, wave that somehow 'guides' quantum particles to their destinations and dynamically 'collapses' when a measurement is made. To regard the wave function in such terms is analogous to regarding the aether of the nineteenth century as a dynamical medium. Just as all experimental attempts to reveal the classical aether led to wilder and weirder attributions of its characteristics, rather than to the simple

---

[11] For an example, see my book review [25].

recognition it did not exist, so too has the strange behavior of quantum systems led to fanciful speculations of the role of the wave function. Unlike the aether, however, the wave function exists *mathematically*; but it is well to keep in mind that it signifies a probability amplitude. As such, the wave function is a construct of the quantum formalism – technically, the projection of a state vector onto a particular basis state – that gives the probability of a particular outcome once the conditions of the physical system are well specified.

It is not known why quantum theory so successfully describes the physical world, in the same sense that we know now why classical mechanics did within its own domain of validity. Nor is there any evidence, experimental or theoretical, to suggest that quantum theory is flawed. Nor is there currently available or on the horizon any comparably successful theory to replace it. Perhaps one day physicists will again devise a nonstatistical fundamental theory of particles and their interactions, although I highly doubt it. In the meantime, quantum mechanics can be understood in much the same way that classical mechanics was previously understood, i.e., through development of an intuitive feeling for the outcome of quantum processes based on frequent and correct application of the quantum formalism.

1.5 Macroscale Objects in Quantum Superpositions

Is it possible for a large object, which Newton's laws of motion would ordinarily describe very well, to manifest a quantum interference of macroscopically distinguishable quantum states? The answer to this question, as discussed briefly in the preceding section, is a qualified 'No', the exceptions being macroscopically coherent systems like a superfluid, superconductor, or Bose–Einstein condensate. Most large systems do not fit this description. Not only is there no way to put something like a Volkswagen Beetle or a real beetle into a coherent linear superposition of macroscopically distinguishable states, but, even if it were possible, no present or foreseeable experiment could distinguish the condition from the corresponding incoherent mixture of quantum states. (I will discuss more thoroughly in a later chapter the differences between ensembles of mixed states and superposition states.) Nevertheless, it is instructive to examine this possibility by means of a mathematically tractable model system.

Consider a system of mass M, which is made up of N particles of mass m. We are not interested here in any internal interactions between the particles, but simply consider the entire system a 'super particle' which can be found in either of two quantum states $|\psi_1\rangle$ or $|\psi_2\rangle$, representing distinguishable spatial locations along the x axis. By varying the particle number N from 1 to some very large number, we will be able to see what the ensuing quantum description leads to for quantum-mechanical scale and classical-mechanical scale objects.

The state vector of the system at time t, which describes a linear superposition of the two states, one ($|\psi_1\rangle$) centered at the coordinate $-L$ and the other ($|\psi_2\rangle$) at coordinate $+L$, with equal probability, can be written as

$$|\Psi_t\rangle = \sqrt{\frac{\eta}{2}}\,\mathrm{e}^{-iHt/\hbar}(|\psi_1\rangle + |\psi_2\rangle)\,, \qquad (1.41)$$

where η is a normalization constant to be determined shortly, and $H = P^2/2M$ is the time-displacement operator (Hamiltonian), representing in the present case the particle kinetic energy. P is the linear momentum operator with a complete set of eigenstates $|p\rangle$ labeled by the momentum eigenvalue. The requirement $\langle\Psi_t|\Psi_t\rangle = 1$, that the probability of finding the particle somewhere in space be 100%, together with the normalization to unity of the two basis states, fixes the overall normalization constant

$$\eta = \frac{1}{1 + \mathrm{Re}\langle\psi_1|\psi_2\rangle}\,, \qquad (1.42)$$

which reduces to unity if $|\psi_1\rangle$ and $|\psi_2\rangle$ are orthogonal states.

In the system under study, the particle is at rest (mean velocity zero), but interacts with a thermal environment. Thus, the two basis states have a finite spatial dispersion, and their overlap, as expressed by the scalar product in (1.42) need not be zero. However, we shall soon consider the case where the initial width of the basis states is much smaller than their separation. In that case η is very close to unity and remains so for all time (since the normalization condition is independent of time).

Projection of (1.41) onto a coordinate basis state $\langle x|$, insertion of a complete set of momentum projection operators equivalent to the unit operator $1 = \int |p\rangle\langle p|\mathrm{d}p$, substitution of the transformation amplitude

$$\langle x|p\rangle = \frac{1}{\sqrt{2\pi\hbar}}\mathrm{e}^{ipx/\hbar}\,, \qquad (1.43)$$

and use of the displacement operator relation

$$\langle p|\psi(L)\rangle = \langle p|\mathrm{e}^{-ipL}|\psi(0)\rangle = \mathrm{e}^{-ipL}\langle p|\psi(0)\rangle = \mathrm{e}^{-ipL}\phi(p)\,, \qquad (1.44)$$

lead to the wave function

$$\Psi(x,t) = \langle x|\Psi_t\rangle \qquad (1.45)$$

$$= \frac{\sqrt{\eta/2}}{\sqrt{2\pi\hbar}}\left[\int \mathrm{e}^{ip(x+L)}\mathrm{e}^{-ip^2t/2M}\phi(p)\mathrm{d}p + \int \mathrm{e}^{ip(x-L)}\mathrm{e}^{-ip^2t/2M}\phi(p)\mathrm{d}p\right].$$

We assume for this model a real-valued momentum wave function $\phi(p)$ which gives rise to a Gaussian momentum distribution function

$$|\phi(p)|^2 = \frac{\mathrm{e}^{-p^2/2\sigma_p^2}}{\sqrt{2\pi}\sigma_p}\,, \qquad (1.46)$$

characteristic of an object classically at rest (mean value $\langle P \rangle = \bar{p} = 0$), but whose momentum is distributed about 0 with standard deviation σ_p as a consequence of thermal fluctuations. I will elaborate on this point, shortly. Substitution of (1.46) into (1.45) and evaluation of the integrals result in a spatial wave function

$$\Psi(x,t) = \frac{\sqrt{\eta/2}\,\sigma}{\sqrt{\sqrt{2\pi}\sigma_p/\hbar}} \left[e^{-(x-L)^2\sigma^2/2} + e^{-(x+L)^2\sigma^2/2} \right], \qquad (1.47)$$

with

$$\sigma^2 = \frac{1}{\dfrac{\hbar^2}{2\sigma_p^2} + \dfrac{i\hbar t}{M}} = \frac{2\sigma_p^2/\hbar^2}{1 + \dfrac{2i\sigma_p^2 t}{\hbar M}}, \qquad (1.48)$$

that is a linear superposition of two Gaussian wave packets separated by a distance $2L$. The function σ^2, which resembles a variance but is complex-valued, comprises an initial quantum uncertainty (to be attributable to thermal fluctuations) and a dynamical uncertainty increasing in time.

For comparison with the superposition state, consider first the special case of the particle initially located at the origin, for which we set $L = 0$ and $\eta = 1/2$ (because the two basis states are now actually the same state). Then the spatial probability distribution deduced from (1.47) and (1.48) is a Gaussian function

$$P_1(x,t) = |\Psi(x,t)|^2 = \frac{e^{-x^2/2\Delta^2}}{\sqrt{2\pi}\Delta}, \qquad (1.49)$$

with variance Δ^2 given by

$$\frac{1}{2\Delta^2} = \mathrm{Re}(\sigma^2) = \frac{2\sigma_p^2/\hbar^2}{1 + (2\sigma_p^2 t/\hbar M)^2}. \qquad (1.50)$$

Thus, a normal momentum probability distribution leads to a normal spatial probability distribution (because the Fourier transform of a Gaussian function is a Gaussian function). Equation (1.50) can be re-expressed in the form of a Heisenberg uncertain relation

$$(\Delta)(\sigma_p) = \frac{\hbar}{2}\left[1 + \left(\frac{2\sigma_p^2 t}{\hbar M} \right)^2 \right]^{1/2}, \qquad (1.51)$$

which reduces to the minimum quantum uncertainty $\hbar/2$ at the initial time ($t = 0$).

In the general case of a particle in a superposition state of two spatially separated Gaussian wave functions, (1.47) leads to a probability distribution function $P(x,t)$ that is a sum of two Gaussian functions like (1.49), but centered at $-L$ and $+L$ respectively, and an interference term

$$\text{Re}(\psi_1^*\psi_2) = \frac{(\eta/2)\hbar|\sigma|^2}{\sqrt{2\pi}\sigma_p}\exp\left[-\frac{2(x^2+L^2)(\sigma_p^2/\hbar^2)}{1+(2\sigma_p^2 t/\hbar M)^2}\right]\cos\frac{(8xL\sigma_p^4/\hbar^3 M)t}{1+(2\sigma_p^2 t/\hbar M)^2}$$

$$= \frac{(\eta/2)\hbar|\sigma|^2}{\sqrt{2\pi}\sigma_p}\exp\left(-\frac{x^2+L^2}{2\Delta^2}\right)\cos\left[K(t)x\right], \qquad (1.52)$$

with time-dependent wave number

$$K(t) \equiv \frac{8L\sigma_p^4 t/\hbar^3 M}{1+(2\sigma_p^2 t/\hbar M)^2} = \frac{8Ltm_p k_B^2 T^2 N^2/\hbar^3}{1+(2tk_B T/\hbar)^2 N}. \qquad (1.53)$$

Figure 1.10 shows three snapshots in time of $P(x,t)$ starting at $t = 0$. As time progresses, the gaussian packets spread, overlap partially and superpose, and eventually merge, producing a pattern very similar in appearance to the Young's two-slit diffraction-interference pattern. Keep in mind, however, that, in the present case, the particle is *not* moving around obstacles or through apertures; it is classically at rest in a superposition state with spatial dispersion.

The calculations for Fig. 1.10 were made for a particle of 1 atomic mass unit (amu), which, for our purposes, is the mass of a proton ($m_p = 1.67 \times 10^{-27}$ kg), in an environment at room temperature (~ 300 K). The two basis states comprising the superposition state are separated spatially by 10 Bohr radii a_0. The dispersion in internal energy of an object in diathermal contact with a heat reservoir, under conditions where volume remains constant, can be shown from thermodynamic fluctuation theory to be given by the expression [26]

$$\text{Var}(E) = \sigma_E^2 = \frac{Nk_B T^2 c_V}{N_{Av}} \implies \sigma_E \sim \sqrt{N}k_B T, \qquad (1.54)$$

in which k_B is Boltzmann's constant, T is the absolute temperature, N_{Av} is Avogadro's number, and c_V is the molar specific heat at constant volume. For most substances at sufficiently high temperature (e.g., room temperature), $c_V \sim R$ (to within a numerical factor of order unity), where $R = N_{Av}k_B$ is the universal gas constant; from this relation follows the approximate equivalence to the right of the arrow in (1.54). Assumption that the internal energy of the particle is exclusively kinetic energy, and use of the statistical properties of Gaussian functions, lead by the following chain of relations

$$\text{Var}(E) = \frac{\text{Var}(P^2)}{(2M)^2} = \frac{1}{(2M)^2}\left(\langle P^4\rangle - \langle P^2\rangle^2\right) = \frac{\sigma_p^4}{2M^2},$$

with

$$\langle P^n\rangle = \int p^n|\phi(p)|^2 dp \implies \begin{cases} \langle P^2\rangle = \sigma_p^2, \\ \langle P^4\rangle = 3\sigma_p^4, \end{cases}$$

to the momentum dispersion parameter

$$\sigma_p^2 = (2M^2\sigma_E^2)^{1/2} \sim N^{3/2}m_p k_B T \,, \qquad (1.55)$$

which is seen to depend on the 3/2 power of the mass number (number of amu constituting the mass).

Figure 1.10 illustrates that complete superposition of the two Gaussians to form a diffraction-interference pattern occurs on a time scale of about

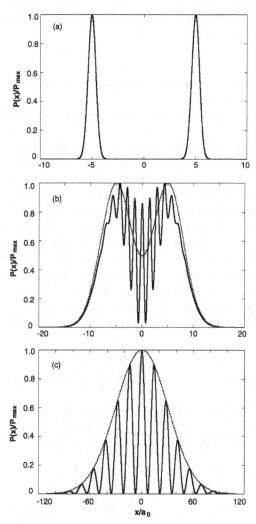

Fig. 1.10. Probability distribution of a 1 amu particle at rest described by a coherent superposition of two Gaussian wave functions initially separated by $2L = 10a_0$. Time t (units of 10^{-13} s) (**a**) 0, (**b**) 1, (**c**) 10. Peak maxima are in the ratio of 1:0.15:0.05. The *dotted line* shows the distribution function of an incoherent mixture. Spatial dispersion is due to interaction with an environment at a temperature of 300 K

10^{-13} s after the particle has been prepared in a coherent superposition of wave packets separated by $10a_0$, or about 5 atomic diameters. What happens, however, after a very long time interval has passed? Is the coherence destroyed and the interference pattern 'washed out'? The answer is 'No'. Figure 1.11 shows the isolated interference term at a time of 1 ms, i.e., 10 billion time units later. The uncertainty in locating the particle has increased enormously, so that the particle is now located somewhere between -10^{11} and $+10^{11}$ Bohr radii, or over a region of about 10.6 m, but the interference pattern is distinct with an oscillation periodicity of measurable size. From (1.53) it follows that the wave number of the oscillatory term

$$K(t) \xrightarrow[t\to\infty \text{ or } N\to\infty]{} \frac{2Lm_\mathrm{p}N}{\hbar t} . \tag{1.56}$$

reduces in either the long-time limit or the large-mass limit to the same quantity in (1.56). The spatial periodicity of the distribution function of a 1 amu particle at 300 K at a time 1 ms after state preparation can be calculated to be about 75 cm.

Now consider the system of a macroscale particle of mass 10^{18} amu (or about 1.7 nanograms) in a superposition of Gaussian basis states separated by 2×10^8 Bohr radii (or about 5.3 mm). While this mass may seem rather small for a macroscopic particle, it will illustrate well enough the foolishness of discussing quantum interference of cats and chickens! In the limit of very large mass number N, one can deduce from (1.48) and (1.55) that the prefactor of the interference term, (1.52), increases as $N^{1/4}$, the argument of the much more rapidly decreasing exponential factor grows as $N^{1/2}$, and the argument of the oscillatory factor increases linearly with N. Thus, the greater the mass of the particle, the smaller should be the amplitude of the interference term and the more rapid should be its spatial variation. Using (1.56), one can show that the periodicity of the interference term of the 10^{18} amu particle,

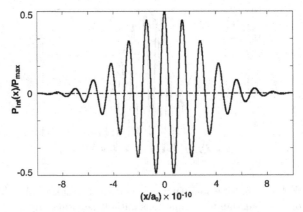

Fig. 1.11. Isolated interference pattern in the probability distribution of the 1 amu particle described in the last figure: $P_\mathrm{int}(x,t) = P(x,t) - P_1(x - L, t) - P_1(x + L, t)$. The pattern is shown for a time $t = 1$ ms after initial preparation. $P_\mathrm{max} = 0.507$

observed 1 s after preparation in a superposition of macroscopically separated basis states, is 3.8×10^{-23} m, which is approximately four tenths of a trillionth the diameter of an atom, or twenty billionths the diameter of a proton.

I have seen it asserted in popularizations (. . . 'sensationalizations') of quantum physics, whereby some macroscopic object like a cat or a chicken is assumed to be in a coherent superposition of two classically distinct states, that the interference pattern, however fine, is 'still there'. This nonsensical conclusion is an outcome of having forgotten what it means physically to produce and observe a quantum interference pattern. One can no more observe, even in principle, the oscillatory structure of a quantum interference pattern of wavelength many powers of ten smaller than an elementary particle, than observe an acoustic interference of wavelength shorter than the interparticle spacing of the medium. From (1.56) one can deduce that the periodicity of the interference pattern becomes comparable to the diameter of an atom after approximately 10^{13} s, or 300 000 years – but that is a long time to wait to do a physics experiment; the system will have become decoherent long before then.

Figure 1.12 shows a series of time snapshots of the probability distribution function of the 10^{18} amu particle, comparable to Fig. 1.10 for a 1 amu particle.

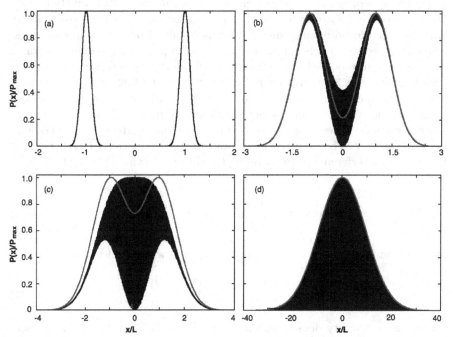

Fig. 1.12. Probability distribution function of a 10^{18} amu particle at rest described by a coherent superposition of two Gaussian wave functions initially separated by $2L = 2 \times 10^8 a_0$. Time t (ms) (**a**) 10, (**b**) 50, (**c**) 75, (**d**) 1000. Peak maxima are in the ratio of 1:0.21:0.20:0.040. *Black regions* are densely filled by oscillations. The *solid red curve* shows the distribution function of an incoherent mixture. Spatial dispersion is due to interaction with an environment at a temperature of 300 K

As time passes, the Gaussian wave packets spread, superpose, and give rise to oscillations so densely distributed that they fill the entire overlap regions at the scale of the figure. Complete overlap occurs at around 1 s. The solid red line shows the probability distribution of an incoherent mixture of two Gaussian packets. Looking at frames (b) and (c) of the figure, one might think that, periodicity notwithstanding, there is a visual – and therefore detectable – difference in the particle probability distribution function, depending on whether the particle is in a coherent superposition state or a mixture. For example, Fig. 1.12c shows that, at 75 ms after preparation, the mixture gives rise to a local minimum, whereas the coherent superposition yields a sort of broad maximum, midway between the positions of the centers of the two macroscopically separated component states. Aside from the fact that the maximum probability density of this figure is about 31 million times smaller than the maximum probability density of the 1 amu particle in Fig. 1.10c, no physically real experiment would ever be able to observe the (coherent) pattern in black, but only the (incoherent) pattern in red.

To appreciate fully the grounds for this assertion, one must recall what such an experiment would entail. The quantum wave function is not a physical entity comprised of matter or energy; it is a mathematical function that leads to a probability distribution. The physical system may comprise a single particle, but the probability distribution is obtained by repeating numerous times a procedure to detect the location of the particle at a specified instant (or, more accurately, within a narrow time window), whereby each subsequent trial is made on a system prepared exactly as in the first trial. Or, equivalently, one can execute a procedure to locate the particle once at the specified instant in each of a large number of identically prepared systems. Either way, one must employ a detector to sample a region of finite extent, which here will be designated R. The detected quantum interference pattern $\overline{P}_{int}(x,t)$ (i.e., the isolated interference corresponding to what is shown in Fig. 1.11) is then the convolution

$$\overline{P}_{int}(x,t) = \int_{-\infty}^{\infty} P_{int}(x',t)F_D(x-x')\mathrm{d}x' \tag{1.57}$$

of the instantaneous probability density $P_{int}(x,t)$ with the detector 'aperture function'

$$F_D(x) = \begin{cases} 1/2R & (|x| \leq R), \\ 0 & (|x| > R). \end{cases} \tag{1.58}$$

Since the probability density in the integrand of (1.57) takes the form [see (1.52)] $P_{int}(x,t) = P_0(x,t)\cos\big[K(t)x\big]$, in which the first factor varies relatively slowly with x compared to the cosine factor, the evaluation of the integral to good approximation leads to the expression

$$\overline{P}_{int}(x,t) \sim P_0(x,t)\cos\big[K(t)x\big]\frac{\sin\big[K(t)R\big]}{K(t)R} . \tag{1.59}$$

For any finite value of R for which $KR \gg 1$, the third factor in (1.59) becomes vanishingly small. The physically observable probability density, therefore, would consist only of the two Gaussian terms like (1.49), and would not be distinguishable from the probability density of an incoherent mixture. In order for the interference pattern to be detectable, the size of the detecting aperture would need to satisfy $KR \sim 1$, which would require a detector window a billion times smaller in size than a proton. No, the interference pattern is not there.

Besides the ultra-small interference wavelength, there is another reason why the probability distribution of a classical mechanical scale object, even if it could be prepared in a superposition of quantum states, would almost certainly be indistinguishable from that of a classical mixture of states. In the model under discussion, the initial momentum dispersion was ascribed to the exchange of thermal energy with the environment, yet no account was taken of the fact that any object at a nonzero absolute temperature will radiate energy. This process of radiation can reduce the degree of coherence of a superposition state, as measured by the contrast or visibility of the resulting interference pattern. Let us examine this point further in the context of the preceding model, for which the geometry is explicitly shown in Fig. 1.13.

The state of the entire system, now including the radiated photon, is described by the entangled state vector

$$|\Psi_t\rangle = \sqrt{\frac{\eta}{2}}\,\mathrm{e}^{-\mathrm{i}Ht/\hbar}\big(|\psi_1\rangle|\phi_1\rangle + |\psi_2\rangle|\phi_2\rangle\big)\,, \qquad (1.60)$$

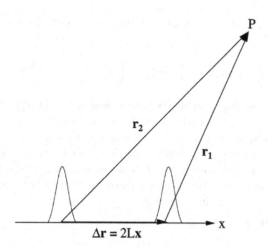

Fig. 1.13. Schematic diagram of a quantum particle, in a superposition state of two spatially separated Gaussian wave packets, radiating a photon received at point P. The photon can be emitted from either of the two particle locations. *Arrows* signify the vectorial displacement of the detector from the centers of the wave packets and of the wave packets from one another

in which the second ket in each term represents the photon state. The state vector is 'entangled' in the sense that it is not factorable into a product of two vectors, one characterizing the state of the particle, the other the state of the radiation. The entanglement of state vectors or wave functions, which will be examined under other circumstances in the next chapter, leads to some of the most fascinating and counter-intuitive phenomena in quantum physics, for which there are no classical counterparts. It is worth emphasizing at this point that, just as there is only one particle, there is only one photon although it can emerge from two different spatial regions.

Projection of the state vector (1.60) onto a product of coordinate bases $\langle x | \langle r |$, locating the particle (still assumed to be a one-dimensional system) and photon (which can be radiated isotropically in three dimensions), leads to the entangled wave function

$$\Psi(x,t;\omega) = \sqrt{\frac{\eta}{2}} \left[\psi_1(x,t) e^{i\boldsymbol{k}\cdot\boldsymbol{r}_1} + \psi_2(x,t) e^{i\boldsymbol{k}\cdot\boldsymbol{r}_2} \right] e^{-i\omega t} , \qquad (1.61)$$

where it has been assumed that the photon is described by a plane-wave state with the familiar dispersion relation $k = |\boldsymbol{k}| = \omega/c$ connecting wave number, wave vector, and frequency. The resulting probability distribution or density (of finding the particle within a differentially small range centered at location x and detecting the radiation within a differentially small range centered at point P) is then given by

$$P(x,t;\omega) = |\Psi(x,t;\omega)|^2 = \frac{1}{2} \left[|\psi_1(x,t)|^2 + |\psi_2(x,t)|^2 + 2\mathrm{Re}\left(\psi_1\psi_2^* e^{-i\boldsymbol{k}\cdot\Delta\boldsymbol{r}}\right) \right] , \qquad (1.62)$$

in which $\Delta\boldsymbol{r} = \boldsymbol{r}_2 - \boldsymbol{r}_1$ and $\boldsymbol{k}\cdot\Delta\boldsymbol{r} = k(2L)\cos\theta$. Only the interference term in relation (1.62),

$$P_{\mathrm{int}}(x,t;\omega) = |\psi_1\psi_2^*| \cos\left(\phi - \frac{2\omega L}{c}\cos\theta\right) , \qquad (1.63)$$

differs from what we have calculated previously in Eq. (1.52), which could be cast into the form of (1.63) without the term containing the radiation angular frequency ω. The quantum phase ϕ is the same in both cases.

Since we are interested primarily in the probability distribution of the particle, and not of the radiation, it is necessary to average the preceding probability density over all the directions into which the photon can be emitted and over all the radiation frequencies:

$$P_{\mathrm{int}}(x,t) = \langle P_{\mathrm{int}}(x,t;\omega)\rangle_\omega \qquad (1.64)$$

$$= |\psi_1\psi_2^*| \frac{1}{I} \int_0^\infty \int_0^\pi \cos\left(\phi - \frac{2\omega L}{c}\cos\theta\right) \cdot$$

$$e^{-\hbar\omega/k_\mathrm{B}T} \frac{2\omega^2}{(2\pi c)^3} 2\pi \sin\theta \mathrm{d}\theta \mathrm{d}\omega .$$

The normalising integral is

$$I = \int\limits_{0}^{\infty}\int\limits_{0}^{\pi} e^{-\hbar\omega/k_{B}T}\frac{8\pi\omega^2 d\omega}{(2\pi c)^3} = \frac{2}{\pi^2}\left(\frac{k_{B}T}{\hbar c}\right)^3 . \tag{1.65}$$

The factors in (1.64) can be understood as follows. First is the Boltzmann factor $\exp(-\hbar\omega/k_{B}T)$ which gives the probability (relative to the ground state) that the particle is in an excited state and therefore can radiate. The statistical factor $2\omega^2/(2\pi c)^3$ is the radiation mode density, i.e., number of modes per volume per frequency within a differential range centered at ω. There is a factor 2 in the numerator to account for two polarization modes at each frequency. The factor 2π in the product of differentials is the integrated photon azimuthal angle. Evaluation of the two integrals in (1.64) is straightforward after a standard variable change $\zeta = \cos\theta$ and recognition of the gamma function $\Gamma(x) = \int_0^\infty u^{x-1}e^{-u}du$, and leads to the result

$$P_{\text{int}}^{(1)}(x,t) = \frac{|\psi_1\psi_2^*|\cos\phi}{(1+\alpha^2)^2} = (1+\alpha^2)^{-2}P_{\text{int}}^{(0)}(x,t) , \tag{1.66}$$

with

$$\alpha = \frac{2k_{B}TL}{\hbar c} \equiv \frac{2L}{\Lambda} . \tag{1.67}$$

The superscripts (1) and (0) in (1.66) show explicitly that the corresponding probability density involves the emission of 1 or 0 photons, respectively.

Equation (1.66) shows that the size of the interference term, upon which the contrast or visibility of an interference pattern depends, is reduced in the case of thermal emission by a factor that depends on the parameter α defined in relation (1.67). There is a straightforward interpretation to this outcome. The quantity $k_{B}T/\hbar$ defines an angular frequency Ω, and therefore a wavelength $\Lambda = 2\pi c/\Omega$, which may be thought of as a characteristic wavelength of the radiation emitted by a particle at temperature T. If Λ is much smaller than the spatial separation of the component states in the superposition, i.e., if $\Lambda \ll 2L$, (or $\alpha \gg 1$), then it would be possible to determine fairly precisely from where the particle emitted a photon. Because the location of an object can be tracked by means of the radiation it emits, once it becomes possible through any means whatever to determine where the object is located, there must have occurred a concomitant loss in the visibility of the associated quantum interference pattern. Otherwise, it would be possible to perform experiments that would violate the uncertainty principle. This is substantiated by (1.67); a large value of α leads to a vanishing $P_{\text{int}}^{(1)}(x,t)$. Conversely, when $\Lambda \gg 2L$, and one cannot tell from which location the photon was emitted, then $\alpha \ll 1$, and $P_{\text{int}}^{(1)}(x,t)$ is hardly changed from $P_{\text{int}}^{(0)}(x,t)$. Figure 1.14 shows a plot of the relative contrast as a function of α.

The loss in coherence (or decoherence) increases in proportion to the temperature of the emitter and the spatial extent of the coherent superposition state. Thus, to use the examples we began with of a 1 amu particle with

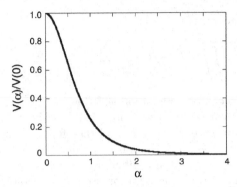

Fig. 1.14. Plot of the relative visibility $V(\alpha)/V(0) = (1 + \alpha^2)^{-2}$ as a function of the parameter $\alpha = 2k_{\mathrm{B}}TL/\hbar c$

approximate coherence length of 5 Bohr radii and a 10^{18} amu particle with approximate coherence length of 10^8 Bohr radii, we find that at room temperature (300 K), the respective decoherence parameters α are 6.96×10^{-5} and 1.39×10^3, leading to interference contrasts of 1.00 and 2.66×10^{-13}. To prevent the decoherence of a 'Schrödinger chicken' ($2L \sim 1$ m), one would have to deep-freeze it at a temperature of about 10^{-4} K.

The occurrence of decoherence is not merely a theoretical proposition, but has been demonstrated in the laboratory [27] by means of a two-slit interference experiment employing a beam of large molecules C_{70} of the fullerene class, i.e., molecules consisting purely of carbon atoms bonded to form closed ball-like structures. A C_{70} molecule made up of ^{12}C atoms has a mass of 840 amu, which is an impressive size for a quantum particle, but nevertheless a very long way from a macroscale one. When the temperature of the molecules, heated by absorption of light from an argon ion laser, was below 1000 K, the molecules exhibited perfect quantum interference upon passing through a free-standing gold diffraction grating with grating periodicity of 991 nm. The contrast in the interference pattern diminished as the temperature was raised, and was entirely lost by 3000 K.

1.6 Quantum Mechanics and Relativity: The 'Wrong-Choice' Experiment

Under circumstances such that the path of a particle through an interferometer cannot be determined, the collective outcome of numerous detections exhibits a quantum interference indicative of wavelike behavior. By contrast, if by some means the path through the interferometer can be determined (even if such a determination is not made), the outcome exhibits no quantum interference. In the early 1980s John Wheeler described a hypothetical modification of the one-particle-at-a-time split-beam experiment in which the decision by the experimenter to configure the interferometer so as to observe an interference pattern or not is made *after* the particle has entered the interferometer, i.e., *after* its wave function is in a superposition state-but before the particle is detected [28]. Actually, to heighten the dramatic impact of the proposal (and possibly also because of contemporary instrumental limitations in realizing an actual table-top experiment), Wheeler framed his 'delayed choice' thought-experiment on a cosmological scale.

Imagine a distant quasar from which an emitted photon on its way to Earth can take a path to one side or the other of a foreground galaxy which, by gravitational lensing, directs a photon along the first path to detector D1 and a photon along the second path to detector D2. A fiber optic delay line of adjustable length inserted before one of the detectors allows data to be collected as a function of the optical path length difference or, equivalently, the relative phase Φ between waves along the two paths. Examination of the results of a large number of counts should show that photons arrive randomly at the detectors one at a time in approximately equal numbers irrespective of the difference in optical path lengths. However, the insertion of a half-silvered mirror at the junction of the two paths before the detectors turns the configuration into a Mach–Zehnder interferometer, as illustrated schematically in Fig. 1.15. The counts at the two detectors plotted as a function of Φ should then produce an oscillatory pattern with one detector registering maximum counts when the other detector ideally registers zero counts. (I analyze various quantum interference experiments with a Mach–Zehnder interferometer in Chap. 3.)

Although the outcomes described above are straightforward and expected, Wheeler's original narrative highlighted the unusual feature of *delayed choice* which cast a new light on the apparent oddness of quantum mechanics. In Wheeler's words:

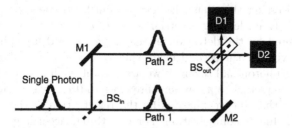

Fig. 1.15. Schematic diagram of a delayed-choice experiment. The wave function of an incoming single photon is split at beam splitter BS$_{in}$ into components corresponding to path 1 and path 2. If there is no output beam splitter BS$_{out}$, then a photon 'taking' path 1 is registered by detector D1 and a photon 'taking' path 2 is registered by detector D2. If BS$_{out}$ is present, the configuration becomes a Mach–Zehnder interferometer with D1 and D2 sensitive to the relative phase corresponding to the difference in optical path lengths. The choice of whether or not to include BS$_{out}$ is made after the photon has passed BS$_{in}$

We get up in the morning and spend the day in meditation whether to observe by 'which route' or to observe interference between 'both routes'. When night comes [...] we leave the half-silvered mirror out or put it in, according to our choice. [...] We may have to wait an hour for the first photon. When it triggers a counter, we discover 'by which route' it came with the one arrangement; or by the other, what the relative phase is of the waves associated with the passage of the photon from source to receptor 'by both routes' – perhaps 50 000 light years apart as they pass the lensing galaxy [...] But the photon has already *passed* that galaxy billions of years before we made our decision.

The strange conclusion Wheeler drew from this imagined experiment has subsequently influenced a number of physicists either to suspect or to proclaim that quantum mechanics somehow violates the special theory of relativity. Again in Wheeler's words:

[...] we are dealing with an elementary act of creation. It reaches into the present from billions of years in the past. It is wrong to think of that past as 'already existing' in all detail. The 'past' is theory. The past has no existence except as it is recorded in the present. By deciding what questions our quantum registering equipment shall put in the present we have an undeniable choice in what we have the right to say about the past.

If the quantum behavior of particles should actually lead to the conclusion that the present determines the past, rather than the reverse, then it would certainly appear that quantum mechanics and relativity are in conflict. Be assured, however, that this is *not* the case. To infer an inconsistency from the Wheeler delayed-choice experiment is an error of thinking attributable to incorrect interpretation of the significance of quantum amplitudes.

Before addressing this issue, however, I must comment on what seems to me a seriously misleading assertion in the above excerpts. Whereas the 'click' engendered by the detection of that long-awaited first photon in the configuration without the second beam splitter could very well inform us of which path the photon had taken (if we are careful to eliminate 'dark counts' by cooling our detectors to a temperature close to 0 K), we certainly will *not* 'discover' what the relative phase is by the registration of that one photon in the Mach–Zehnder configuration. For one thing, according to Wheeler's description of his thought experiment, we already know what the relative phase will be because we have adjusted the length of the delay line. More to the point, however, is to recognize that in general the determination of relative phase requires numerous detection events, as demonstrated dramatically with the build-up of an electron interference pattern in the experiment I proposed to the Hitachi electron microscopy group. The registration of one electron or one photon in a split-beam experiment conveys no information about phase. Indeed there is a kind of uncertainty relation between particle number and phase.[12] The lesson here is significant: quantum mechanics is an irreducibly statistical theory.

Various attempts to implement the delayed-choice experiment in the laboratory have been made since Wheeler's original proposal. Surprisingly, the claim to having achieved for the first time a relativistic spacelike separation between the entry of single photons into the interferometer and the choice between open and closed interferometer configurations was reported only shortly before this book was completed in 2007 [30]. Figure 1.15 depicts the experimental configuration. Single-photon pulses, obtained from photoluminescence of a single nitrogen-vacancy color center in a diamond nanocrystal, were directed by input beam splitter BS_{in} into path 1 or path 2 with orthogonal linear polarizations. The output beam splitter BS_{out} comprised in part an electro-optic modulator (EOM) which, when unexcited, allowed the orthogonally polarized photons to pass so that photons of one polarization were received by detector D1 and photons of the other polarization were received by detector D2. Thus, path identification was possible, and the configuration corresponded to the *absence* of the half-silvered mirror. Upon application of a voltage, the EOM became equivalent to a half-wave plate that rotated the input polarizations by 45°, whereupon amplitudes along the two paths could superpose and interfere, corresponding to the *presence* of a half-silvered mirror. In the 40 ns time interval to switch the EOM between states, a light signal could propagate about 12 m, whereas the path length through the interferometer was 48 m.

In implementing the delayed-choice protocol for each incoming photon, the decision of whether to excite the EOM or not was made after the photon passed BS_{in} by a binary quantum random number generator. For each photon passing BS_{in}, the experimenters recorded the configuration choice ('which route' or 'both routes'), the detection outcome (D1 or D2), and the position of a piezoelectric actuator in BS_{out} which determined the phase shift Φ between the two arms of the interferometer. Figure 1.16 illustrates the results which, as stated earlier, are precisely in accord with the predictions of quantum mechanics. Like Wheeler however, the authors remark upon this agreement "even in surprising situations where a tension with relativity seems to appear".

Is there any 'tension' with special relativity? No, none that I know of from any quantum mechanical experiment. Indeed, no experimental implementation of Wheeler's version of delayed choice, however ingeniously devised and carefully executed, can ever show the kind of inconsistency between relativity and quantum physics, i.e., the 'present' determines the 'past', that has caused

---

[12] In contrast to the uncertainty relation between particle coordinate and linear momentum, the widely used uncertainty relation between particle number and relative phase does not derive from quantum commutation conditions because there is no operator rigorously corresponding to phase. Nevertheless, there are various ways to express the experimental fact that a single interferometer configuration cannot reveal both the exact number of particles and the relative phase of the divided particle wave function. See, for example, [29].

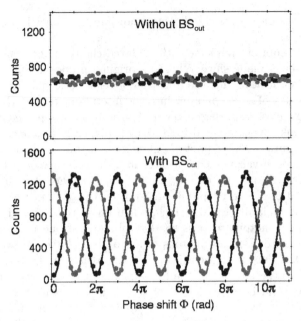

Fig. 1.16. Experimental outcome of the delayed-choice experiment of Jacques et al. [30]. *Blue (red) points* correspond to detections at D1 (D2) minus dark counts

some physicists anxiety. Why? Because Wheeler's delayed choice is the *wrong* choice.

Quantum mechanical amplitudes (which is perhaps better terminology than waves or wave functions) do not represent parcels of matter or energy. They represent information or states of knowledge about a physical system. Failure to understand this point has often given rise to problematic circumstances which some physicists then try to resolve when, in fact, there is no problem to begin with. The 'collapse of the wave function' is one such case whereby the conclusion of a measurement creates a discontinuous event because the outcome provides new information which changes the quantum description of a system and the probabilities of subsequent events. Yet to search for a physical mechanism of the 'collapse' is tantamount, in my opinion, to searching for some unknown physical force 'causing' the Lorentz contraction or time dilatation, rather than realizing, as nearly all physicists now do, that these are intrinsic to the process of making and comparing measurements.

In the delayed-choice configuration depicted in Fig. 1.15, quantum amplitudes for propagation along path 1 or path 2 are created at the moment photons encounter the first beam splitter. That act of creation is the one relevant 'reality' of the past, and whatever choices are made subsequently take place at a later time. Nothing the experimenter decides to do can alter the past or what can be said about the past. If, subsequent to the entry of the photon into the interferometer, the experimenter decides to insert the second beam splitter, then the two already extant coherent amplitudes are 'available' for superposition and interference. If the experimenter decides not to insert the second beam splitter, then the amplitudes are not superposed in the calculation of detection probabilities. In no case is there a tension with, or violation of, special relativity because neither choice of what to do with beam splitter BS_{out} influences the fact that the photon wave function was split earlier at BS_{in}.

The situation calls to mind (as an analogy, not an equivalence) a dramatic classroom demonstration of Faraday's law of induction. One has a conducting ring containing a small incandescent bulb and a long iron rod whose base is inserted into a solenoid carrying alternating current. The ring itself is not connected by any wires to a source of energy, yet the incandescent bulb glows brightly as the ring is lowered over the rod. We infer from this that the oscillating magnetic flux within the iron core gives rise to an electric field of cylindrical symmetry centered on the bar, and that this electric field drives a current within the ring. If the ring were made of plastic or some other nonconducting material, the bulb would not light, but no physicist would conclude that the induced electric field was not present. The glowing of the light merely makes the presence of the induced electric field manifest. In a heuristically similar way, the presence of the second beam splitter makes the presence of coherent quantum amplitudes manifest through quantum superposition and interference – but removal of that beam splitter with resulting absence of interference does not imply that the amplitudes do not exist.

To look for an inconsistency between special relativity and quantum mechanics by means of a delayed-choice experiment, it is the *first* beam splitter BS_{in}, *not* the second beam splitter BS_{out}, that one must choose to remove or insert. If BS_{in} is present when the photon (or other kind of particle) arrives, then amplitudes are created for a nonvanishing probability of propagation along path 1 or path 2, and the presence of BS_{out} can manifest quantum interference. If BS_{in} is not present when the photon arrives, but is inserted *after* the photon passes this location but *before* reaching BS_{out}, there can be only a single quantum amplitude, and therefore no manifestation of quantum interference is expected. Should the *post-passage insertion* of BS_{in} lead to quantum interference (i.e., phase-dependent detection probabilities) in the registrations of D1 and D2, one could legitimately proclaim a conflict with relativity.

Personally, I am not worried about the outcome.

Correlations and Entanglements I: Fluctuations of Light and Particles

2.1 Ghostly Correlations of Entangled States

In a 1935 paper [31] that has since become a classic in the literature regarding conceptual implications of quantum mechanics, Einstein and his colleagues Boris Podolsky and Nathan Rosen (to be designated EPR) raised in one of its starkest forms the issue of nonlocality – that is, the occurrence of interactions instantaneously at a distance in violation of physicists' intuitive sense of cause and effect as embodied in the principles of special relativity. Actually, Einstein's primary focus of concern was the completeness of quantum mechanics as a self-consistent theory of individual particles (as opposed to a purely statistical theory of ensembles of particles), but the Gedankenexperiment proposed by EPR illustrated what many physicists throughout the ensuing years have considered to be one of the strangest features of quantum mechanics.

In its barest essentials, the EPR experiment concerns the correlation of coordinates and momenta of two particles that have interacted at some time in the past and then separated to such an extent that for all practical purposes (from the perspective of classical physics) they would appear to be distinctly independent systems at the time a measurement was to be performed on them. Although it is not possible to measure both the position and momentum of each particle simultaneously – to do so would violate the uncertainty principle – one could in principle, according to EPR, measure one of these variables for one particle of the pair and, *without in any way disturbing the state of the second particle*, deduce the corresponding variable with 100% certainty. However, since the choice of whether to measure coordinate or momentum is a decision to be made by the experimenter-and this decision can even be made after the two particles have separated-the remote unprobed particle can not 'know' *which* measurement was made on the examined particle. A different experimental configuration is required to measure position than to measure momentum, and, depending upon which configuration is employed, the unprobed particle would be expected to manifest sharp values of either one or the other – but not both – of two canonically conjugate variables. Can the fact that a measurement was made on one particle be transmitted instantaneously to the other? If not, then do the particles have well-defined positions and momenta even though both properties are not simultaneously measurable? This enigma of completeness versus locality constitutes in part what has been termed the EPR paradox.

The quantitative features of the one-dimensional model system analyzed by EPR can be summarized briefly in the following way (as first shown by Niels Bohr). From the momenta (p_1, p_2) and coordinates (q_1, q_2) of the two separated particles, which satisfy the usual quantum commutation conditions (with $i, j = 1, 2$)

$$[q_i, q_j] = 0 , \qquad [p_i, p_j] = 0 , \tag{2.1}$$

$$[q_i, p_j] = i\hbar \delta_{ij} , \tag{2.2}$$

one can define new pairs of conjugate variables (Q_1, P_1) and (Q_2, P_2) by means of a rotational transformation with parameter θ, viz.,

$$\begin{pmatrix} Q_1 \\ Q_2 \end{pmatrix} = \begin{pmatrix} \cos\theta & \sin\theta \\ -\sin\theta & \cos\theta \end{pmatrix} \begin{pmatrix} q_1 \\ q_2 \end{pmatrix} , \tag{2.3}$$

$$\begin{pmatrix} P_1 \\ P_2 \end{pmatrix} = \begin{pmatrix} \cos\theta & \sin\theta \\ -\sin\theta & \cos\theta \end{pmatrix} \begin{pmatrix} p_1 \\ p_2 \end{pmatrix} . \tag{2.4}$$

The transformed variables satisfy commutation rules identical in form to (2.1) and (2.2). Although it is not possible to assign definite numerical values to both Q_1 and P_1 (since they do not commute), one could, however, prepare the two-particle system in a state such that Q_1 and P_2 (which *do* commute) have known, sharp values. Then, since

$$Q_1 = q_1 \cos\theta + q_2 \sin\theta , \qquad P_2 = -p_1 \sin\theta + p_2 \cos\theta , \tag{2.5}$$

a measurement of either q_1 or p_1 will allow one to predict respectively the corresponding quantity q_2 or p_2.

Although admittedly arbitrary, EPR adopted as a reasonable definition of reality the criterion that: "If, without in any way disturbing a system, we can predict with certainty (i.e., with probability equal to unity) the value of a physical quantity, then there exists an element of physical reality corresponding to this physical quantity." By this criterion, then, the coordinate and momentum of particle 2 must be real – since it can be predicted with certainty by measurements made on distant particle 1. Yet, according to quantum mechanics, a complete description of the system does not permit simultaneous knowledge of the coordinate and momentum of a particle. Thus, quantum mechanics is incomplete, or the properties of a particle have no physical reality until measured, an implication unacceptable to any theory that purports to make objective sense of the world according to EPR.

Published replies to the EPR paper followed quickly from Bohr [32] and many others. In fact, as soon as the EPR paper was published, Einstein personally received a large number of letters from physicists pointing out to him where the argument failed. He found it considerably amusing that "while all the scientists were quite positive that the argument was wrong, they all gave different reasons for their belief!" [33]. In the ensuing years to the present time, entire conferences devoted to the conceptual foundations of quantum theory testify still to the undiminished fascination that many physicists and laymen alike have with the issues raised by EPR. Numerous experimental tests of the EPR correlations have since been performed, although not by measuring the position and momentum of massive particles, but rather the polarization of correlated photons.[1]

Because entanglement of photon polarizations provides an instructive system for investigating the counter-intuitive nature of quantum correlations, it is worth examining this consequences in some detail. Moreover, the model to be discussed here prepares the groundwork for examining the phenomenon of correlated quantum beats in chapter 4. Consider a process (such as parametric down-conversion by a nonlinear optical crystal) which gives rise at the origin of the z axis to pairs of momentum- and polarization-correlated photons of angular frequencies ω_α and ω_β in a rotationally symmetric state (i.e., symmetric with respect to rotation of the polarization basis vectors about the propagation axis) (see Fig. 2.1). The correlations are such that, if photon α, propagating to the left along the z axis, is observed to be in one of two orthogonal polarization states x or y (representing vertical or horizontal linear polarization, respectively), then photon β, propagating to the right along the z axis, will be observed to be in the same polarization state (x or y) as α. Whether photon α goes to the left or right is entirely random, but the

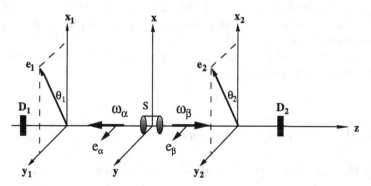

Fig. 2.1. Schematic diagram of a counter-propagating entangled pair of photons emitted from source S and received at detectors D_1 and D_2. The transmission axes of polarizers 1 and 2 are represented, respectively, by unit vectors e_1 and e_2. The polarizations e_α, e_β of the two photons are randomly oriented from emission to emission, but correlated with one another

---

[1] See, for example, the comprehensive technical discussion by Clauser and Shimony [34] and the popular articles by d'Espagnat [35], Heppenheimer [36], Shimony [37], and Ghirardi [38].

correlation is 100% that photon β propagates in the opposite direction with the same polarization as α. The state vector describing such a state can be written in the form

$$|\Psi\rangle = \frac{1}{\sqrt{2}}\big(|\boldsymbol{x}_1\boldsymbol{x}_2\rangle + |\boldsymbol{y}_1\boldsymbol{y}_2\rangle\big) , \tag{2.6}$$

where

$$|\boldsymbol{x}_1\boldsymbol{x}_2\rangle = \frac{1}{\sqrt{2}}\Big[|\boldsymbol{x}_\alpha\rangle|\boldsymbol{x}_\beta\rangle e^{i(\phi_{\alpha,1}+\phi_{\beta,2})} + |\boldsymbol{x}_\beta\rangle|\boldsymbol{x}_\alpha\rangle e^{i(\phi_{\beta,1}+\phi_{\alpha,2})}\Big] , \tag{2.7}$$

with corresponding definition for $|\boldsymbol{y}_1\boldsymbol{y}_2\rangle$. The subscript 1 or 2 in the two-photon ket respectively designates the left-going or right-going photon. Likewise, in the product of two single-photon kets in (2.7), the first ket represents the left-going photon and the second ket the right-going photon. The single-photon polarization basis states are represented by orthonormalized kets:

$$\langle\boldsymbol{x}|\boldsymbol{x}\rangle = \langle\boldsymbol{y}|\boldsymbol{y}\rangle = 1 , \qquad \langle\boldsymbol{x}|\boldsymbol{y}\rangle = \langle\boldsymbol{y}|\boldsymbol{x}\rangle = 0 . \tag{2.8}$$

For simplicity, I have assumed the space-time dependence of plane waves with phases of the form

$$\phi_{\mu,i} = \omega_\mu\left(t_i - \frac{z_i}{c}\right) , \qquad \mu = \alpha, \beta , \quad i = 1, 2 , \tag{2.9}$$

but the significant attributes of entanglement that follow do not require this assumption. The symmetric form of (2.7) takes account of the fact that photons are bosons; therefore, under particle exchange, the photon state vector does not change sign.

The state vector of (2.6) represents a pure state; the density matrix (or density operator) of the system

$$\rho = |\Psi\rangle\langle\Psi| \tag{2.10}$$

$$= \frac{1}{2}\big(|\boldsymbol{x}_1\boldsymbol{x}_2\rangle\langle\boldsymbol{x}_1\boldsymbol{x}_2| + |\boldsymbol{x}_1\boldsymbol{x}_2\rangle\langle\boldsymbol{y}_1\boldsymbol{y}_2| + |\boldsymbol{y}_1\boldsymbol{y}_2\rangle\langle\boldsymbol{x}_1\boldsymbol{x}_2| + |\boldsymbol{y}_1\boldsymbol{y}_2\rangle\langle\boldsymbol{y}_1\boldsymbol{y}_2|\big)$$

satisfies the condition of being an idempotent operator

$$\rho = \rho^2 , \tag{2.11}$$

as well as the condition for conservation of probability

$$\mathrm{Tr}(\rho) = 1 . \tag{2.12}$$

Tr symbolizes the operation of taking a trace, i.e., summing the diagonal elements, of a matrix. One must not misinterpret the state vector (2.6) to represent a system containing vertically and horizontally polarized pairs of photons – or, in other words, a statistical mixture of states of different, but well-defined, polarizations. A mixture of this kind can not be represented by a state vector, but by a density matrix of the form

$$\rho_{\text{mix}} = \frac{1}{2}\Big(|\boldsymbol{x}_1\boldsymbol{x}_2\rangle\langle\boldsymbol{x}_1\boldsymbol{x}_2| + |\boldsymbol{y}_1\boldsymbol{y}_2\rangle\langle\boldsymbol{y}_1\boldsymbol{y}_2|\Big)\,, \tag{2.13}$$

which is not idempotent and which leads to significantly different physical properties than the pure-state density matrix (2.10).

We can now ask the question: What is the probability that photon 1 (i.e., the photon propagating to the left) passes a polarizing filter with transmission axis at an angle θ_1 to the vertical (represented by unit polarization vector $\boldsymbol{e}_1$) and photon 2 (propagating to the right) passes a polarizing filter with transmission axis at angle θ_2 to the vertical (represented by unit polarization vector $\boldsymbol{e}_2$)? In order not to complicate the discussion unnecessarily, let us assume the polarizers and detectors are 100% efficient, and there are no reflective losses or other processes irrelevant to the issue of entanglement. The filtering action of the two polarizers is formally represented by projection operators $P_1 = |\boldsymbol{e}_1\rangle\langle\boldsymbol{e}_1|$ and $P_2 = |\boldsymbol{e}_2\rangle\langle\boldsymbol{e}_2|$ upon which the density matrix is projected and traced as follows:

$$P_{12}(\theta_1, \theta_2) = \text{Tr}(P_1\rho P_2) = \cos^2(\theta_1 - \theta_2)\cos^2\left[\frac{1}{2}\omega_{\alpha\beta}\left(\Delta t - \frac{\Delta z}{c}\right)\right]\,, \tag{2.14}$$

where $\omega_{\alpha\beta} = \omega_\alpha - \omega_\beta$, and $\Delta t = t_1 - t_2$ and $\Delta z = z_1 - z_2$ mark the difference in propagation time and detector separation. The angular dependence of (2.14) is ultimately a consequence of the scalar products: $\langle\boldsymbol{e}_i|\boldsymbol{x}\rangle = \cos\theta_i$ and $\langle\boldsymbol{e}_i|\boldsymbol{y}\rangle = \sin\theta_i$ for $i = 1, 2$. (The effect of spectral width on the correlation function (2.14) is examined in Appendix 2A.)

If, for the moment, we dispense with the space-time dependence of the signal by assuming that photons 1 and 2 have identical frequencies, then (2.14) tells us that the detection probability depends only on the *difference* in angular orientation of the two polarizers and not on their individual orientations. Thus, if photon 1 is $\boldsymbol{x}$-polarized, then the probability is 100% that photon 2 is also $\boldsymbol{x}$-polarized, in accordance with the polarization correlation initially 'built in' to the state vector; likewise for a left-detected photon with $\boldsymbol{y}$-polarization. As just described, the formal outcome (2.14) is a direct consequence of the symmetry (particle-exchange and rotational) of the state vector and may not seem too surprising, but the phenomenological properties of the entangled state raise some subtle issues when looked at more closely, particularly from the perspective of the individual observers at each detector.

It is to be borne in mind that the light source is periodically emitting pairs of counter-propagating photons, not classical waves which comprise large numbers of photons. A single photon either passes a polarizing filter or it does not pass; one never observes a fraction of a photon. The probability that observer 1 detects a photon, irrespective of what observer 2 may find, is obtained by summing relation (2.14) over all possible orientations of the polarizer for photon 2:

$$
P_1(\theta_1) = \begin{cases} \dfrac{1}{2\pi} \displaystyle\int_0^{2\pi} P_{12}(\theta_1, \theta_2)\,d\theta_2 = \dfrac{1}{2}\,, \\[1em] \text{or} \\[0.5em] \dfrac{1}{2}\Big[P_{12}(\theta_1, \theta_2 = 0) + P_{12}(\theta_1, \theta_2 = \pi/2)\Big] = \dfrac{1}{2}\,, \end{cases} \tag{2.15}
$$

with an identical result for the probability $P_2(\theta_2)$. Thus, from the perspective of each observer, the regular arrival of single photons generates a random binary sequence of transmissions (50%) and absorptions (50%) like the outcomes (H, T) of tossing a fair coin. Each observer may be inclined to believe he is receiving photons from an unpolarized source. However, if the two observers subsequently compare their detection sequences, they will find, for parallel transmission axes of the two polarizers, the sequences to be identical

Observer 1: H H T T T H ... ,
Observer 2: H H T T T H ... ,

or, completely opposite

Observer 1: H H T T T H ... ,
Observer 2: T T H H H T ... ,

for the two transmission axes perpendicular to one another.

Although the state vector in (2.6) was written in terms of vertical and horizontal polarization states, it could well have been written in terms of any two orthogonally polarized basis states. This, in fact, is what is meant by referring to the entangled state as rotationally symmetric. A rotational transformation of the basis states from $(\boldsymbol{x}, \boldsymbol{y})$ to $(\boldsymbol{e}_a, \boldsymbol{e}_b)$ with $\boldsymbol{e}_a \perp \boldsymbol{e}_b$ leads to the same probability correlation given by (2.14). This property markedly distinguishes the system of photons with entangled polarization states from a system comprising an equal mixture of vertically and horizontally polarized photons, as described by the density matrix of (2.13). In the latter case, the correlation function corresponding to (2.14) becomes

$$
P_{12}^{\text{mix}}(\theta_1, \theta_2) = \text{Tr}(P_1 \rho_{\text{mix}} P_2) = \frac{1}{2}\Big(\cos^2\theta_1 \cos^2\theta_2 + \sin^2\theta_1 \sin^2\theta_2 \Big)\,, \tag{2.16}
$$

and depends on the individual angular settings of the two polarizers. (2.16) is not rotationally invariant. Furthermore, the absence of a space-time dependent function in (2.16) is *not* a consequence of assuming for the sake of simplicity that the frequencies of photons 1 and 2 are identical; rather there occurs in the analysis leading to (2.16) a sum of the 'squares' (i.e., absolute magnitude squared), and not a 'square' of the sum, of plane-wave phase factors, with the result that the space-time dependence vanishes from the expression for probability. As an illustration of the different predictions made by (2.14) and (2.16), consider the polarizer settings and results shown in Table 2.1.

To some physicists, the preceding experimental configuration with entangled photons leads to philosophical difficulties analogous to those raised in the EPR paradox. The kernel of the problem goes something like this: Prior to determining whether a photon, let us say photon 1, passes a polarizer, we cannot think of the photons in a correlated pair as possessing the attribute of polarization, since the state vector in (2.6) can give rise to photons of any polarization; there is no orientation of polarizer 1 or polarizer 2 for which 100% of the incident single photons pass. The photons of the pair, however, are correlated. Thus, should an observation reveal photon 1 to be vertically polarized, then we know with 100% certainty that photon 2 will also be found to be vertically polarized when it is detected. Some physicists argue, therefore, that the state vector of the system has instantaneously 'collapsed' or 'reduced' to the form $|\Psi'\rangle = |x_1 x_2\rangle/\sqrt{2}$, in which case now the two photons *do* possess the attribute of vertical polarization. Moreover, the reduction to a form $|\Psi'\rangle = |e_1 e_2\rangle/\sqrt{2}$ occurs whenever the first-observed photon is found to have a polarization e. [Note: the orientation of the polarization vector is specified by e, *not* by e_1 (or e_2); the numerals 1 and 2 signify respectively the direction of propagation as left and right.] Since the two photons of a correlated pair can be arbitrarily far apart when the first one is observed, a measurement on one photon appears to have instantaneously affected the outcome of a measurement to be performed on the second, in violation of our sense of propriety that physical systems should be influenced only by local interactions. Hence the alleged paradox:

- either photon 2 (and therefore photon 1) had from the outset the polarization it revealed upon measurement, in which case quantum mechanics is an incomplete theory because it does not permit this polarization to be predicted prior to the detection of photon 1;
- or else the polarization of photon 2 was engendered by an instantaneous action at a distance, in conflict with our sense of causality, if not specifically with the special theory of special relativity.

Table 2.1. Joint detection probabilities of entangled and mixed states

		Joint probability P_{12}	
Polarizer 1 θ_1	Polarizer 2 θ_2	Entangled states	Mixed states
0	0	1.00	0.50
$\pi/2$	$\pi/2$	1.00	0.50
0	$\pi/2$	0.00	0.00
$\pi/2$	0	0.00	0.00
$\pi/4$	$\pi/4$	1.00	0.25

Frankly, I believe that the preceding perspective, embracing notions like the collapse of a state vector and instantaneous action at a distance, is *not* a satisfactory way to consider the correlations inherent in entangled states. To begin with, the expression 'collapse of the state vector' evokes a kind of mechanical transformation that is inappropriate to the function which a state vector serves in the quantum formalism, viz., as a theoretical construct for determining probabilities of allowable outcomes. As such, the projection of the state vector (as employed above) onto basis states corresponding to the measurement outcome whose probability is sought, is not a physical interaction like the collapse of a bridge or collapse of an ocean wave reaching the shore, but the exercise of a standard mathematical procedure linking the quantum formalism to experimental results. The state vector (2.6) gives a statistical description of a *system* of pairs of entangled photons, and, provided the properties of the light source do not change over the course of the experiment, the same state vector (2.6) [or density operator (2.10)] continues to describe the correlated pairs irrespective of measurements made by observer 1 or 2. By interposing a polarizer between incident photons and the detector, however, observer 1 or 2 has selected out of the original ensemble of correlated photons a subset of single photons that now manifests a well-defined polarization. Subsequent statistical predictions concerning this subset must be based on the appropriately modified ('collapsed') state vector or density operator. The so-called collapse reflects the loss of phase coherence (i.e., correlation) between the two detected photons.

Whether or not each photon of a correlated pair 'had' a polarization before detection – an issue that deeply concerns some who see a paradox here – is, in my opinion, not a physically significant, point. Theoretically, i.e., according to (2.6) or (2.10), only the pair, and not the indistinguishable photons that comprise it, have sharp properties: zero total linear momentum, zero total angular momentum. Quantum theory no more permits us to enquire how the polarization of the first- or second-observed photon came to be what it is, than it permits us to enquire through which of two slits a single electron passed on its way to creating a pattern of interference fringes at a distant screen. Experimentally, there is no way to probe the pair of correlated photons to determine the properties of its constituents without destroying the correlations. I see little point in a philosophical debate over nonobservable things.

I also believe there is no basis for claiming that the polarization correlation of widely separately photons is evidence of an instantaneous interaction at a distance. For one thing, because the sequence of polarizations recorded by each observer is random, the correlations provide no means by which the observers could send a superluminal signal to one another. As pointed out previously, the two observers would not even recognize that the outcomes of their polarization measurements are correlated until they compare them- and this cannot be done instantaneously. Thus, the quantum phenomenon of entanglement does not violate special relativity. The issue is more subtle, however, for according to special relativity it is not even possible to specify uniquely which polarization measurement – that of photon 1 or of photon 2 – instantaneously affected the other.

Suppose, for example, that observer 1 is farther from the light source than observer 2, as judged by observers at rest with respect to the laboratory (including the light source, detectors, polarizers, and observers 1 and 2). They will agree, therefore, that observer 2 makes the first observation. If there is an instantaneous interaction at a distance, then it is the observation of the polarization of photon 2 that 'collapses' the state vector and 'forces' photon 1 into the same polarization state. However, suppose this experiment is being viewed by an observer (observer 3) traveling at uniform velocity $-v$ towards the left in the direction of propagation of photon 1. To observer 3, the entire experimental set-up is traveling at a uniform velocity $+v$ to the right. By a simple application of the Lorentz transformation, it follows that, according to observer 3, the time interval $t_2 - t_1$ between detections of photon 2 and photon 1 is given by

$$t_2 - t_1 = \gamma \frac{L_2 + L_1}{c} \left(\frac{L_2 - L_1}{L_2 + L_1} + \frac{v}{c} \right) , \qquad (2.17)$$

in which L_1 and L_2 are, respectively, the locations of detectors 1 and 2 relative to the light source at the origin of the laboratory reference frame. Thus, according to (2.17), if $L_1 > L_2$, it is nevertheless possible for $t_2 > t_1$ if the speed of observer 3 satisfies the inequality $v/c > (L_1 - L_2)/(L_1 + L_2)$. Observations made in two different inertial reference frames therefore lead to opposite conclusions regarding which photon, 1 or 2, was detected first and, through the agency of 'state-vector collapse', instantaneously modified the polarization state of the other photon.

None of what I have written in the preceding paragraphs need minimize the fascinating consequences of entanglement, which is one of the features of quantum mechanics that sets it apart most sharply from classical mechanics. Strange and counter-intuitive as long-range correlations of entangled states are, the EPR paradox is *not* a real paradox. There is no internal inconsistency in the quantum formalism and no failure – at least as of the moment I write this sentence – in the capacity of quantum theory to account successfully for the outcomes of experiments on quantum systems. The 'paradox' is primarily one of unfulfilled expectations of philosophical preferences ('objective reality', 'locality') and deceptive physical images evoked by semantically poor labels ('state-vector collapse', 'instantaneous action at a distance'). In its present form – and most likely for any future incarnation – quantum theory does not describe single events, but only the statistical properties (count rates, correlations, cross-sections, etc.) of numerous events. The assertion, first enunciated by EPR, that quantum theory is incomplete may some day be experimentally vindicated – but that discovery, in my opinion, will have to await entirely new types of experiments or observations in hitherto unprobed regions of the scientific wilderness.

2.2 A Dance of Correlated Fluctuations.
The 'Hanbury Brown Twiss'

There is a second type of multiparticle correlation that finds its origin, not in the entanglement of wave functions resulting from special state preparation as in an EPR configuration, but as an indirect consequence of spin through the spin–statistics connection discussed earlier. Although neglect of the bosonic nature of photons in the preceding section would have reduced the magnitude of the correlation by a factor of two, it did not modify the essential feature of an angular dependence on the *difference* in settings of the polarizers, i.e., the experimental signature that distinguished the system of entangled photon pairs from a mixture of photon pairs. Now we will consider a type of multiparticle interference that depends entirely on the quantum statistics of the particles.

In the mid-1950s two astronomers, R. Hanbury Brown and R.Q. Twiss (to be designated HBT) developed a new type of interferometer [39] whose underlying explanation was eventually to have as profound an impact on quantum physics – in particular quantum optics – as did the EPR paradox and the AB effect. Known as an intensity interferometer, the apparatus operated by correlating (i.e., multiplying together and time-averaging) the output currents from two photodetectors illuminated by light from a thermal source such as a star (see Fig. 2.2). Since it is the light *intensity* to which each detector responds and to which the output current is proportional, the resulting oscillatory correlation as a function of detector separation was regarded by many as highly surprising. All physicists know that wave *amplitudes*, not intensities, interfere. HBT took pains to point out that the phenomenon did not actually involve the interference of light intensities and could, in fact, be understood without difficulty by a radio engineer within the framework of the classical wave theory of radiation. Nevertheless, physicists probing the quantum implications of the HBT effect were even more surprised, if not altogether incredulous, to learn that photons, emitted apparently randomly from a thermal source, were correlated in their arrivals at the two detectors. In the amusing description of Hanbury Brown, later reminiscing about this period [40]:

> Now to a surprising number of people, this idea seemed not only heretical but patently absurd and they told us so in person, by letter, in publications, and by actually doing experiments which claimed to show that we were wrong. At the most basic level they asked how, if photons are emitted at random in a thermal source, can they appear in pairs at two detectors. At a more sophisticated level the enraged physicist would brandish some sacred text, usually by Heitler, and point out that [...] our analysis was invalidated by the uncertainty relation [...].

The book by Heitler [41], *The Quantum Theory of Radiation*, was very well known to generations of physicists before more modern treatments of radiation theory, like Feynman's *Quantum Electrodynamics* [42], became available, and together with Dirac's *Principles of Quantum Mechanics* [43] constituted a quantum physicist's 'New Testament' and 'Old Testament'. Nevertheless, unfolding developments were to show that HBT were not only *not* wrong, but their work-although originally conceived for the purpose of measuring stellar angular diameters – actually laid the experimental foundation for what has become contemporary quantum optics and provided new methods for fundamental tests of quantum mechanics.

A heuristic understanding of the intensity interferometer, schematically diagramed in Fig. 2.2, can be gained by examining the superposition at each detector of the classical electromagnetic waves emitted from two different locations in an extended optical source. Of the broad range of frequencies radiated by a thermal source, consider just two Fourier components of the same linear polarization, $E_1 \sin(\omega_1 t + \phi_1)$ from point P_1 and $E_2 \sin(\omega_2 t + \phi_2)$ from point P_2, that reach detector D_1. The phases ϕ_1 and ϕ_2 in the argument vary randomly from one emitted wave front to another. The instantaneous intensity at D_1, to which the output current i_1 is proportional (with detector proportionality constant K_1) then takes the form

$$i_1 = K_1 \Big[E_1 \sin(\omega_1 t + \phi_1) + E_2 \sin(\omega_2 t + \phi_2) \Big]^2 . \tag{2.18}$$

A similar expression

$$i_2 = K_2 \Big\{ E_1 \sin \big[\omega_1(t + d_1/c) + \phi_1 \big] + E_2 \sin \big[\omega_2(t + d_2/c) + \phi_2 \big] \Big\}^2 \tag{2.19}$$

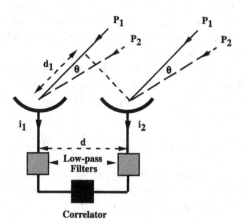

Fig. 2.2. Hanbury Brown–Twiss (HBT) intensity interferometer. Wave fronts issuing from two points (P_1,P_2) of an extended source are received at two photodetectors whose output currents are first filtered to let pass low-frequency harmonics and then correlated. d_1 is the greater distance traveled by the wave front from P_1 to detector 1 than to detector 2. (The corresponding interval d_2 for the wave front from P_2 is not shown)

gives the output photocurrent of detector D_2 where d_i $(i = 1, 2)$ is the difference in optical path lengths between source point i and the two detectors. By computing the squares in (2.18) and (2.19) and using trigonometric identities, one finds that the resulting photocurrents contain Fourier components with frequencies ω_1, ω_2, $2\omega_1$, $2\omega_2$, $\omega_1 + \omega_2$, and $\omega_1 - \omega_2$. A low-frequency filter (e.g., with pass-band of approximately 1–100 MHz) that allows only the Fourier component at the difference frequency $\omega_1 - \omega_2$ to pass to the correlator gives rise to the following photocurrents from D_1 and D_2

$$i_1 = K_1 E_1 E_2 \cos \left[(\omega_1 - \omega_2)t + (\phi_1 - \phi_2) \right] , \qquad (2.20)$$

$$i_2 = K_2 E_1 E_2 \cos \left[(\omega_1 - \omega_2)t + (\phi_1 - \phi_2) + \frac{\omega_1 d_1 - \omega_2 d_2}{c} \right] . \qquad (2.21)$$

Although the initial phases are random functions of time, the *difference* in these phases appears in the expressions for *both* photocurrents. These two photocurrents (and therefore the incident light intensities) are clearly correlated since at any instant they have the same frequency and differ in phase only by a constant term for a given interferometer configuration. Upon multiplying relations (2.20) and (2.21) and time-averaging over one or more periods, one obtains the correlation function

$$\langle i_1(t) i_2(t) \rangle \approx K_1 K_2 E_1^2 E_2^2 \cos \left[\frac{\omega}{c} (d_1 - d_2) \right] , \qquad (2.22)$$

where, within the narrow pass-band, it is adequate to set $\omega = \omega_1 \approx \omega_2$. A complete analysis of the intensity interferometer, which we do not need here, would require that (2.22) be integrated over all pairs of points on the source and that all appropriate Fourier components be included. For our present purposes, however, the above expression is sufficient to show that the origin of the intensity 'interference' can be readily understood on the basis of amplitude interference in classical wave theory.

It should be noted that technically HBT did not measure directly a correlation in the output photocurrents, but rather a correlation in the *fluctuations* of these currents. Representing each current as the sum of a stationary average term and a fluctuating term

$$i_k(t) = \bar{i}_k + \Delta i_k(t) \qquad (k = 1, 2) , \qquad (2.23)$$

one can express the correlation in the fluctuations of the two beams by means of the so-called second-order correlation function (i.e., second-order in the intensities or fourth order in the amplitudes)

$$g_{12}^{(2)} = \frac{\langle i_1(t_1) i_2(t_2) \rangle}{\bar{i}_1 \bar{i}_2} = 1 + \frac{\langle \Delta i_1(t_1) \Delta i_2(t_2) \rangle}{\bar{i}_1 \bar{i}_2} . \qquad (2.24)$$

For the generally nonperiodic signals $i_1(t)$ and $i_2(t)$ the angular bracket signifies an average over a time interval long compared with the coherence time t_c of the light beam, where t_c is the reciprocal of the bandwidth in analogy to (1.6). In essence, HBT demonstrated that the fluctuation term, the second term of the second equality in (2.24), is greater than zero.

If the incident light beams are sufficiently weak, then, rather than multiplying the fluctuations in the output currents of the two detectors, one could in principle count the detected photons individually and establish the number of coincident arrivals as a function of detector separation. The nature of the controversial *quantum* effects revealed by this version of intensity interferometry can be made clearer by considering first an experimental configuration in which photons incident upon a *single* detector are counted repeatedly within a prescribed time interval T. (Afterward, we will re-examine intensity interferometry as a split-beam experiment where photons are incident upon two correlated detectors.) The mean number of photoelectrons in the detector output is proportional to the mean light intensity and T. Now if the incident photons, all of which are presumed to have the same polarization, could be thought of as randomly arriving particles, then the variance in count rate would be predicted to be

$$\overline{(\Delta n)^2} \equiv \overline{\left(n - \overline{n}\right)^2} = \overline{n^2} - \overline{n}^2 = \overline{n} \,, \tag{2.25}$$

where the final equality follows from Poisson statistics. A correct theoretical analysis [44] leads, however, to a variance larger by an amount proportional to the mean number of photons counted

$$\overline{(\Delta n)^2} = \overline{n} \left(1 + \overline{n}\frac{t_c}{T}\right) \,, \tag{2.26}$$

under the circumstances (assumed here) that the spectral density of the light is uniform over the optical bandwidth $\Delta\nu = 1/t_c$. If the light source is unpolarized, then the supplemental term is to be multiplied by one half, since fluctuations in orthogonal polarizations are independent. What is the origin of this additional spread in count rate?

Purcell, has given a simple, visualizable explanation of relation (2.26) in the above-cited reference which cannot be expressed any clearer than in his own words:

> If one insists on representing photons by wave packets and demands an explanation in those terms of the extra fluctuation, such an explanation can be given. But I shall have to use language which ought, as a rule, to be used warily. Think, then, of a stream of wave packets, each about $c/\Delta\nu$ long, in a random sequence. There is a certain probability that two such trains accidentally overlap. When this occurs they interfere and one may find (to speak rather loosely) four photons, or none, or something in between as a result. It is proper to speak of interference in this situation because the conditions of the experiment are just such as will ensure that these photons are in the same quantum state. To such interference one may ascribe the 'abnormal' density fluctuations in any assemblage of bosons.

We will see later the wisdom of Purcell's words and the difficulties to which a too cavalier use of the imagery of wave packets can lead. In any event, the broader variance reflects a quantum interference effect deriving from the statistical properties of light as a system of Bose–Einstein particles.

Figure 2.3 shows an intensity interferometer configuration where the counts – or more precisely, the fluctuation in counts – at two detectors are correlated. Let n_1 photons be received at detector D_1 and n_2 photons at detector D_2 within a time interval T. For simplicity all photons will be assumed to have identical polarization. The variance in counts at each detector takes the form of relation (2.26)

$$\overline{(\Delta n_k)^2} = \overline{n}_k \left(1 + \overline{n}_k \frac{t_c}{T}\right) \qquad (k = 1, 2) . \tag{2.27}$$

By correlating the two photodetector outputs, i.e., by linking the two outputs together, one has effectively a single-detector configuration again, but with total count rate $n = n_1 + n_2$ and variance

$$\overline{(\Delta n)^2} = \overline{(\Delta n_1 + \Delta n_2)^2} . \tag{2.28}$$

Expansion of the right hand side of expression (2.28) and comparison with relation (2.26) leads to the positive cross-correlation

$$\overline{\Delta n_1 \Delta n_2} = \overline{n}_1 \overline{n}_2 \frac{t_c}{T} , \tag{2.29}$$

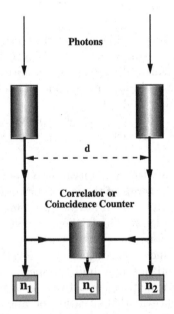

Fig. 2.3. Intensity interferometer based on photon counting. The number of photons received at each detector is individually determined (n_1, n_2) as well as the number (n_c) that arrive coincidentally within a specified time window

observed by HBT. [For an unpolarized light source multiply the right-hand side of (2.29) by 1/2.]

From the standpoint of quantum physics, the nonvanishing correlation between the two components of a split light beam is a consequence of, in Purcell's words, the 'clumping' of the photons. The contemporary term, 'photon bunching', refers to the tendency of a sequence of photon arrival times (registered at a single photodetector) to be more narrowly spaced than that predicted on the basis of Poisson statistics for randomly occurring events (e.g., the arrival of raindrops), as illustrated in Fig. 2.4. The same effect shows up in the number of joint photon arrivals at two detectors, such as in the configuration of Fig. 2.3 with the correlator replaced by a coincidence counter. In the case of polarized light coherent over both detecting surfaces, the number of coincident detections n_c within a time window T much longer than the longitudinal coherence time t_c takes the form

$$n_c = n_c^0 \left(1 + \frac{t_c}{T} \right) , \tag{2.30}$$

where the coincidence count for two uncorrelated beams of light is proportional to the number of counts (n_1, n_2) received at each detector. The supplementary positive term in relation (2.30) represents photon bunching, and is seen to have the same form (to within a proportionality factor) as the cross-correlation (2.29). It is worth stressing, however, that – as demonstrated by Hanbury Brown's classical wave argument summarized above – the explanation of this phenomenon does not require the introduction of photons.

Indeed, Einstein had examined the same problem, albeit in a different form, in a 1909 summary of the "present state of the problem of radiation" [45], and

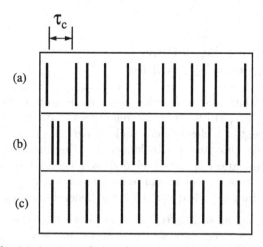

Fig. 2.4. Distribution in time of particle counts illustrating (a) antibunched, (b) bunched, and (c) random arrivals. The type of clustering is gauged against the coherence time τ_c of the source

it is instructive to look at his results. Concerned with the properties of black-body radiation, Einstein showed that the variance in dE_ν – the energy of thermal radiation in the spectral range between ν and $d\nu$ and volume V – can be written as the sum of two terms as follows

$$\overline{(\Delta dE_\nu)^2} = h\nu\overline{dE_\nu} + \frac{c^3}{8\pi\nu^2 V d\nu}\ \overline{(dE_\nu)^2}\ . \tag{2.31}$$

The first term corresponds to shot noise, the fluctuations in energy resulting from fluctuations in the number of particles, each of energy $h\nu$. This was the exciting part at the time of Einstein's report, for it reflected the grainy or particulate nature of the radiation field and was inexplicable on the basis of classical electromagnetic theory. The first term supported the notion of light quanta and indicated that these quanta (the photons) were subject to the same statistical laws as were the molecules of an ideal gas. On the other hand, the second term – referred to as wave noise by HBT – is a purely classical term representing energy fluctuations deriving from fluctuations in the amplitudes of interfering waves. From the standpoint of optical intensity interferometry, the quantum shot noise at one detector is uncorrelated with that at the other, and so their product averages away leaving only the correlations due to classical wave noise.

From the foregoing remarks, which pertain largely to the chaotic sources[2] employed by HBT – namely, thermal light (e.g., starlight) or light from arc lamps – one should not infer that quantum effects play no intrinsic role in the correlations of all light fields. Quite the contrary. States of light, such as so-called squeezed light, have been predicted and observed that give rise to correlation effects totally inexplicable within the framework of classical wave theory [47]. The study of these effects falls within the province of quantum optics.

2.3 Measurable Distinctions Between Quantum Ensembles

Statistical uncertainty and probability have a role to play in both classical and quantum physics, although at different levels of fundamentality. An elementary constituent of a classical ensemble of particles can be endowed with dynamical variables that may not be known, but that are, in principle, knowable; their determination is not precluded by physical law. An elementary constituent (e.g., an atom) of a quantum ensemble cannot ordinarily be endowed with definite dynamical variables; the general quantum description –

---

[2] For a comprehensive description the nature of chaotic light, which includes black-body radiation as a special case, see [46]. A significant feature is that the density or statistical operator of a chaotic radiation field is diagonal in a basis of photon number states. For example, in the case of a single optical mode, it would have the form $\hat{\rho} = \sum_{(n)} \rho_{n,n}|n\rangle\langle n|$.

even of a single particle – is not in terms of the observable attributes themselves (e.g., energy or momentum), but in terms of probability amplitudes
for the potential realization of these attributes. For a multiparticle quantum
system, there may also be a statistical distribution of probability amplitudes
over the elementary constituents, whereupon the use of statistics as in classical
physics becomes necessary.

Suppose one has in the laboratory two containers, one with an ensemble
of atoms (or some other collection of quantum 'particles') each of which is
endowed with definite, although statistically distributed, energy values, and
the other in which each atom is in a linear superposition of energy eigenstates
encompassing the same energy values as the first ensemble. Are the two ensembles experimentally distinguishable? Always, never, or only under certain
conditions?

To make the example concrete, consider an ensemble of atoms with effectively just two excited states with energy eigenvalues E_1 and E_2. Each atom,
initially in its ground state, is subjected to an impulsive excitation at time t_0
that prepares it in a linear superposition described by the state vector

$$|\Psi(t_0)\rangle = a_1|1\rangle + a_2|2\rangle .\tag{2.32}$$

The atoms are said to be coherently excited. The system constitutes a pure
state whose density matrix or operator $\rho_0 = |\Psi(t_0)\rangle\langle\Psi(t_0)|$ in an energy rep-
resentation has elements

$$\rho_{ii}^0 = |a_i|^2 \quad (i = 1, 2) ,\tag{2.33}$$

$$\rho_{12}^0 = |a_1 a_2|e^{-i\phi} , \qquad \phi = \arg(a_2) - \arg(a_1) .\tag{2.34}$$

A measurement of the energy of randomly chosen atoms at t_0 would yield
either E_1 or E_2 with probability $|a_1|^2$ or $|a_2|^2$, respectively. This outcome is
exactly the same as if the system consisted of a mixture of atoms distributed
with probabilities $|a_1|^2$ and $|a_2|^2$ over definite states, as characterized by a den-
sity matrix with elements

$$\rho_{ij}^0 = |a_i|^2 \delta_{ij} \quad (i, j = 1, 2) ,\tag{2.35}$$

where the Kronecker delta function δ_{ij} equals unity for $i = j$, and zero otherwise. Nevertheless, the two ensembles are quite different; quantum correlations
that distinguish a superposition state from a mixture of definite states can
be manifested by quantum interference effects in atomic fluorescence. The
subject of quantum interference in the time domain will be taken up more
thoroughly in Chap. 4. However, it is instructive to examine in the present
section some aspects of this matter closely tied to HBT-type correlations [48].

The excitation of individual atoms into a superposition state like that of
(2.32) does not necessarily mean that the ensemble as a whole will manifest

quantum interference effects if subpopulations of atoms are in linear superpositions of eigenstates with different relative phases. It is of both practical
and conceptual significance to ascertain to what extent the incursion of phase
disorder extinguishes the coherence observable in an ensemble of what one
might call mixed superposition states.

As an experimental problem, the distinction between theoretically different quantum ensembles may be difficult. The experimenter is not presented
in the laboratory with the elements of a density matrix, but, at best, with the
ensemble of atoms and the source of excitation. For example, the theoretical
possibility of quantum interference in light emission from coherently excited
atoms was pointed out by Breit in the 1930s [49]. When observed in atomic
beam experiments in the 1960s, the modulated fluorescence ('quantum beats'),
was incorrectly attributed to electric-field-induced Stark mixing of degenerate
states, rather than being correctly attributed to field-free decay of nondegenerate states [50]. In an astrophysical setting, the experimenter has control over
neither the ensemble of atoms nor the source of excitation. Not long after the
correct interpretation of quantum beats in spontaneous emission, the question
was raised of whether atomic excitation by an optical pulsar could engender
observable quantum interference among Zeeman states of atoms in the Crab
Nebula [51]. If such were the case, astronomers could determine the magnitude of the magnetic field in an interstellar medium by a spectroscopic method
largely free of Doppler broadening. The atoms of the nebula, however, do not
comprise a pure superposition state; constraints on the direct observation of
a modulated fluorescent intensity are severe. An alternative way to detect
quantum coherence of an ensemble of mixed superposition states would be
useful.

In a certain sense, the answer to the question raised above (are the two
ensembles experimentally distinguishable?) is trivially simple. A phase average of the density matrix (2.33), (2.34) eliminates the off-diagonal elements
and leads to density matrix (2.35). So the two ensembles are experimentally
indistinguishable with respect to any process that averages over the quantum phases. The core of the issue here, however, is what it means to 'average
over phases'. For example, it has been demonstrated that a fluctuating phase
still leads to manifestations of quantum interference in the power spectrum
of intensity fluctuations engendered by *absorption* of a stationary light beam
transmitted through an atomic gas [52]. At any instant in time, the absorption
of radiation produced a well-defined phase for the off-diagonal density matrix
elements of the ensemble. Similarly, it has been shown that the correlation
of intensity fluctuations in *spontaneous emission*, received at two detectors,
also manifests quantum interference of atomic states if, again, the quantum
phases, even though they fluctuate, are common to all constituents of the sample at each instant in time [53]. Thus, there may be information in intensity
interferometry that is not accessible in the measurement of intensity alone.

Before considering this point further, it is useful to examine the relationship between the density matrix and the degree of disorder (or entropy) of

a system. For an ensemble of two-state atoms in a pure state such as (2.32), the elements of the density matrix (2.33), (2.34) can be parameterized by a polar angle θ and a phase angle ϕ in the following way

$$\rho_{11} = \cos^2\theta , \qquad \rho_{22} = \sin^2\theta , \qquad (2.36)$$

$$\rho_{12} = \rho_{21}^* = \frac{1}{2}\sin 2\theta e^{-i\phi} . \qquad (2.37)$$

The conservation of probability, $\mathrm{Tr}(\rho) = 1$, for a closed system is reflected in (2.36). The determinant of ρ is 0, which, together with (2.36), reveals immediately that the two eigenvalues of ρ are 1 and 0. The diagonal elements ('populations') can range from 0 to 1, and the magnitude of the off-diagonal elements ('coherences') can range from 0 to 1/2. For a mixture of states with no quantum correlations in a designated representation, the off-diagonal elements vanish.

In general, i.e., for mixed as well as pure states, the density matrix can be expressed in terms of the elements of the mean population P_0 and a polarization vector $\boldsymbol{P}$

$$\rho = P_0(1 + \boldsymbol{P}\cdot\boldsymbol{\sigma}) , \qquad (2.38)$$

where $\boldsymbol{\sigma}$ is a vector whose components are the Pauli spin matrices.[3] P_0, which would be 1/2 for a closed system, and the components of $\boldsymbol{P}$, given by

$$P_0 = \frac{1}{2}(\rho_{11} + \rho_{22}) , \qquad (2.39)$$

$$P_0P_1 = \frac{1}{2}(\rho_{12} + \rho_{21}) , \qquad (2.40)$$

$$P_0P_2 = \frac{1}{2i}(\rho_{21} - \rho_{12}) , \qquad (2.41)$$

$$P_0P_3 = \frac{1}{2}(\rho_{11} - \rho_{22}) , \qquad (2.42)$$

are analogous to the Stokes parameters familiar in optics [7].

The disorder or information content of a system can be specified by the statistical entropy function S

$$S = -\mathrm{Tr}(\rho\ln\rho) = -\sum_{i=1,2}\rho_i\ln\rho_i , \qquad (2.43)$$

where ρ_i ($i = 1, 2$) are the eigenvalues of ρ given by

$$\rho_{1,2} = \frac{1}{2}\left[(\rho_{11} + \rho_{22}) \pm \sqrt{(\rho_{11} - \rho_{22})^2 + 4|\rho_{12}|^2}\right]$$

$$= \frac{1}{2}\left\{\mathrm{Tr}(\rho) \pm \sqrt{\left[\mathrm{Tr}(\rho)\right]^2 - \det(\rho)}\right\} , \qquad (2.44)$$

---

[3] The Pauli spin matrices are defined by invariant algebraic relations which have given rise to a number of standard representations. The explicit form of the spin matrices used in this book is given in Chap. 4.

or equivalently, in terms of the polarization, by

$$\rho_{1,2} = P_0\left(1 \pm |\boldsymbol{P}|\right), \tag{2.45}$$

with

$$|\boldsymbol{P}| = \frac{\sqrt{(\rho_{11} - \rho_{22})^2 + 4|\rho_{12}|^2}}{\rho_{11} + \rho_{22}} = \frac{\sqrt{[\mathrm{Tr}(\rho)]^2 - \det(\rho)}}{\mathrm{Tr}(\rho)}. \tag{2.46}$$

The second relation in (2.44) and (2.46) respectively expresses the eigenvalues and magnitude of the polarization in terms of the trace and determinant of the density matrix, which are properties of a matrix invariant under changes of representation.

From (2.43) and (2.44)–(2.46), one can relate the disorder and polarization by

$$S = -2P_0 \ln P_0 - P_0\left[\left(1 + |\boldsymbol{P}|\right) \ln\left(1 + |\boldsymbol{P}|\right) + \left(1 - |\boldsymbol{P}|\right) \ln\left(1 - |\boldsymbol{P}|\right)\right]. \tag{2.47}$$

Thus, a pure state (eigenvalues $\rho_{1,2} = 1, 0$) corresponds to a completely polarized ensemble ($P_0 = 1/2$, $|\boldsymbol{P}| = 1$) with $S = 0$. For a weakly polarized ensemble ($|\boldsymbol{P}| \ll 1$), (2.47) reduces approximately to $S \sim -2P_0 \ln P_0 - P_0|\boldsymbol{P}|^2$, which yields $S = \ln 2$ for a completely unpolarized ensemble.

For the system under consideration, each atom, prepared in a linear super-position of excited states, is presumed to decay radiatively to a lower state. Strictly speaking, therefore, we do not have a two-state system. The decay can be included phenomenologically in the equation of motion of the excited state density matrix by means of a suitable relaxation term. The subsystem of excited atomic states is then effectively an open one, and after the initial excitation, the condition $\mathrm{Tr}(\rho) = 2P_0 \neq 1$ reflects the fact that probability (within the manifold of excited states) is no longer conserved.

The time-evolution of the excited atoms is governed by a Hamiltonian $H = H_0 + \mathrm{i}\hbar\Gamma/2$ comprising a Hermitian operator H_0 with eigenvalues $E_i = \hbar\omega_i$ ($i = 1, 2$) and an anti-Hermitian decay operator $\mathrm{i}\hbar\Gamma/2$ with eigenvalue $\mathrm{i}\hbar\gamma/2$ taken to be the same for both excited states.[4] The density matrix equation of motion is then

$$\frac{\mathrm{d}\rho}{\mathrm{d}t} = -\frac{\mathrm{i}}{\hbar}[H, \rho] - \frac{1}{2}\{\Gamma, \rho\}, \tag{2.48}$$

in which the square brackets and curly brackets designate, respectively, the operations of commutation and anticommutation. (We discuss the origin of

---

[4] The definition of the decay operator includes a factor $1/2$ because the eigenvalue γ represents the decay rate (inverse lifetime) of a state population. Thus, if the occupation number of a state decreases exponentially in time as $e^{-\gamma t}$, then the corresponding quantum amplitude must decrease as $e^{-\gamma t/2}$.

this relation in Chap. 5.) The phenomenological relaxation term leads to the exponential decay of the excited state populations, as expressed by

$$\frac{d\mathrm{Tr}(\rho)}{dt} = -\mathrm{Tr}(\rho\Gamma) = -\gamma\mathrm{Tr}(\rho) . \tag{2.49}$$

It then follows from (2.43) and (2.48), (2.49) that the excited-state disorder evolves in time according to

$$\frac{dS}{dt} = -\mathrm{Tr}(\rho\Gamma) + \mathrm{Tr}(\Gamma\rho\ln\rho) , \tag{2.50}$$

which, under the present circumstances where Γ is proportional to the unit two-dimensional matrix, leads to the simpler expression

$$\frac{dS}{dt} + \gamma S = \gamma\mathrm{Tr}(\rho_0)e^{-\gamma t} . \tag{2.51}$$

The time-evolution of S is therefore independent of H. Integration of (2.51) with imposition of the initial condition $\mathrm{Tr}(\rho_0) = 1$ at $t = 0$, gives the result

$$S(t) = (S_0 + \gamma t)e^{-\gamma t} , \tag{2.52}$$

where S_0 is the initial measure of disorder. In the absence of decay, the disorder remains constant, $S(t) = S_0$.

2.4 Correlated Emission from Coherently Excited Atoms

Experimentally, the impulsive preparation of an ensemble of two-state atoms, in which the phase ϕ is common to all the atoms at the interaction time t_0, can be achieved in a variety of ways, such as electron bombardment [54], fast ion-beam passage through thin foils [55, 56], optical excitation by means of shuttered spectral lamps [57], and pulsed laser excitation [58]. Integration of the equation of motion (2.48) leads to the density matrix elements

$$\rho_{ij}(t, t_0) = \rho_{ij}^0 e^{-\omega_{ij}(t-t_0)}e^{-\gamma(t-t_0)} , \tag{2.53}$$

where $\omega_{ij} = \omega_i - \omega_j$. The initial polarization is $|P| = 1$; the initial disorder is $S_0 = 0$; and the evolution of these quantities takes the form

$$|P(t, t_0)| = e^{-\gamma(t-t_0)} , \tag{2.54}$$

$$S(t, t_0) = \gamma(t - t_0)e^{-\gamma(t-t_0)} . \tag{2.55}$$

The disorder within the system begins to increase as the two excited states are depopulated; maximum disorder is reached at a time γ^{-1} after excitation.

There is, again, total order at an infinite time after excitation when all atoms have returned to the (nondegenerate) ground state.

The spontaneous emission from the ensemble of atoms is represented theoretically by the expectation value of a detection operator X

$$I(t, t_0) = \text{Tr}(\rho X) , \tag{2.56}$$

where

$$X = e \cdot D|0\rangle\langle 0|e \cdot D \tag{2.57}$$

is defined in terms of the light polarization vector e, the atomic electric dipole operator D, and the operator $|0\rangle\langle 0|$ projecting onto the lower state (ground state) to which the excited states decay. It is assumed for simplicity that the matrix elements of $e \cdot D$ are real. The signal (emission intensity) derived from (2.56), (2.57) is

$$I(t, t_0) = \left\{ A + B \cos \left[\omega_{21}(t - t_0) - \phi \right] \right\} e^{-\gamma(t-t_0)} , \tag{2.58}$$

where

$$A = D_{10}^2 \rho_{11}^0 + D_{20}^2 \rho_{22}^0 , \tag{2.59}$$

$$B = 2D_{10}D_{20} \left| \rho_{12}^0 \right| . \tag{2.60}$$

The exponentially damped fluorescence from the ensemble of atoms is modulated at the Bohr frequency ω_{21}, with a modulation depth B/A. If the transition matrix elements from states $|1\rangle$ and $|2\rangle$ to $|0\rangle$ are equal, and the excitation parameter $\theta = \pi/4$, the contrast B/A will be 100%. The total unpolarized fluorescent emission in a given direction is not modulated; oscillations in intensity are observable only on a polarized component. This can be demonstrated by the Wigner–Eckart theorem. There is no Doppler spread in the beat frequency (to first order in v/c) because the interference effect is produced by coherence in individual atoms, not between different atoms.

For an incoherently prepared state, $B = 0$. The fluorescent emission then follows the simple exponential damping

$$I(t, t_0) = Ae^{-\gamma(t-t_0)} , \tag{2.61}$$

which is characteristic of incoherent decay in a wide variety of processes (e.g., the disintegration of radioactive nuclei).

If the system of atoms is subjected to impulsive excitations occurring over an indefinitely long period of time at a rate given by

$$r(t) = W(1 + m \cos \Omega t) , \tag{2.62}$$

the density matrix $\rho(t)$ is obtained by averaging $\rho(t, t_0)$ over the excitation time

$$\rho(t) = \int_{-\infty}^{t} \rho(t, t_0) r(t_0) dt_0 , \tag{2.63}$$

to yield the elements

$$\rho_{ij}(t) = \frac{W}{\gamma} \left[\frac{\gamma}{\gamma + i\omega_{ij}} + \frac{m}{2} \frac{\gamma e^{i\Omega t}}{\gamma + i(\omega_{ij} + \Omega)} + \frac{m}{2} \frac{\gamma e^{-i\Omega t}}{\gamma + i(\omega_{ij} - \Omega)} \right] \rho_{ij}^0 . \quad (2.64)$$

The intensity of the emitted light following from (2.56) is then

$$I(t) = \frac{W}{\gamma} \left\{ A \left[1 + \frac{m\gamma(\gamma \cos \Omega t + \Omega \sin \Omega t)}{\gamma^2 + \Omega^2} \right] + B \frac{\gamma(\gamma \cos \phi + \omega_{21} \sin \phi)}{\gamma^2 + \omega_{21}^2} \right.$$

$$+ \frac{B}{2} \frac{m\gamma[\gamma \cos(\Omega t + \phi) + (\omega_{21} + \Omega) \sin(\Omega t + \phi)]}{\gamma^2 + (\omega_{21} + \Omega)^2}$$

$$\left. + \frac{B}{2} \frac{m\gamma[\gamma \cos(\Omega t - \phi) - (\omega_{21} - \Omega) \sin(\Omega t - \phi)]}{\gamma^2 + (\omega_{21} - \Omega)^2} \right\} . \quad (2.65)$$

The rate parameter W is assumed small compared to the atomic decay rate γ, and one may then neglect the occurrence of cycles of stimulated emission and absorption over the period of the excitation pulse (a phenomenon that will be taken up in Chap. 4).

In the case of incoherent excitation leading to an ensemble of mixed definite states, the populations oscillate in time simply as a trivial consequence of turning the excitation on and off at frequency Ω. For $\Omega \gg \gamma$, the populations respond sluggishly to the time-variation of the excitation; the modulated components of ρ_{11} and ρ_{22} fall off as Ω^{-1}. In effect, the system is described by a density matrix with elements

$$\rho_{ij} = \frac{W}{\gamma} \rho_{ii}^0 \delta_{ij} \quad (\Omega \gg \gamma) , \quad (2.66)$$

which lead to a polarization

$$|\boldsymbol{P}| = \left| \rho_{11}^0 - \rho_{22}^0 \right| \quad (2.67)$$

and disorder

$$S = -\frac{W}{\gamma} \ln \frac{W}{\gamma} - \frac{W}{\gamma} \left(\rho_{11}^0 \ln \rho_{11}^0 + \rho_{22}^0 \ln \rho_{22}^0 \right) . \quad (2.68)$$

The first term of (2.68),

$$S_{\mathrm{min}} \equiv -\frac{W}{\gamma} \ln \frac{W}{\gamma} ,$$

is the minimum disorder engendered by the excitation and decay processes. The fluorescent intensity

$$I(t) = \frac{WA}{\gamma} \quad (2.69)$$

is constant.

For excitation into nondegenerate states ($\omega_{21} > \gamma$) at a resonant frequency ($\Omega = \omega_{21}$), the off-diagonal density matrix element is

$$\rho_{12}(t) = \frac{Wm}{2\gamma}\rho_{12}^0 e^{i\Omega t} \qquad (\Omega = \omega_{21}) , \tag{2.70}$$

and the diagonal elements are as given in (2.66). The system now displays a greater polarization

$$|\boldsymbol{P}| = \sqrt{\left(\rho_{11}^0 - \rho_{22}^0\right)^2 + m^2\left|\rho_{12}^0\right|^2} , \tag{2.71}$$

which results in the maximum value $|\boldsymbol{P}| = 1$ and minimum disorder $S_{\min}$ for modulation amplitude $m = 2$. The resonant intensity,

$$I(t) = \frac{W}{\gamma}\left[A + \frac{1}{2}mB\cos(\Omega t - \phi)\right] \qquad (\Omega = \omega_{21}) , \tag{2.72}$$

characterizes a steady state, not transient, optical emission. Under the condition of minimum disorder ($m = 2$), the contrast is B/A, which is the same as in the case of quantum beats from a pure state.

Far from resonance, $\Omega \gg (\omega_{21}$ or $\gamma)$, one can ignore the modulation to good approximation and consider the system of atoms prepared by a random excitation occurring at a constant rate W. The off-diagonal density matrix element then becomes independent of time

$$\rho_{12} = \frac{W}{\gamma + i\omega_{21}}\rho_{12}^0 . \tag{2.73}$$

The polarization and fluorescent intensity depend on the relative size of the two (angular) frequencies ω_{21} and γ

$$|\boldsymbol{P}| = \sqrt{\left(\rho_{11}^0 - \rho_{22}^0\right)^2 + 4\frac{\gamma^2}{\gamma^2 + \omega_{21}^2}\left|\rho_{12}^0\right|^2} , \tag{2.74}$$

$$I(t) = \frac{W}{\gamma}\left[A + B\frac{\gamma(\gamma\cos\phi + \omega_{21}\sin\phi)}{\gamma^2 + \omega_{21}^2}\right] . \tag{2.75}$$

For degenerate states ($\omega_{21} = 0$), $|\boldsymbol{P}|$ reduces to the maximum value of 1; the disorder of the system is again $S_{\min}$, the same minimum value as in the case of resonant excitation of nondegenerate states. In both cases, the quantum coherence in the excitation of single atoms is maximally preserved in the entire ensemble. For resonant excitation of nondegenerate states, however, the source of excitation forces the entire system into phased emission. In the case of degenerate excited states, the quantum oscillations occur at zero frequency for each atom, and the emissions from different atoms consequently do not become dephased in time. The fluorescent intensity is

$$I(t) = \frac{W}{\gamma}(A + B\cos\phi) \qquad (\omega_{21} = 0) , \qquad (2.76)$$

and depends on the atomic coherence.

If the states are nondegenerate ($\omega_{21} > \gamma$), the polarization is not very sensitive to $|\rho_{12}^0|^2$ unless the state populations are nearly equal. The dependence of the fluorescent intensity

$$I(t) = \frac{W}{\gamma}\left(A + B\frac{\gamma}{\omega_{21}}\sin\phi\right) \qquad (2.77)$$

on the atomic coherence is scaled by the ratio of the decay rate to Bohr frequency.

Let us consider now the situation in which the source of excitation may produce superpositions of states with randomly distributed relative phase ϕ. As stated earlier, the resulting phase-averaged density matrix is indistinguishable from the density matrix of a mixture of definite energy states. The immediate consequence of this is that all terms manifesting a quantum interference will vanish from the expression for the intensity (2.65). This does not mean, however, that the optical signal from the ensemble of atoms will provide no retrievable information concerning the atomic energy level structure.

From the perspective of classical optics, the instantaneous intensity can be represented by an expression similar to (2.23)

$$I(t) = \langle I(t) \rangle + \Delta I(t) , \qquad (2.78)$$

where the first term is the phase-averaged intensity and the second term is the fluctuation about the average. A measure of the intensity correlation is provided by the degree of second-order coherence, defined by the relation [analogous to (2.24)]

$$g^{(2)}(t) \equiv g_{11}^{(2)}(t) = \frac{\langle I(0)I(t)\rangle}{\langle I(0)\rangle\langle I(t)\rangle} = 1 + \frac{\langle \Delta I(0)\Delta I(t)\rangle}{\langle I(0)\rangle\langle I(t)\rangle} . \qquad (2.79)$$

Only the time delay t, and not the origin of time, is of consequence here.

From (2.58)–(2.60) and (2.79), it follows that

$$g^{(2)}(t) = 1 + \frac{1}{2}\left(\frac{B}{A}\right)^2 \cos(\omega_{21}t) \qquad (2.80)$$

for free radiative decay from a mixture of superposition states impulsively generated by the same excitation event. For excitations occurring at constant rate $W \ll \gamma$ over an indefinitely long time, (2.65) and (2.79), with modulation amplitude $m = 0$, lead to

$$g^{(2)}(t) = 1 + \frac{1}{2} \left(\frac{B}{A} \right)^2 \frac{\gamma^2}{\gamma^2 + \omega_{21}^2} . \tag{2.81}$$

The second term in (2.80), (2.81) represents the correlation within the light field from individual coherently excited atoms. That $g^{(2)}(t)$ contains a term dependent upon $|\rho_{12}^0|$ is not unreasonable because the atomic polarization [see (2.46)] of each subgroup of atoms of given phase depends only on the magnitude, not on the phase, of the off-diagonal matrix elements.

2.5 The Quantum Optical Perspective

Throughout the preceding section, the radiation emitted from the system of excited atoms was treated as comprising classical waves. However, from a quantum mechanical perspective, an atom in a coherent superposition of excited states emits, at each de-excitation event, only one photon, and it is the interference between the amplitudes of two undecidable decay pathways of each atom that leads to quantum beats; photons emitted from different atoms are not supposed to interfere. To examine more closely the influence of atomic coherence on the properties of the detected radiation, it is necessary to calculate the coherence function for a quantized radiation field. This is interesting to do because the procedure differs radically from that in classical optics, and can be applied to field states for which there are no classical counterparts.

In classical physical optics, the measurable properties of light (e.g., the intensity, degree of linear polarization, and degree of circular polarization[5]) emitted from a particular source can be determined completely from knowledge of the electric and magnetic field vectors that comprise the associated electromagnetic wave. Briefly summarized, the electric and magnetic fields together *are* the light wave. Moreover, it is the fields of a detected wave that act upon the charged particles (electrons, 'holes', ions, etc.) of the detector. However, in the logical structure of quantum optics, a complete exposition of which must be left to references [59], the electric and magnetic fields are operators, and the state of the field upon which they act is a ket (or bra) vector labeled by the quantum numbers of a suitable representation. As is often the case in both classical and quantum optics, the magnetic field is of secondary importance and will be of no further concern here.

Based on the analogy of the electromagnetic field to a system of harmonic oscillators, the electric field operators are expanded in a Fourier series of its modes (i.e., allowed frequencies and polarizations), the expansion coefficients

---

[5] See, for example, [7].

of which are annihilation (a) and creation $(a^\dagger)$ operators satisfying the familiar commutation relations for each mode

$$[a, a^\dagger] = 1 , \qquad [a, a] = [a^\dagger, a^\dagger] = 0 . \tag{2.82}$$

Creation and annihilation operators of different modes commute with one another. The action of these operators on states of well-defined photon number $|n\rangle$ is as follows:

$$a|n\rangle = \sqrt{n-1}|n-1\rangle , \qquad a^\dagger|n\rangle = \sqrt{n+1}|n+1\rangle . \tag{2.83}$$

From the preceding relations, one can readily show that the number operator $\hat{n} = a^\dagger a$ counts the number of photons in a number state, $\hat{n}|n\rangle = n|n\rangle$, and satisfies the commutation relations[6]

$$[\hat{n}, a] = -a , \qquad [\hat{n}, a^\dagger] = a^\dagger . \tag{2.84}$$

The electric field operator of a linearly polarized bimodal field can be decomposed into a 'positive-frequency' component containing annihilation operators

$$E^+(t) \sim -i \left(\omega_1 a_1 e^{-i\omega_1 t} + \omega_2 a_2 e^{-i\omega_2 t} \right) \tag{2.85}$$

and 'negative-frequency' components containing creation operators

$$E^-(t) \sim -i \left(\omega_1 a_1^\dagger e^{i\omega_1 t} + \omega_2 a_2^\dagger e^{i\omega_2 t} \right) , \tag{2.86}$$

the terminology regarding frequency being based on the physicist's convention of representing the time dependence of a harmonic wave (of positive frequency) by the factor $e^{-i\omega t}$. The equivalence sign, rather than equal sign, appearing in (2.85), (2.86) indicates that constant factors irrelevant to the present discussion have been omitted. Likewise, since spatial retardation plays no role in the discussion to follow, the creation and annihilation operators, as well as the photon states, will not be labeled by the wave vector (which is equivalent to photon momentum). Also dropped from the notation is the unit vector expressing photon polarization, since all photons will have the same polarization in the example to be worked out. Finally, the slight dispersion in frequency resulting from the finite lifetime of the atomic states has been ignored, and the

---

[6] These commutation relations are special cases of more general relations that find wide usage throughout quantum mechanics and quantum optics. If $f(a, a^\dagger)$ is some general, well-behaved function of the annihilation and creation operators, then

$$[a, f(a^\dagger)] = \frac{df(a^\dagger)}{da^\dagger} , \qquad [a^\dagger, f(a)] = \frac{df(a)}{da} ,$$

$$[\hat{n}, f(a)] = -a\frac{df(a)}{da} , \qquad [\hat{n}, f(a^\dagger)] = a^\dagger\frac{df(a^\dagger)}{da^\dagger} .$$

optical frequencies ω_1 and ω_2 will be assumed equal in expressions containing their product or sum. What is significant here is the frequency difference.

The quantum optical expression for intensity, comparable to the classical-field relation (2.56), is

$$I(t) = \langle \Psi | E^-(t) E^+(t) | \Psi \rangle \,, \tag{2.87}$$

in which the order of the fields is important. Similarly, the quantum optical counterpart to (2.79) for intensity correlation, as expressed by the degree of second-order coherence, is

$$g^{(2)}(t) = \frac{\langle \Psi | E^-(0) E^-(t) E^+(t) E^+(0) | \Psi \rangle}{\langle \Psi | E^-(0) E^+(0) | \Psi \rangle \langle \Psi | E^-(t) E^+(t) | \Psi \rangle} \,, \tag{2.88}$$

where, again, the order of the electric field operators is crucial to obtaining correct results. From the properties of the annihilation and creation operators and the conditions articulated above, one can show that the quadratic and quartic product of field operators, apart from unimportant constant factors, take the forms

$$E^-(t) E^+(t) \sim \hat{n}_1 + \hat{n}_2 + a_1^\dagger a_2 e^{-i\omega_{21}t} + a_1 a_2^\dagger e^{i\omega_{21}t} \,, \tag{2.89}$$

$$
\begin{aligned}
E^-(0) E^-(t) E^+(t) E^+(0) \sim\ & \left(\hat{n}_1^2 - \hat{n}_1 \right) + \left(\hat{n}_2^2 - \hat{n}_2 \right) + 2 \hat{n}_1 \hat{n}_2 (1 + \cos \omega_{21} t) \\
& + \left[(a_1^\dagger)^2 (a_2)^2 e^{-i\omega_{21}t} + (a_1)^2 (a_2^\dagger)^2 e^{i\omega_{21}t} \right] \\
& + \left[a_1^\dagger \hat{n}_1 a_2 (1 + e^{-i\omega_{21}t}) + \hat{n}_1 a_1 a_2^\dagger (1 + e^{i\omega_{21}t}) \right] \\
& + \left[a_1^\dagger \hat{n}_2 a_2 (1 + e^{-i\omega_{21}t}) + a_1 a_2^\dagger \hat{n}_2 (1 + e^{i\omega_{21}t}) \right] \,,
\end{aligned}
\tag{2.90}
$$

where square brackets enclose Hermitian conjugate pairs in the latter equation.

The state vector of the atom-radiation field, whose coherence properties we are interested in here, is adapted from the well-known Wigner–Weisskopf theory of spontaneous emission, as applied to a two-state atom decaying to a nondegenerate ground state [60]. This state vector, derived on the basis of single-photon electric dipole processes only, consists in essence of a linear superposition of the quantum electrodynamic vacuum state $|00\rangle$, and the single-photon states $|10\rangle$, $|01\rangle$, where $|n_1 n_2\rangle$ designates a state with n_i photons of frequency ω_i ($i = 1, 2$). We will consider, however, the more general state

$$
\begin{aligned}
|\Psi\rangle = \ & A_0 |n_1, n_2\rangle + A_1 |n_1 + 1, n_2\rangle + A_2 |n_1, n_2 + 1\rangle \\
& + B_1 |n_1 - 1, n_2\rangle + B_2 |n_1, n_2 - 1\rangle \,,
\end{aligned}
\tag{2.91}
$$

which allows for single-photon stimulated absorption and emission processes, as well as spontaneous emission. The amplitudes A_i, B_i can be related to atomic dipole transition and density matrix elements. Of particular interest ultimately, are the relative phases between the amplitudes. The expectation values on $|\Psi\rangle$ indicated in (2.87), (2.88) lead to

$$
\begin{aligned}
\langle E^-(t)E^+(t)\rangle = {}& |A_0|^2(n_1 + n_2) + \left(|A_1|^2 + |A_2|^2\right)(n_1 + n_2 + 1) \\
& + \left(|B_1|^2 + |B_2|^2\right)(n_1 + n_2 - 1) \\
& + 2\Big[\sqrt{(n_1 + 1)(n_2 + 1)}|A_1 A_2|\cos(\omega_{21}t - \phi_A) \\
& \qquad + \sqrt{n_1 n_2}|B_1 B_2|\cos(\omega_{21}t + \phi_B)\Big] ,
\end{aligned}
\tag{2.92}
$$

$$
\begin{aligned}
\langle E^-(0) & E^-(t)E^+(t)E^+(0)\rangle \\
= {}& |A_0|^2(n_1 + n_2)(n_1 + n_2 - 1) \\
& + \left(|A_1|^2 + |A_2|^2\right)(n_1 + n_2)(n_1 + n_2 + 1) \\
& + \left(|B_1|^2 + |B_2|^2\right)(n_1 + n_2 - 1)(n_1 + n_2 - 2) \\
& + 2\Big[|A_0^2|n_1 n_2 + |A_1|^2(n_1 + 1)n_2 + |A_2|^2 n_1(n_2 + 1) \\
& \qquad + |B_1|^2(n_1 - 1)n_2 + |B_2^2|n_1(n_2 - 1)\Big]\cos\omega_{21}t \\
& + 2|A_1 A_2|\sqrt{(n_1 + 1)(n_2 + 1)}\Big[(n_1 + n_2)\cos\phi_A + n_1\cos(\omega_{21}t - \phi_A) \\
& \qquad\qquad\qquad\qquad + n_2\cos(\omega_{21}t + \phi_A)\Big] \\
& + 2|B_1 B_2|\sqrt{n_1 n_2}\Big[(n_1 + n_2 - 2)\cos\phi_B + (n_1 - 1)\cos(\omega_{21}t + \phi_B) \\
& \qquad\qquad\qquad\qquad + (n_2 - 1)\cos(\omega_{21}t - \phi_B)\Big] ,
\end{aligned}
\tag{2.93}
$$

where

$$
\phi_A = \arg(A_2) - \arg(A_1) , \qquad \phi_B = \arg(B_2) - \arg(B_1) .
\tag{2.94}
$$

For the case of spontaneous emission into the vacuum (in essence, the Wigner–Weisskopf state), $n_1 = n_2 = 0$ and $B_1 = B_2 = 0$. Equation (2.92) then displays a modulated intensity, which vanishes for random phase ϕ_A. Equation (2.93) results in an intensity correlation that is identically null independent of relative phase. There is no contribution from the coherent excitation and decay of individual atoms. The distinction between the classical and quantum optical results may be understood simply. In the classical picture, the light field produced by an atom consists of a superposition of two waves of different frequencies that may interfere at the detector. In the quantum picture, however, the emission of one photon of energy $\hbar\omega_1$ or $\hbar\omega_2$ precludes the emission of the other by the same atom for a given excitation event. There can be no correlation between two photons from the same atom under the above circumstances.

If, however, the atoms are immersed in a sea of photons with $(n_1, n_2) \gg 1$ and $|A_0| \sim 1 \gg |A_1|, |A_2|, |B_1|, |B_2|$, the intensity and intensity correlations become

$$\langle E^-(t)E^+(t)\rangle \sim |A_0|^2(n_1 + n_2) + 2\sqrt{n_1 n_2}\Big[|A_1 A_2|\cos(\omega_{12}t - \phi_A) \qquad (2.95)$$

$$+ |B_1 B_2|\cos(\omega_{12}t + \phi_B)\Big] ,$$

$$\langle E^-(0)E^-(t)E^+(t)E^+(0)\rangle$$

$$= |A_0|^2\Big[(n_1 + n_2)^2 + 2n_1 n_2 \cos\omega_{21}t\Big]$$

$$+ 2|A_1 A_2|\sqrt{n_1 n_2}\Big[(n_1 + n_2)\cos\phi_A + n_1 \cos(\omega_{21}t - \phi_A)$$

$$+ n_2 \cos(\omega_{21}t + \phi_A)\Big]$$

$$+ 2|B_1 B_2|\sqrt{n_1 n_2}\Big[(n_1 + n_2)\cos\phi_B + n_1 \cos(\omega_{21}t + \phi_B)$$

$$+ n_2 \cos(\omega_{21}t - \phi_B)\Big] . \qquad (2.96)$$

For fixed relative phases, the intensity (2.95) shows quantum interference at the Bohr frequency, resulting from the two undecidable pathways by which a photon can be absorbed from one field mode (ω_1 or ω_2) and emitted into the other. A random distribution of phases results in an unmodulated intensity. The intensity correlation (2.96), however, displays a modulated term independent of phase. This is analogous to the classical wave noise of the Hanbury Brown–Twiss experiment, since it represents correlations in the light field from different atoms. There is no violation of Dirac's dictum that [61]:

> Each photon [...] interferes only with itself. Interference between two different photons never occurs.

This is because, in a multiphoton state such as represented by (2.91), there is no way to determine which atom emitted or absorbed which photon. (We will consider this point again shortly when examining correlations of massive particles.) In the quantum description of light, the modulated term in (2.96) characterizes the bosonic nature of the photon.

Substitution of relations (2.95) and (2.96) into (2.88) leads to the second-order correlation function

$$g^{(2)}(t) = 1 + \frac{2n_1 n_2}{(n_1 + n_2)^2} \cos(\omega_{21} t) \ . \tag{2.97}$$

A derivation based on a classical light field gives rise to the same expression, but with photon number n_i replaced by intensity I_i. Since atoms are usually moving relative to the observer, (2.97) must be averaged over the Doppler profile. Photons emitted with frequency ω are perceived by the detector to have frequency $\omega' = \omega(1 - \beta)$, where $\beta = v_x/c$ is the ratio of the line-of-sight velocity to the speed of light. (β is assumed to be nonrelativistic, so that only the first-order Doppler effect need be considered.) The distribution of atomic velocities is given by Maxwell's distribution law, which is a Gaussian distribution function, $f(v_x) \propto e^{-mv_x^2/2k_B T}$, where m is the atomic mass, k_B is the Boltzmann constant, and T the absolute temperature. The velocity-average of the cosine factor in (2.97) is then

$$\langle \cos \omega_{21} t \rangle_\beta = \int_{-1}^{1} \cos \left[\omega_{21}^0 t (1 - \beta) \right] e^{-\beta^2/2\sigma^2} \frac{d\beta}{\sqrt{2\pi}\sigma} \ , \tag{2.98}$$

where $\omega_{21}^0 = \omega_2^0 - \omega_1^0$ is the excited level separation (in angular frequency units) in the atomic rest frame, and the Doppler width $\sigma = \sqrt{k_B T/mc^2} \ll 1$. The integration variable can be be transformed to $u = \beta/\sigma\sqrt{2}$ which ranges, for all practical purposes, from $-\infty$ to $+\infty$, leading to the result

$$\left\langle g^{(2)}(t) \right\rangle_\beta = 1 + \frac{2n_1 n_2}{(n_1 + n_2)^2} e^{-t^2/2\tau^2} \cos(\omega_{21}^0 t) \ , \tag{2.99}$$

with coherence time parameter τ defined by $\omega_{21}^0 \tau = 1/\sigma = \sqrt{mc^2/k_B T}$. Thus the degree of second-order coherence contains information concerning the internal level structure of the emitting atoms that is not provided by direct measurement of the intensity. The information can be accessed by measuring the $\left\langle g^{(2)}(t) \right\rangle_\beta$ as a function of delay time t within an approximate range $2\pi/\omega_{21}^0 \leq t \leq 2\tau$. The lower limit ensures that at least one oscillation occurs, whereas the upper limit ensures that the correlation term has not decayed to insignificance. Figure 2.5 illustrates the variation in $\left\langle g^{(2)}(t) \right\rangle_\beta$ as a function of the dimensionless ratio t/τ for a system parameter $\omega_{21}^0 \tau = 20$. One can also extract structural information from the Fourier transform

$$G(\omega) = \int_0^\infty \left\langle g^{(2)}(t) \right\rangle_\beta e^{-i\omega t} \tag{2.100}$$

$$= \pi\delta(\omega) + \frac{\sqrt{\pi}}{4} \frac{2n_1 n_2}{n_1 + n_2} \left[e^{-(\omega - \omega_{21}^0)^2 \tau^2/2} + e^{-(\omega + \omega_{21}^0)^2 \tau^2/2} \right] \ ,$$

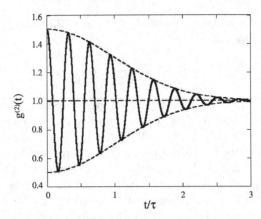

Fig. 2.5. Plot of second-order coherence as a function of t/τ for system parameter $\omega_{21}^0 \tau = 20$. *Dashed lines* show the Gaussian decay envelope. At $t = 0$, it is seen that $g^{(2)}(0) > 1$, indicative of photon bunching

which, apart from the delta spike at $\omega = 0$, shows peaks at $\omega = \pm\omega_{21}^0$.

As an example, Fig. 2.6 illustrates the Doppler profile

$$F(\omega) = \exp\left[-\frac{(\omega - \omega_1)^2/\omega_1^2}{2\sigma^2}\right] + \exp\left[-\frac{(\omega - \omega_2)^2/\omega_2^2}{2\sigma^2}\right]$$

as a function of temperature for a 30 amu atom with two resonances at optical frequencies separated by ω_{21}^0, which falls in the radiofrequency range. The two peaks are resolved at the lowest temperature shown (30 K), for which $\tau = 6.1 \times 10^{-4}$ s, but are broadened into a single peak at the higher temperatures 300 K

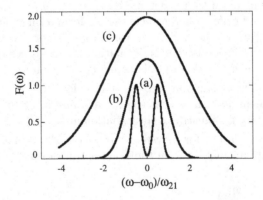

Fig. 2.6. Doppler profiles of a 30 amu atom with two absorption resonances at temperatures T: (**a**) 30 K, (**b**) 300 K, (**c**) 3000 K. The mean absorption frequency is $\omega_0 = (\omega_1 + \omega_2)/2 = 2\pi \times (5.1 \times 10^{14} \text{ Hz})$, and the frequency difference is $\omega_{21} = \omega_2 - \omega_1 = 2\pi \times (865 \text{ MHz})$

and 3000 K, for which $\tau = 1.9 \times 10^{-4}$ s and $\tau = 6.1 \times 10^{-5}$ s, respectively. At a temperature of 3000 K, therefore, where the peaks in the directly measured spontaneous emission spectrum completely overlap, one could determine the energy difference of the two excited states from the second-order correlation of the light field for delay times within the range (1.2 ns $< t <$ 1.2 μs).

To recapitulate briefly the seminal results of this section, I note that the quantum interference that distinguishes an ensemble of atoms coherently excited into a linear superposition of energy eigenstates from an ensemble of atoms in definite energy states persists in the spontaneous emission intensity, even if the excitation occurs over an indefinite period of time, provided that the relative quantum phases are well-defined within the ensemble. Under conditions where the experimenter has access to the atoms and source of excitation, ensembles with maximum polarization $|\boldsymbol{P}| = 1$, and minimum disorder $S = -(W/\gamma)\ln(W/\gamma)$ can be produced by resonant amplitude modulation of the rate of excitation of nondegenerate states or by external field-induced degeneracy of states excited at a constant rate. Quantum interference in stimulated emission contributes to the correlation of intensity fluctuations of the light field and also might serve to indicate coherent excitation of individual atoms under conditions where the observer has access only to the radiation and not to the atoms or source of excitation.

If the quantum phases vary randomly at each instant over the ensemble, quantum interference effects do not appear directly in the fluorescent intensity. In both classical and quantum optics, interference terms may occur in the second-order coherence resulting from correlations between the light from different atoms. Unlike the quantum beats from coherently excited individual atoms, the latter beats are affected by the distribution of atomic velocities.

2.6 Coherence of Thermal Electrons

Although still to be worked out more completely and tested experimentally, the quantum optics of particles with mass is an emerging counterpart to that of light, yet embraces processes for which no light optical analogue actually exists [62]. Thermal electrons for example, like thermal photons, may also give the impression of being randomly emitted, yet arrive correlated in space and time at one or more detectors. There are marked differences, however, in the clustering behavior of thermal electrons and photons owing to differences in particle number conservation, the tensorial character (vector vs. spinor) of the basic fields, and the applicable quantum statistics [63]. The correlations of electrons, or more generally any fermionic system, in contrast to those of photons, requires a quantum description at the very outset, for with only one fermion per quantum state, as required by the Pauli exclusion principle, there is simply no classical wave theory of fermions.

It is instructive to examine the coherence of thermal electrons, which is a system as fundamental to the study of particles as black-body radiation is to the study of light. Massive particles like electrons and neutrons, although subject to a different quantum statistics than that of photons, give rise under comparable conditions to diffraction-interference patterns that are indistinguishable, except for wavelength scale, from those of light. Dennis Gabor, who received the Nobel Prize in physics for his researches in electron microscopy leading primarily, however, to the development of optical holography, commented on this point some thirty-five years after quantum mechanics was created, in referring to "the almost complete identity of light optics and electron optics" [64]. The identity is only an apparent one, arising from the circumstance that quantum interference effects produced with free beams of massive particles have been characterized, until relatively recently, only by the first-order correlation function (in effect, the intensity) of the particle field. Such experiments involve the self-interference of single-particle wave packets and are insensitive to multiparticle effects governed by quantum statistics.

The coherence properties of black-body radiation were studied during the 1960s by Bourret [65] and by Wolf and his colleagues [66]. Bourret has pointed out that optical coherence functions depend in general upon both the properties of the source and the geometrical relations between source and detector. Only in the case of black-body radiation, however, do the coherence functions depend solely upon the intrinsic statistical properties of the photons, and are completely determined by the single physical parameter, temperature, apart from relative coordinates. Thermal electrons constitute a physical system of similar theoretical significance. In the general case, the coherence functions are parameterized by both temperature and chemical potential, the latter quantity being an implicit function of temperature and particle density.

One principal distinction between thermal radiation and thermal electrons is that the chemical potential of photons is identically null. This reflects the important fact that the number of black-body photons within an enclosure is not conserved. For a nondegenerate electron system, the chemical potential is small compared to the thermal energy, and the coherence functions are then parameterized primarily by temperature alone. In this case, the analogy to black-body radiation is close; the coherence time is basically the same for both the electron and photon systems at a given temperature. There still remain, of course, explicit quantum-statistical distinctions arising from a sign difference in the particle distribution function. For a highly degenerate electron system, the chemical potential is an explicit function of the particle density alone; the coherence time is then essentially independent of temperature and determined only by the chemical potential. These points will be elucidated shortly.

A second principal distinction between the coherence functions of thermal radiation and thermal electrons lies in their tensorial character. The basic fields $(\boldsymbol{E}, \boldsymbol{H})$ of electromagnetism are vector-valued functions; the resulting two-point second-order correlation functions are second-order tensors. The basic electron field is a spinor-valued function. The electron second-order correlation function, however, can be effectively considered a scalar quantity because electrons of opposite spin orientation are uncorrelated.

Consider now the matter of spectral distribution and coherence times. As in the case of thermal radiation, an examination of the energy distribution of thermal electrons leads to order-of-magnitude estimates of the characteristic coherence parameters. These parameters also emerge naturally in a field-theoretical analysis of the second-order coherence of a thermal-electron field. The coherence time of an electron field is, in essence, the reciprocal of the energy spread, or

$$\tau_c = \frac{\hbar}{\Delta E} = \frac{1}{\Delta \omega} , \tag{2.101}$$

where, for nonrelativistic electrons of mass m, the relation $\varepsilon_p = p^2/2m$ between the single-electron kinetic energy and momentum leads to the dispersion relation

$$\omega_k = \frac{\varepsilon_p}{\hbar} = \frac{\hbar k^2}{2m} \tag{2.102}$$

between angular frequency and wave number.

The average internal energy U per volume V of a system of thermal electrons is given by the quantum statistical expression

$$u \equiv \frac{U}{V} = \int \frac{\varepsilon g(\varepsilon) \mathrm{d}\varepsilon}{\mathrm{e}^{\beta(\varepsilon - \mu)} + 1} = C \int_0^\infty \mathcal{E}(x) \mathrm{d}x , \tag{2.103}$$

in which

$$g(\varepsilon) = \frac{(2m)^{3/2} \varepsilon^{1/2}}{2\pi^2 \hbar^3} \tag{2.104}$$

is the mode density, including a degeneracy factor $2S + 1 = 2$ because electrons have spin quantum number $S = 1/2$. The Fermi–Dirac statistical factor $[\mathrm{e}^{\beta(\varepsilon - \mu)} + 1]^{-1}$ is the mean particle number per mode, and is distinguished from the comparable function for bosons by the plus sign in the denominator. The second equality in (2.103) re-expresses the energy density u as the product of a constant

$$C = \frac{(2m)^{3/2}}{2\pi^2 \hbar^3 \beta^{5/2}} = \sqrt{\frac{2}{\pi}} \frac{k_B T}{\lambda_T^3} , \tag{2.105}$$

with the dimension of energy ($k_B T \equiv 1/\beta$) per volume (λ_T^3) and a dimensionless integral whose integrand

$$\mathcal{E}(x) = \frac{x^{3/2}}{\mathrm{e}^{x - \beta\mu} + 1} \tag{2.106}$$

is a function of the dimensionless parameter $x = \beta\varepsilon$. (Regrettably, the symbol β is traditionally employed by physicists to represent both a thermal parameter inversely proportional to temperature and a dynamical parameter proportional to velocity; the context should make it clear, however, how β is used in this chapter.) The characteristic length in (2.105) is the electron thermal wavelength

$$\lambda_T \equiv \frac{h}{\sqrt{2\pi m k_B T}} = \left(\frac{2\pi \hbar^2 \beta}{m}\right)^{1/2}, \tag{2.107}$$

which sets the scale at which multiparticle quantum effects become important. This is the case when the particle density is sufficiently high that more than one particle is likely to be found within a region of volume λ_T^3.

Figures 2.7 and 2.8 show the evolution in form of the density (2.106) as a function of x as the electron system goes from low degeneracy (or high temperature), $\beta\mu < 1$, to high degeneracy (or low temperature), $\beta\mu \gg 1$. For the low-degeneracy system, one can verify either analytically or by examining the plots, that the maximum value of $\mathcal{E}(x)$ occurs in the vicinity of $x = 2$,

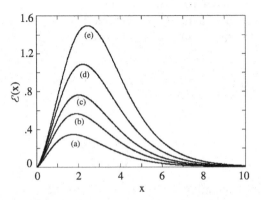

Fig. 2.7. Spectral distribution of a low-degeneracy thermal-electron system; x is the dimensionless variable $\beta\varepsilon$ with $\beta = (k_B T)^{-1}$. The curves are labeled by the dimensionless parameter $\beta\mu$: (**a**) 0, (**b**) 0.6, (**c**) 1.0, (**d**) 1.5, (**e**) 2.0

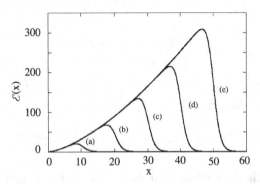

Fig. 2.8. Spectral distribution of a degenerate thermal-electron system. The curves are labeled by the dimensionless parameter $\beta\mu$: (**a**) 10, (**b**) 20, (**c**) 30, (**d**) 40, (**e**) 50

and that the spread between half-maximum points is approximately $\Delta x \sim 3$. It then follows from (2.101) that the coherence time of the field is adequately estimated by

$$\tau_{\mathrm{c}}^{(\mathrm{nd})} = \frac{\hbar\beta}{\Delta x} \sim \frac{1}{3}\hbar\beta \ . \tag{2.108}$$

Thus, for a nondegenerate thermal-electron system, the coherence time is of the order of $\hbar\beta$, which is also the case for a system of thermal radiation in an equilibrium enclosure.

The situation is quite different for a system of degenerate thermal electrons, as shown in Fig. 2.8. To be noted are the considerably larger ordinate and abscissa values in comparison with Fig. 2.7, and especially the very rapid descent of $\mathcal{E}(x)$ with x in the region beyond the point $x_{\mathrm{max}} \sim \beta\mu$ at which $\mathcal{E}(x)$ reaches its maximum value. Since the point below $\beta\mu$ at which $\mathcal{E}(x)$ reaches half its maximum value is approximately $0.63\beta\mu$, it follows that $\Delta x \sim 0.4\beta\mu$, and the coherence time of the degenerate system is then

$$\tau_{\mathrm{c}}^{(\mathrm{deg})} = \frac{\hbar\beta}{\Delta x} \sim \frac{\hbar}{0.4\mu} \ . \tag{2.109}$$

Thus, the coherence time of a highly degenerate thermal-electron system is of the order $\hbar/\mu$ and independent of temperature, since the chemical potential depends in this case only on particle density. Currently available sources of thermal electrons almost certainly fall in the nondegenerate category. To produce coherence times on the order of the response times of the fastest particle detectors (order of picoseconds) would require source temperatures of a few kelvin $[\tau_{\mathrm{c}}^{(\mathrm{nd})}(1\ \mathrm{K}) \sim 2.5\ \mathrm{ps}]$.

Consider next the second-order correlation function of thermal electrons. For a spin-polarized electron field characterized by a density matrix

$$\rho\big((k_1, s_1), \ldots, (k_n, s_n)\big) = \prod_{i=1}^{n} f_{s_i}(k_i) \tag{2.110}$$

that factors into a product of single-particle momentum (or energy) states, the second-order correlation function can be shown to take the form

$$g_{\mathrm{pol}}^{(2)}(\xi, \tau) = 1 - F(\xi, \tau) \ , \tag{2.111}$$

where ξ, τ are respectively the relative spatial and temporal coordinates of two points in the field. The coherence function

$$F_{(d)}(\xi, \tau) = \left| g^{(1)}(\xi, \tau) \right|^2 = \left| \frac{\int f(k) k \mathrm{e}^{\mathrm{i}(k\xi \cos\theta_k + \omega_k \tau)} \mathrm{d}V_k^{(d)}}{\int f(k) k \mathrm{d}V_k^{(d)}} \right|^2 , \tag{2.112}$$

which is the squared modulus of the first-order correlation function, is a quantum interference exchange term arising from the indistinguishability of the electrons. For an idealized one-dimensional $(d = 1)$ electron beam, the volume element is $dV_k^{(1)} = dk$ and $\theta_k = 0$ in the exponent, whereas for a three-dimensional $(d = 3)$ electron gas, the volume element is $dV_k^{(3)} = 2\pi \sin \theta_k k^2 dk d\theta_k$. Since electrons of opposite spin are uncorrelated, the second-order correlation function of an unpolarized electron field is

$$g_{\text{unp}}^{(2)}(\xi, \tau) = 1 - \frac{1}{2}F(\xi, \tau) . \tag{2.113}$$

In view of the preceding remarks on spin orientation and correlation and the trivial modification of $g^{(2)}$ for a spin-unpolarized field, it is permissible in the following discussion to neglect spin further (or, equivalently, to consider the system perfectly polarized). The basic field is then the scalar-valued electron wave function, and the second-order correlation function is likewise a scalar. Under these circumstances, the single-particle momentum weighting function $f_s(k)$ in (2.110) and (2.112) is independent of spin orientation and given simply by the Fermi–Dirac statistical factor $[e^{\beta(\varepsilon_k - \mu)} + 1]^{-1}$ with ε_k expressed in terms of k by the nonrelativistic dispersion relation (2.102). One sees immediately from (2.111), (2.112) that $g^{(2)}(0,0) = 0$, which expresses the antibunching of fermionic particles. For ξ or τ large in comparison to the coherence parameters of the system, F approaches 0 and $g^{(2)}$ approaches 1, indicative of uncorrelated particles.

Consider first the temporal correlation function obtained by evaluating (2.112) in the case of $\xi = 0$. The solid angle within which the electrons are confined drops out of the expression, and one is left with an integral of the form (for $d = 1$ or 3)

$$I^{(d)}(\tau) = \int \frac{k^d e^{i\omega_k \tau} dk}{e^{\beta(\hbar\omega_k - \mu)} + 1} .$$

Integrals like this one are frequently encountered in the quantum statistics of fermionic systems. Although in general they cannot be reduced to closed-form analytical expressions, procedures exist for transforming them into expressions more convenient for analytical, numerical, and graphical evaluation. These details in the present case, while straightforward, are tedious and must be left to the original literature [67]. In brief, the integration variable is changed from momentum to energy (i.e., from k to ω), and the resulting integral of the form

$$J = \int_0^\infty \frac{F(\omega)d\omega}{e^{\beta(\omega - \mu)} + 1}$$

is transformed into

$$J = \int_0^\mu F(\omega)d\omega + \frac{1}{\beta}\left[\int_0^\infty \frac{F(\mu + \beta^{-1}x)dx}{e^x + 1} - \int_0^{\beta\mu} \frac{F(\mu - \beta^{-1}x)dx}{e^x + 1}\right] .$$

The factor $(e^x + 1)^{-1}$ can be rewritten as $1 - (1 - e^x)^{-1}$ and expanded in the infinite series $\sum_{n=0}^\infty (-1)^n x^n$, whereupon the integrations over x can then be performed, giving rise to expressions recognizable as, or bearing a resemblance to, generalized Riemann zeta functions

$$\zeta(u, v) \equiv \sum_{n=0}^{\infty} (v + n)^{-u} , \qquad \mathrm{Re}(u) > 1, \; v \neq 0, -1, -2, \ldots . \qquad (2.114)$$

Application of the foregoing procedure to $I^{(d)}(\tau)$ leads to an intermediate expression

$$J = \int_0^{\mu} e^{i\omega\tau} d\omega + \frac{e^{i\omega(\tau/\tau_c^{\mathrm{deg}})}}{\beta} \left[\int_0^{\infty} \frac{e^{ix(\tau/\tau_c^{\mathrm{nd}})} dx}{e^x + 1} - \int_0^{\beta\mu} \frac{e^{-ix(\tau/\tau_c^{\mathrm{nd}})} dx}{e^x + 1} \right] ,$$

in which the ratio of the correlation time delay τ to the nondegenerate coherence time $\tau_c^{\mathrm{nd}} = \hbar\beta$ and to the degenerate coherence time $\tau_c^{\mathrm{deg}} = \hbar/\mu$ both naturally appear. In the case of a nondegenerate thermal electron system, the analysis leads to the approximate analytical expressions

$$F_{(1)}(\tau) = \left| \frac{\Phi_1(\tau/\hbar\beta)}{\ln 2} \right|^2 , \qquad (2.115)$$

$$F_{(3)}(\tau) = \left| \frac{12}{\pi^2} \Phi_2(\tau/\hbar\beta) \right|^2 \qquad (2.116)$$

for the coherence functions, in which

$$\Phi_1(t) = \int_0^{\infty} \frac{e^{itx} dx}{e^x + 1} = \sum_{n=1}^{\infty} (-1)^{n-1} (n - it)^{-1} , \qquad (2.117)$$

$$\Phi_2(t) = \sum_{n=1}^{\infty} (-1)^{n-1} (n - it)^{-2} = \zeta(2, -it) - \frac{1}{2}\zeta(2, -it/2) . \qquad (2.118)$$

At $t = 0$, the phi functions reduce to $\Phi_1(0) = \ln 2$ and $\Phi_2(0) = \pi^2/12$. The corresponding coherence functions for the case of a highly degenerate ($\beta\mu \gg 1$) electron system take the form

$$F_{(1)}(\tau) = \frac{\sin^2(\tau\mu/2\hbar)}{(\tau\mu/2\hbar)^2} , \qquad (2.119)$$

$$F_{(3)}(\tau) = \frac{8}{(\tau\mu/\hbar)^4} \left[1 + \frac{1}{2} \left(\frac{\tau\mu}{\hbar} \right)^2 - \left(\cos\frac{\tau\mu}{\hbar} + \frac{\tau\mu}{\hbar} \sin\frac{\tau\mu}{\hbar} \right) \right] . \qquad (2.120)$$

The expressions for intermediate degeneracy are cumbersome and will not be given here.

Figures 2.9 and 2.10 show the variation in $g_{\text{pol}}^{(2)}(0,\tau) = 1 - F(0,\tau)$ as a function of $\tau/\hbar\beta$ for one- and three-dimensional systems, respectively, for increasing values of $\beta\mu$ that lead from nondegeneracy to high degeneracy. The plots were obtained by direct numerical integration of the integrals defined by (2.112). As already noted, the correlation functions are null at zero delay time, in marked contrast to the behavior of thermal radiation for which the correlation function is 2 at $\tau = 0$, indicative of the property of bosons to bunch in phase space. For delay times long in comparison to the coherence time, the correlation functions of both electrons and photons approach 1, the value expected for uncorrelated particles. At a fixed temperature (β),

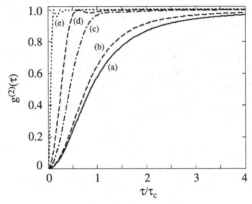

Fig. 2.9. Temporal variation of the second-order correlation function for a one-dimensional thermal-electron beam as a function of the dimensionless ratio of delay time to (nondegenerate) coherence time $\tau_c = \hbar\beta$. The plots correspond to different values of $\beta\mu$: (a) 0, (b) 1.0, (c) 5.0, (d) 10, (e) 50

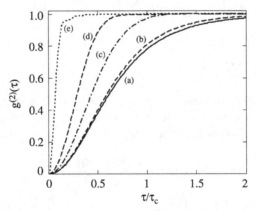

Fig. 2.10. Temporal variation of the second-order correlation function for a three-dimensional thermal-electron gas as a function of the dimensionless ratio of delay time to (nondegenerate) coherence time $\tau_c = \hbar\beta$. The plots correspond to different values of $\beta\mu$: (a) 0, (b) 1.0, (c) 5.0, (d) 10, (e) 50

the greater the chemical potential, the faster the correlation function rises from the temporal region where quantum effects of particle correlation are significant to the region where the correlations are entirely random.

Consider next the spatial correlation function obtained by evaluating (2.112) in the case of $\tau = 0$. The defining integrals, after integrating over the angular variable in the three-dimensional case, can be cast into the forms

$$I^{(1)}(\xi) = \int_0^\infty \frac{e^{i(\xi/\xi_c)x}dx}{e^{x-\beta\mu}} , \tag{2.121}$$

$$I^{(3)}(\xi) = \left(\frac{\xi}{\xi_c}\right)^{-1} \int_0^\infty \frac{\sin\left[(\xi/\xi_c)x^{1/2}\right]x^{1/2}dx}{e^{x-\beta\mu}} , \tag{2.122}$$

in which

$$\xi_c = \frac{\beta\hbar^2}{2m} \tag{2.123}$$

is the spatial coherence length $l_c^{(nd)}$ for a nondegenerate system, as can be seen by rewriting (2.123) as $l_c^{(nd)} = \bar{v}\tau_c^{(nd)}$ where $\bar{v} = \sqrt{k_B T/2m}$ is a measure of the root-mean-square thermal speed. For a highly degenerate system, the integrals in (2.121), (2.122) can be evaluated approximately by the procedure previously outlined, and lead to expressions (which will not be given here) that depend on the variable $\xi/l_c^{(deg)}$, where

$$l_c^{(deg)} = \frac{\hbar}{\sqrt{2m\mu}} = v_F \tau_c^{(deg)} \tag{2.124}$$

is the spatial coherence length for a degenerate system and $v_F = \sqrt{2\mu/m}$ is the speed of electrons at the Fermi energy μ.

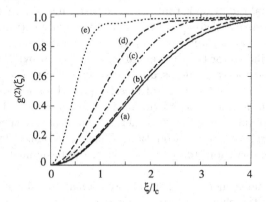

Fig. 2.11. Spatial variation of the second-order correlation function for a one-dimensional thermal electron beam as a function of the dimensionless ratio of retardation to (nondegenerate) coherence length $l_c = \sqrt{\hbar^2\beta/2m}$. The plots correspond to different values of $\beta\mu$: (a) 0, (b) 1.0, (c) 5.0, (d) 10, (e) 50

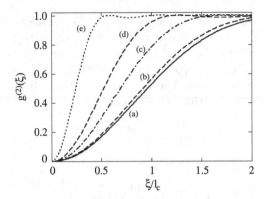

Fig. 2.12. Spatial variation of the second-order correlation function for a three-dimensional thermal electron gas as a function of the dimensionless ratio of retardation to (nondegenerate) coherence length $l_c = \sqrt{\hbar^2\beta/2m}$. The plots correspond to different values of $\beta\mu$: (**a**) 0, (**b**) 1.0, (**c**) 5.0, (**d**) 10, (**e**) 50

Figures 2.11 and 2.12 show the variation of $g_{\mathrm{pol}}^{(2)}(\xi,0) = 1 - F(\xi,0)$ as a function of ξ/ξ_c for one- and three-dimensional systems, respectively, for increasing values of $\beta\mu$. The plots were again obtained by numerical integration of the defining integrals, and show spatial properties similar to the temporal properties represented in Figs. 2.9 and 2.10.

2.7 Comparison of Thermal Electrons and Thermal Radiation

Past researches have shown that thermal radiation exhibits coherence in a sufficiently small space-time region. One early verification of the second-order coherence of thermal radiation is that of the intensity interferometry of starlight by Hanbury Brown and Twiss [68]. A system of thermal electrons is a fermionic analogue to black-body radiation, and its second-order coherence properties are likewise significant for the theoretical insights they provide and the potential applications to electron microscopy and various electron devices. Although the first-order coherence functions of massive particles and radiation are virtually identical in mathematical form for a given experimental configuration, there are marked differences between thermal electrons and thermal radiation with regard to second-order correlations. These differences derive primarily from particle conservation, tensor character of the basic fields, and quantum statistics.

Because the mean energy of thermal radiation at equilibrium depends only on temperature, the chemical potential $\mu = (\partial U/\partial N)_{S,V}$ of a photon gas is identically zero, and photons can be emitted or absorbed at the walls of the enclosure at no cost of energy. Thus, the number of photons is not conserved. By contrast, for a system of electrons in thermal equilibrium, particle conservation is tantamount to the conservation of electric charge, a law that has never been known to be violated. Apart from relative coordinates, the electron correlation functions depend on both temperature and chemical potential. For a nondegenerate system, $\beta\mu \ll 1$, the coherence time is effectively $\hbar\beta$, and the coherence length is essentially $\bar{v}\hbar\beta$, where $\bar{v}$ is the r.m.s. thermal speed. In a highly degenerate system, the coherence time is $\hbar/\mu$, and the coherence length is essentially $v_F\hbar/\mu$, where v_F is the mean electron speed at the Fermi energy. Both parameters depend only on particle density, independent of temperature. There is no radiation analogue to this case.

The basic electromagnetic fields are vector-valued functions and lead to two-point correlation functions that are elements of a second-order tensor. The basic electron field is a spinor-valued function, which leads to scalar correlation functions because electrons of opposite spin orientation are uncorrelated.

The second-order correlation function is sensitive to the quantum statistics that govern the aggregational behavior of the particles for temporal or spatial intervals short compared with the appropriate coherence times or lengths. Thermal radiation manifests the phenomenon of boson bunching, i.e., the correlation function is greater than unity for time intervals short compared with the coherence time and falls to unity for long time intervals. Thermal electrons manifest the phenomenon of antibunching, i.e., the correlation function is less than unity for time intervals short compared with the coherence time, and rises to unity for long intervals.

Experimental verification of the correlations discussed in this section would provide new kinds of tests of the wavelike properties, exchange antisymmetry, and quantum statistics of a multifermion system. Such experiments, however, have been very difficult to do because of the unavailability of suitable fermion (primarily electron or neutron) sources. A number of the predictions from the quantum analyses I first made in the 1980s, which will be examined in the following chapter, have yet to be tested, while laboratory confirmation of others have only relatively recently been achieved.

As a final matter of consideration, it is worthwhile to note that the correlations exhibited by a particular class of particles are not necessarily intrinsic to those particles. Quantum statistics requires that a boson state vector be symmetric under particle exchange, but photon bunching is not a general property of photons, since radiation states can also be constructed for which photons are uncorrelated or anticorrelated. This fact has long been known. Much less widely known, however, is that a similar circumstance prevails for fermions. Although thermal electrons – and, in general, 'chaotic' fermionic systems, i.e., systems characterized by a density matrix, like (2.110), diagonal in a momentum–spin basis – manifest antibunching, other fermion ensembles are conceivable that can exhibit other types of clustering behavior [69].

2.8 Brighter Than a Million Suns:
Electron Beams from Atom-Size Sources

The statistical properties of optical fields have been an active area of investigation since first stimulated by the pioneering experiments of Hanbury Brown and Twiss in the 1950s. As already pointed out, these experiments demonstrated correlations in intensity fluctuations and in photon arrival times of partially coherent light beams. In the parlance of optical coherence theory, the HBT experiments manifest the second-order coherence characteristics of light deriving from wave noise (in the imagery of classical optics), or photon clumping (in the imagery of quantum optics). This property of light is to be contrasted with first-order coherence which, both classically and quantum mechanically, refers to fringe contrast in an interference pattern.

The wave–particle duality intrinsic to all particles, suggests that what can be done with light should also be feasible with massive particles, and, indeed, intensity interferometry has been used in nuclear physics to investigate, for example, the geometry of the emission region in high energy nuclear reactions [70]. The relativistic heavy ion beams employed in these experiments are basically incoherent, and the hadronic interactions to which they give rise are not sufficiently understood to permit theoretical calculation of the two-particle correlation functions. Such a source would not be suitable for quantum interference experiments of the kind to be described in the following chapter. What is needed is a bright, coherent, charged or neutral (depending on particular experiment) fermion source whose physical interactions are well understood. Practically speaking, this means an electron source. Weak interaction phenomena aside, the behavior of electrons (and positrons) is completely specified by quantum electrodynamics. What are the prospects, then, of observing quantum correlations of electrons with currently available beams of electrons?

It is well beyond the scope of this chapter to consider the details of all experimental difficulties that must be surmounted, but one conceptually important issue merits close attention. First and foremost, it is necessary to have a source for which there is a reasonable probability of obtaining correlated particles. Loosely speaking, the wave functions of two particles should in some sense overlap at the time of production (although it must be stressed again that classical images can be misleading, especially when employed to explain quantum effects of multiparticle states). All familiar electron sources – including the most coherent, such as the field-emission electron source – produce for the most part single-particle states. This is not simply a matter of beam *intensity* – i.e., a large electron accelerator in place of an electron microscope would not necessarily solve the problem – but a question of beam *brightness*, which is related to the concept of degeneracy, referred to earlier, but not yet precisely defined. It is worth examining more quantitatively the connections between the brightness, coherence, and degeneracy of a particle source.

Although the concept of brightness can be rigorously delineated [71], I adopt here the widely employed experimental definition of mean brightness B

as the current density j (current per unit area normal to the beam) emitted into solid angle Ω. Thus, the number of particles received in a time interval Δt within a solid angle $\Delta\Omega$ through a detecting surface ΔA can be written as

$$\Delta n = \frac{B}{e}\Delta A\Delta t\Delta\Omega\,, \qquad (2.125)$$

where e is the particle charge. The quantities preceded by Δ are to be regarded as sufficiently small (strictly speaking, infinitesimally small) that particle direction, location, and arrival time are reasonably well defined. The beam degeneracy δ is defined in the equivalent, expression based on the flux of unpolarized spin-1/2 particles with momentum $\boldsymbol{p}$ and speed v in phase space

$\Delta n = $ (mean number of particles per cell of phase space)

$\qquad \times$ (number of occupied cells)

$$= \delta\left(\frac{2\Delta p_x\Delta p_y\Delta p_z\Delta x\Delta y\Delta z}{h^3}\right) = \delta\left(\frac{2(p^2\Delta p\Delta\Omega)(v\Delta t\Delta A)}{h^3}\right)\,. \quad (2.126)$$

It is the degeneracy parameter that governs the performance of electron interferometers and fundamentally determines the magnitude of quantum interference effects involving electron correlations [72]. The factor 2 in (2.126) takes account of the two spin degrees of freedom. Comparison of (2.125) and (2.126) leads to the expression

$$B = \delta B_{\max}\,, \qquad (2.127)$$

where the maximum brightness

$$B_{\max} = \frac{2evp^2\Delta p}{h^3} = \frac{2ep^2\Delta\varepsilon}{h^3} = \frac{2e\Delta\varepsilon}{h\lambda^2} \qquad (2.128)$$

occurs for degeneracy parameter $\delta = 1$. Conversion of the momentum dispersion Δp into energy dispersion $\Delta\varepsilon$ by means of the relativistically exact relation $\varepsilon^2 = p^2c^2 + m^2c^4$ leads to the second equality in (2.128). The third equality follows from substitution of the de Broglie relation $p = h/\lambda$ specifying the electron wavelength. For nonrelativistic electrons ($\Delta\varepsilon = p\Delta p/m$), the maximum brightness can be conveniently expressed in terms of kinetic energy and energy dispersion

$$B_{\max}^{(\mathrm{nr})} = \frac{4me\varepsilon\Delta\varepsilon}{h^3}\,. \qquad (2.129)$$

From (2.127) and the various expressions for $B_{\max}$, the essential dependences of the degeneracy parameter on beam characteristics can be summarized as follows:

$$\delta = B\frac{h}{2e}\frac{\lambda^2}{\Delta E} = B\frac{m}{2e}\frac{\lambda^3}{\Delta p} \sim B\frac{h^3}{4me}\frac{1}{\varepsilon\Delta\varepsilon}\,. \qquad (2.130)$$

Since the components of a split wave packet interfere to the extent that they occupy the same region of phase space, one can express the volume of a cell in phase space in terms of the (longitudinal) coherence length $l_c \sim v t_c$ and (transverse) coherence area $A_c \sim l_t^2$ of the beam by the relation $(p^2 \Delta p \Delta \Omega)(l_c A_c) = h^3$ in which the coherence time and coherence lengths were defined in the previous chapter. Use of (2.127), (2.128) and the definition of B then leads to an equivalent expression for degeneracy

$$\delta \sim \frac{j}{e} A_c t_c , \tag{2.131}$$

interpretable as the mean number of particles per coherence time traversing a coherence area normal to the beam.

To understand better the connections between occupation probability, brightness, and degeneracy, it is useful to examine the practical case of thermionic emission of electrons from a metal surface (Richardson effect), schematically represented in Fig. 2.13. In the absence of an external electric field, the electrons fill all states up to the Fermi level ε_F of a square-well potential of depth W, an approximation suitable for purposes of illustration. The work function ϕ is the energy required to escape the surface, which electrons acquire through thermal interaction with the environment as in the evaporation of molecules from the surface of a liquid. The rate of electron emission per unit surface area is given by the expression

$$\frac{d^2N}{dtdA} = \frac{2(2\pi)}{h^3} \int_{\sqrt{2mW}}^{\infty} dp_z \frac{p_z}{m} \int_0^{\infty} p_t dp_t \frac{1}{e^{\beta(\varepsilon_n + \varepsilon_t - \mu)} + 1} , \tag{2.132}$$

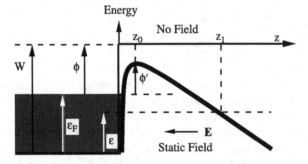

Fig. 2.13. Schematic diagram of the potential energy of electrons in a metal whose surface is located in the plane $z = 0$. Electrons fill all levels up to the Fermi level ε_F. In the absence of an electric field, the work function ϕ is the energy required to escape the surface of a square-well potential of depth W. An external electric field $\boldsymbol{E} = -E\hat{\boldsymbol{z}}$ and electron image field reduces the work function to ϕ' and tapers the walls of the potential, thereby increasing the probability of electron evaporation (thermionic emission) or quantum tunneling (field-emission). Coordinates z_0 and z_1 denote, respectively, the point of maximum potential and the second classical turning point at which the total energy (ε) of an electron equals the potential energy

in which the first factor of 2 is the spin degeneracy, the product

$$\frac{\mathrm{d}A\dfrac{p_z\mathrm{d}t}{m}(2\pi p_z p_t\mathrm{d}p_t)}{h^3}$$

is the number of occupied cells in phase space expressed in terms of electron momentum normal to the surface (p_z) and transverse to the surface (p_t), and

$$\frac{1}{e^{\beta(\varepsilon_n+\varepsilon_t-\mu)}+1}$$

is the Fermi–Dirac mean occupation per mode of total energy $\varepsilon = \varepsilon_n + \varepsilon_t$, where $\varepsilon_n = p_z^2/2m$, $\varepsilon_t = p_t^2/2m$, and $\beta = 1/k_BT$. The integral can be re-expressed exclusively in terms of the normal and transverse energies to give the current density

$$j = e\frac{\mathrm{d}^2N}{\mathrm{d}t\mathrm{d}A} = e\int_W^\infty \mathrm{d}\varepsilon_n \int_0^\infty \mathrm{d}\varepsilon_t P(\varepsilon_n,\varepsilon_t) = e\int_0^\infty \mathrm{d}\varepsilon_n' \int_0^\infty \mathrm{d}\varepsilon_t P(\varepsilon_n',\varepsilon_t) \qquad (2.133)$$

as an integral over a joint probability density with

$$P(\varepsilon_n,\varepsilon_t) = \frac{4\pi m}{h^3}\frac{1}{e^{\beta(\varepsilon_n+\varepsilon_t-\mu)}+1} \quad \text{or} \quad P(\varepsilon_n',\varepsilon_t) = \frac{4\pi m}{h^3}e^{-\beta(\varepsilon_n+\varepsilon_t+\phi)} . \tag{2.134}$$

To go from the first relation to the second in (2.134) it is useful to recall that Boltzmann's constant corresponds to about 1 eV per 12 000 K, and the chemical potential of metals is ordinarily a few electron volts. Thus, the temperature of the electron gas within the solid phase of a metal is effectively at 0 K, and one can equate the Fermi energy ε_F to the chemical potential μ. (An expression for the chemical potential of the electron gas within a metal at low temperature, or equivalently at high degeneracy, is derived in Appendix 2B.) Next, define $\varepsilon_n' = \varepsilon_n - W$, which shifts the origin of energy so that the normal kinetic energy is measured from 0, and recognize that $W - \mu = \phi$. Since the argument $\beta(\varepsilon_n' + \varepsilon_t + \phi)$ of the exponential in the first relation of (2.134) is much greater than unity, we can neglect the 1 in the denominator to obtain the second form of the probability density which looks exactly like what one would get by using Maxwell–Boltzmann (MB) statistics instead of Fermi–Dirac (FD) statistics. However, MB statistics are valid for the condition $\varepsilon_F \ll k_BT$, i.e., opposite that pertaining to the electron gas in a metal, and do not lead to the correct prediction of the current density, which from (2.133), (2.134), is

$$j = \frac{4\pi me}{h^3}(k_BT)^2 e^{-\phi/k_BT} . \tag{2.135}$$

The refractory metal tungsten, with work function $\phi = 4.5$ eV, chemical potential $\mu \sim 8$ eV, and melting point of nearly 3700 K, is frequently employed

as a thermionic emitter. Applied to tungsten at temperatures $T = 2500$ K and
3000 K, (2.135) leads to current densities 0.65 and 30.2 A/cm^2, respectively.

The axial brightness of a source is the current density per solid angle into
which the particles are emitted, in the limit of vanishing transverse energy,
$B_c = (\mathrm{d}j_c/\mathrm{d}\Omega)_{\varepsilon_t \to 0}$, where the subscript c signifies evaluation at the location
of the source (cathode). To implement this derivative, note that the solid
angle defined by a cone of apex angle θ is $\Omega(\theta) = 2\pi(1 - \cos\theta) \sim \pi\theta^2$, where
the approximation holds for $\theta \ll 1$ (as is required for a well-defined electron
beam). Thus the differential solid angle

$$\mathrm{d}\Omega \sim 2\pi\theta\mathrm{d}\theta \sim 2\pi\frac{p_t}{p_z}\frac{\mathrm{d}p_t}{p_z} \sim \pi\frac{\mathrm{d}\varepsilon_t}{\varepsilon'_n}$$

can be expressed in terms of the differential transverse energy $\mathrm{d}\varepsilon_t$, leading to
$\mathrm{d}\varepsilon_t = \varepsilon'_n \mathrm{d}\Omega/\pi$ which is substituted in the integral (2.133) defining the current
density j. It then follows that the axial brightness of a thermionic emitter is

$$B_c = \frac{e}{\pi} \int_0^\infty P(\varepsilon'_n, 0)\varepsilon'_n \mathrm{d}\varepsilon'_n = \frac{4\pi me}{\pi h^3}(k_B T)^2 e^{-\beta/k_B T} = \frac{j_c}{\pi}. \qquad (2.136)$$

Figure 2.14 shows the variation in B_c for a thermionic emitter with $\phi = 4.5$ eV as a function of temperature. Within the range of temperatures (2500 K
to 3000 K) at which tungsten can be thus employed, the brightness varies
between approximately 1 and 10 A/cm^{2}sr.

To estimate the theoretical maximum brightness of an electron source
from (2.129), and ultimately the degeneracy, we need the mean energy and

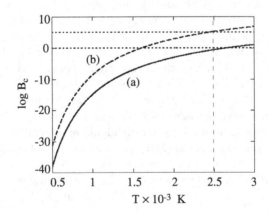

Fig. 2.14. Brightness of a tungsten thermionic source ($\phi = 4.5$ eV) as a function of
emitter temperature for conditions of (**a**) no accelerating voltage, and (**b**) an accelerating voltage of 150 kV. *Dotted lines* mark brightnesses of 1 and 10^5 A cm^{-2}sr^{-1};
the *dashed line* marks a lower bound to the temperatures at which the source would
likely be used

energy dispersion. The joint probability density (2.134) leads to the marginal probability densities

$$P_n(\varepsilon_n') = \frac{\int\limits_0^\infty P(\varepsilon_n', \varepsilon_t) d\varepsilon_t}{\int\int P(\varepsilon_n', \varepsilon_t) d\varepsilon_t d\varepsilon_n'} = \beta e^{-\beta \varepsilon_n'} , \tag{2.137}$$

$$P_t(\varepsilon_t) = \frac{\int\limits_0^\infty P(\varepsilon_n', \varepsilon_t) d\varepsilon_n'}{\int\int P(\varepsilon_n', \varepsilon_t) d\varepsilon_t d\varepsilon_n'} = \beta e^{-\beta \varepsilon_t} , \tag{2.138}$$

and to the probability density for the total energy $\varepsilon = \varepsilon_n' + \varepsilon_t$

$$P_\varepsilon(\varepsilon) = \int\limits_0^\varepsilon P_n(\varepsilon - \varepsilon_t) P_t(\varepsilon_t) d\varepsilon_t = \beta^2 \varepsilon e^{-\beta \varepsilon} , \tag{2.139}$$

which is the convolution of the two marginal densities. (The probability density of the sum of two random variables is derived in Appendix 2C.) Using (2.139), one can show that the mean and mean-square electron energies are

$$\langle \varepsilon \rangle = \int\limits_0^\infty \varepsilon P_\varepsilon(\varepsilon) d\varepsilon = 2\beta^{-1} , \tag{2.140}$$

$$\langle \varepsilon^2 \rangle = \int\limits_0^\infty \varepsilon^2 P_\varepsilon(\varepsilon) d\varepsilon = 6\beta^{-2} , \tag{2.141}$$

and the corresponding energy dispersion is

$$\Delta\varepsilon = \sqrt{\langle \varepsilon^2 \rangle - \langle \varepsilon \rangle^2} = \sqrt{2}\beta . \tag{2.142}$$

The degeneracy parameter calculated from (2.130) is then

$$\delta = \frac{e^{-\beta\phi}}{2\pi\sqrt{2}} \sim \frac{e^{-\beta\phi}}{10} . \tag{2.143}$$

Figure 2.15 shows the variation in δ as a function of temperature. At the upper end of the useful temperature range for a tungsten emitter, the degeneracy reaches a maximum of the order of 10^{-8}.

The presence of a static electric field can enhance brightness in two distinctly different ways by (1) changing the shape of the potential energy well, thereby increasing the rate of electron emission, and (2) modifying the factors

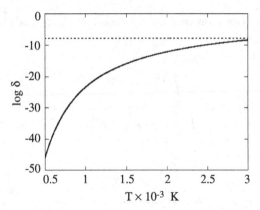

Fig. 2.15. Field-free degeneracy as a function of temperature for a thermionic electron source with work function $\phi = 4.5$ eV. The *dotted line* marks a maximum degeneracy of about 10^{-8} at the upper limit of the temperature range of the source

of the phase-space density of electrons accelerated after emission. Consider the second process first.

Liouville's theorem, applied to a beam of electrons emitted along the z axis with mean speed v from a transverse area dS, maintains that the phase-space density

$$(d^3x d^3p) = (dSvdt)(p^2 dpd\Omega) \propto \varepsilon'_n dS d\varepsilon'_n d\Omega$$

is constant along the axis. Thus, comparing the phase-space density at the cathode (c) to the corresponding density at an axial location (a) where electrons have been accelerated to a potential V, one finds

$$\varepsilon'_n dS_c d\Omega_c = (\varepsilon'_n + eV) dS_a d\Omega_a \ ,$$

which, by (2.136), leads to the relation

$$B_a = B_c(1 + \beta eV) \tag{2.144}$$

between the brightnesses at the two locations. For thermionic electrons emitted from a 3000 K cathode and accelerated to 150 kV, the brightness enhancement is approximately $eV/k_BT = 6 \times 10^5$.

In the second process, a static electric field perpendicular to the metal surface, as shown in Fig. 2.13, contributes a term $-eEz$ to the potential energy outside the surface, reshaping the wall of the potential well from a horizontal 'step' to a triangular 'wedge'. Moreover, an emitted electron at a distance z outside the surface interacts with its positive image charge within the metal a distance z from the surface, contributing an additional term $-(e^2/4\pi\varepsilon_0)/4z$ to the potential energy. The net result is a potential energy function in the form of a 'rounded' wedge with maximum value $-e^{3/2}E^{1/2}/\sqrt{4\pi\varepsilon_0}$ at a distance

$$z_0 = \sqrt{\frac{e/4\pi\varepsilon_0}{4E}} \ .$$

The dynamical effect of these two contributions is a reduced work function

$$\phi' = \phi - \frac{e^{3/2} E^{1/2}}{\sqrt{4\pi\varepsilon_0}} \ ,$$

which leads, by an analysis identical to the one resulting in (2.135), to a larger current density

$$j(E) = j(0) \exp \frac{e^{3/2} E^{1/2}}{\sqrt{4\pi\varepsilon_0}} \tag{2.145}$$

and corresponding brightness. At a field of 10^8 Vm^{-1}, beyond which (2.145) is no longer likely to be valid, the brightness of a field-assisted thermionic beam would be enhanced by a factor of about 5.81 at $T = 2500$ K and 4.33 at $T = 3000$ K. This is helpful, but not a huge increase.

Beyond 10^8 Vm^{-1}, however, a different field-induced process predominates, as schematically indicated in Fig. 2.13. Instead of a vertical transition out of the potential well, corresponding to thermally-induced electron evaporation, an electron can make a horizontal transition through the potential by quantum tunneling. This is the process of field-emission, the theory of which was first developed by Fowler and Nordheim for the case of the triangular wedge potential [73] and subsequently generalized by Nordheim to include the electron image potential [74]. The Fowler–Nordheim (FN) theory of field-emission, which entails the solution of the Schrödinger equation in the WKB (Wentzel–Kramers–Brillouin) approximation will not be examined in detail here, but it will be instructive to consider the theoretical predictions to which it leads.

As a baseline for comparison with thermionic emission, we consider the fundamental case of cold-cathode emission, in which the temperature of the filament is effectively at 0 K. In contrast to the previous case of thermionic emission, in which an emitted electron has energy in excess of $\mu + \phi$, an electron tunneling through the potential barrier will have energy less than μ. According to the FN theory, the joint probability density that an electron has kinetic energy of motion normal to the surface between ε_n and $\varepsilon_n + d\varepsilon_n$ and kinetic energy of motion transverse to the surface between ε_t and $\varepsilon_t + d\varepsilon_t$ takes the form [75]

$$P(\varepsilon_n, \varepsilon_t) = \frac{4\pi m}{h^3} \frac{e^{-c} e^{-(\mu - \varepsilon_n)/d}}{1 + e^{-\beta(\mu - \varepsilon_n - \varepsilon_t)}} \xrightarrow[T \to 0]{} \frac{4\pi m}{h^3} e^{-c} e^{-(\mu - \varepsilon_n)/d} \ , \tag{2.146}$$

where $\mu \geq \varepsilon_n \geq 0$ and $\mu - \varepsilon_n \geq \varepsilon_t \geq 0$, and

$$c = \frac{4}{3} \left(\frac{\phi}{eE\xi} \right)^{3/2} \ , \qquad d = \frac{1}{2} \frac{(eE\xi)^{3/2}}{\phi^{1/2}} \ , \qquad \xi = \left(\frac{\hbar^2}{2meE} \right)^{1/3} \ . \tag{2.147}$$

The field-dependent length ξ is defined by the relation $eE\xi = \hbar^2/2m\xi^2$, which is interpretable as an application of the work–energy theorem to an electron which has penetrated a distance ξ through the potential barrier.

From (2.146) one can calculate by the same procedure used in the case of thermionic emission, the current density

$$j_{FN} = \frac{4\pi m e}{h^3} d^2 e^{-c} \tag{2.148}$$

and brightness

$$B = \frac{1}{\pi} j_{FN} \frac{\mu}{d} = \frac{4me}{h^3} \mu d e^{-c} . \tag{2.149}$$

Comparison of (2.148) and (2.135) suggests that the quantity d plays the role in field-emission that $k_B T$ plays in thermionic emission with an analogous correspondence between c and $\phi/k_B T$. Equation (2.146) leads to the marginal probability density for ε_n

$$P_n(\varepsilon_n) = \frac{\mu - \varepsilon_n}{d^2} e^{-(\mu - \varepsilon_n)/d} \tag{2.150}$$

and to the probability density for total energy

$$P_\varepsilon(\varepsilon) = \frac{1}{d} e^{-(\mu - \varepsilon)/d} , \tag{2.151}$$

from which follow the mean energies

$$\langle \varepsilon \rangle = \mu - d , \qquad \langle \varepsilon^2 \rangle = \mu^2 - 2\mu d + 2d^2 , \qquad \langle \varepsilon_n \rangle = \mu - 2d , \tag{2.152}$$

and total energy dispersion

$$\Delta \varepsilon = \sqrt{\langle \varepsilon^2 \rangle - \langle \varepsilon \rangle^2} = d . \tag{2.153}$$

The quantity d is therefore interpretable as either the mean transverse energy $\langle \varepsilon_t \rangle = \langle \varepsilon \rangle - \langle \varepsilon_n \rangle$, or the dispersion in total energy. In the calculation of the preceding mean values, the inequality $\mu/d \gg 1$ was employed to eliminate terms containing the exponential $e^{-\mu/d}$. For cold-cathode field emission from tungsten, for example, d is a few tenths of an electron volt, and $\mu/d \sim 20$.

Substitution of the above values of $\langle \varepsilon \rangle$ and $\Delta \varepsilon$ into expression (2.129) for the maximum brightness leads to a degeneracy parameter

$$\delta_{FN} \sim e^{-c} , \tag{2.154}$$

to be compared with relation (2.143). Figure 2.16 shows the variation in brightness and degeneracy as a function of the external electric field strength. Field emission begins to be significant at fields of about 2×10^9 V/m or 0.2 V/angstrom. At the very high, but still achievable, field of 1 V/angstrom, a brightness of $\sim 2 \times 10^7$ A/cm^{2}sr and degeneracy of $\sim 8 \times 10^{-4}$ are predicted. Cold-cathode field-emission leads to the brightest and most coherent electron beams currently available. Experimentally, the brightness, and therefore the degeneracy, can be obtained directly from measurements of current density and solid angle or indirectly from coherence parameters inferred from an interference pattern [76]. In the case of a coherent beam such as required for interferometry, the latter procedure is to be preferred.

Of all the known types of particle sources, field-emission electron sources are the brightest known, considerably brighter even than the Sun or linear

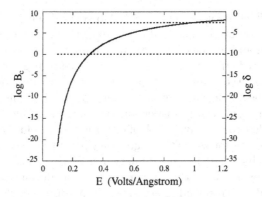

Fig. 2.16. Variation in brightness B (in $A\,cm^{-2}sr^{-1}$) and degeneracy δ of an electron field-emission source as a function of electric field ($1\ V/Å = 10^{10}\ V/m$) as calculated from the Fowler–Nordheim theory. On the *left scale*, the *dotted lines* mark unit brightness and $B = 10^{7.3}$ which occurs at $1\ V/Å$. On the *right scale*, the *dotted lines* mark degeneracies of 10^{-10} (corresponding to a tungsten thermionic emitter at 2500 K) and 8.4×10^{-4} (corresponding to a tungsten field emitter at $1\ V/Å$)

Table 2.2. Characteristics of a high-voltage field-emission electron beam

Energy ε	10^5 eV
Energy dispersion $\Delta\varepsilon$	0.3 eV
Wavelength λ	4×10^{-10} cm
Coherence time t_c	1.4×10^{-14} s
Current density j	3.3×10^{13} e cm^{-2}s^{-1}
Angular divergence α	1.3×10^{-7} rad
Transverse coherence length l_t	1.5×10^{-3} cm
Coherence area A_c	2.2×10^{-6} cm^{-2}
Brightness	10^8 A cm^{-2}sr^{-1}
	6×10^{26} e cm^{-2}s^{-1}sr^{-1}
Degeneracy δ	10^{-6}

accelerators or synchrotron/wiggler/undulator systems. For example, from knowledge of the total solar irradiance ($\sim 1367\ W/m^2$ measured at the top of the Earth's atmosphere) [77], the solar radius ($\sim 7 \times 10^8$ m), and the Earth–Sun distance ($\sim 1.5 \times 10^{11}$ m), one can estimate a solar brightness of approximately $10^{10}\ erg\,s^{-1}cm^{-2}sr^{-1}$, which translates into roughly 10^{21} photons s^{-1}cm^{-2}sr^{-1}, assuming a photon wavelength of 500 nm (peak of the solar emission spectrum). By contrast, as seen from Table 2.2, the brightness of a standard 100 kV field-emission microscope[7] is of the order of 10^8 A cm^{-2}sr^{-1}

---

[7] It is to be noted that the 100 kV is the accelerating potential, not the extraction potential. Because the actual cathode in a field-emission microscope is a very

or about 6×10^{26} electrons $s^{-1}cm^{-2}sr^{-1}$. Moreover, from (2.128), the maximum brightness potentially achievable for an electron source of the specified energy dispersion and wavelength is about a million times greater.

Nevertheless, a higher brightness does not insure a greater degeneracy in comparisons between different types of sources. Examination of (2.130) shows that the beam degeneracy is also proportional to the square of the wavelength for a fixed energy dispersion-and optical wavelengths are much longer than the de Broglie wavelength of electrons from field-emission sources. Unlike photon sources for which the degeneracy parameter can be many orders of magnitude – about 10^4 for an ordinary 1 mW HeNe laser, and in excess of 10^{16} for a ruby laser[8] – the maximum fermion degeneracy is unity. In practice, thermal neutron and thermal electron sources have a degeneracy parameter many orders of magnitude smaller than one, and even in the case of a 150 kV field-emission source the degeneracy is only about 10^{-6}–10^{-4}. It is the low fermion degeneracy that prompted Gabor to comment on the "almost complete identity of light optics and electron optics", as previously noted. Still, the situation is improving.

An area of development that augurs well for the interferometry of correlated electrons is the fabrication of ultrasharp field-emission tips emitting electrons from one or at most a few surface atoms [79]. Such 'nanotips', illustrated schematically in Fig. 2.17, are expected to produce brighter, and therefore more degenerate, beams than previous field-emission tips – perhaps by two or more orders of magnitude. The nanotip consists of a small protrusion or 'teton' of nanometer size produced at the end of a sharp tungsten needle of a kind generally employed in scanning tunneling microscopes. In the method initially developed for making single-atom emitters, a tungsten atom is deposited onto a trimer of atoms in the (111) plane at the end of a tungsten needle previously prepared by controlled field evaporation of the apex. Other methods have been subsequently developed that employ heat treatment and diffusion of atoms at the apex. Interestingly, as I was completing this chapter, a research group at the National Institute for Nanotechnology in Edmonton, Canada, reported the fabrication of the sharpest needle ever made – radius of curvature less than 1 nm – by a new method employing the controlled reaction of nitrogen gas with atoms at the tungsten tip [80]. A striking image of the end of such

---

sharp pointed tip, a much lower potential can generate the requisite high electric fields needed for extraction. Also, the Fowler–Nordheim theory, which treats an idealization of the field-emission process, actually understates the achievable current densities for a given field strength.

[8] The degeneracy of a quasi-monochromatic laser source of power P, frequency ν, and bandwidth $\Delta\nu$ or pulse width τ is effectively the number of photons $\delta = Pt_c/h\nu$ emitted in a coherence time $t_c = 1/\Delta\nu$ or τ. Thus, a continuous-wave HeNe beam of wavelength 633 nm and spectral width 0.2 nm can be shown to have a degeneracy of approximately 2.14×10^4. A ruby laser producing a train of 5 mW pulses each of 1 μs duration at 694 nm emits about 1.8×10^{16} photons per pulse. See, for example, [78].

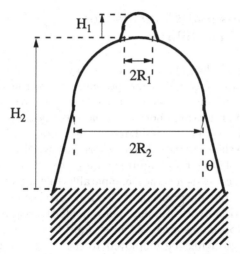

Fig. 2.17. Idealized model of a field-emission electron nanotip consisting of a nanometer-size protrusion or 'teton' on top of a larger supporting cylindrical tip similar to that designed for scanning tunneling microscopes. The dimensions of the nanotip are typically the following: teton height $H_1 \sim 3$ nm and radius $R_1 \sim 1$ nm; base length H_2 and radius $R_2 \sim 100$ nm, and shank angle $\theta \sim 10°$

a sub-nanotip, as recorded by field ion microscopy, is shown in Fig. 2.18. The image was part of a motion picture made by the group, which captured the displacement of surface atoms during the one-second period of recording.

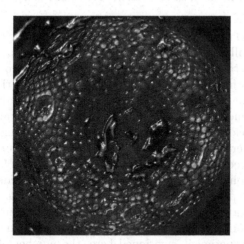

Fig. 2.18. Atomic-resolution field ion microscope image of the tip of the sharpest tungsten needle ever made at the time of writing. Small spheres show individual atoms; smearing at the tip reveals atomic displacement during the imaging time of approximately 1 s. The tip comprises sequential layers, separated by 0.22 nm, of stacked spheres, beginning with a single atom of radius 0.15 nm at the apex. (Courtesy of R. Wolkow)

2.9 Correlations and Coincidences: Experimental Possibilities

All things being equal, the greater the degeneracy of the source, the more suitable is the source for use in experiments probing higher-order (than one) coherence effects of matter. The degeneracy of the arc light source employed by HBT in experiments demonstrating photon correlations was about 10^{-3}. Although low, sufficient statistics were accumulated in approximately 10 hours of counting to demonstrate the sought-for bosonic correlations. With nanometer-sized field-emission tips of brightness approximately 10^{10} A cm^{-2}sr^{-1}, one would have electron sources of degeneracy comparable to that of the light source employed by HBT. The observation of electron correlations may be difficult, but, Gabor's remark notwithstanding, not outside the realm of possibility.

Given an appropriate source, at least four types of experimental approaches could in principle be employed to manifest electron correlations. One could count electrons arriving at a single detector to determine:

- the variance in the number of counts about the mean, or
- the conditional probability of receiving a second electron at a predetermined time interval after having detected a first.

Alternatively, one could count electrons at two detectors to determine:

- the correlation in fluctuations in the numbers of counts received within the same time interval, or
- the number of coincidences as a function of a time delay in one of the input channels.

There are practical differences in implementation of the different approaches, and the theoretically expected ratio of signal (i.e., fermionic deviation from random particle statistics) to noise (the random background counts), for a given total counting time can differ substantially among them.

Consider, as an illustration, a simple, if somewhat idealized, experiment involving coincidence counting at two detectors, as illustrated in Fig. 2.19. Suppose that a field-emission source emits spin-polarized electrons at a rate of R_S particles per second. Two detectors (D1, D2), assumed to be 100% efficient, are located symmetrically about the optic axis of the source, and each subtends a solid angle Ω at the source, leading to a count rate of $R_1 = R_2 = R_S \Omega \equiv R$ particles/sec. A coincidence results whenever one of the detectors receives a particle within a specified interval of time – the detector resolution or response time t_r – of the prior arrival of a particle at the other detector. In the experiment envisioned here, the number of coincidences is recorded as a function of the time interval $\tau = t_1 - t_2$ between the arrival of a 'start' particle at D1 and a 'stop' particle at D2.

If particles were emitted randomly and without any correlation, one would nevertheless expect a certain number of accidental coincidences. With particles arriving at the rate R_1 at detector D1, the counting circuitry will be active

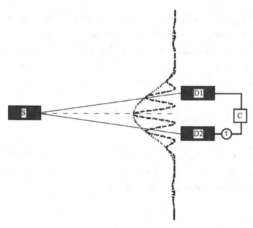

Fig. 2.19. Schematic diagram of coincidence-counting experiment. Electrons emitted from a source S arrive at detectors D1 and D2. The number of coincident arrivals within the resolution time t_r of the apparatus is registered by a correlating device C as a function of delay time τ. The counters will be coherently illuminated by the source if they are positioned well within a transverse coherence length (as determined by a separate two-slit interference experiment indicated by the dashed diffraction pattern)

for a fraction $R_1 t_r$ of the total counting time t. Thus, the fraction of the R_2 particles/sec that will accidentally arrive at detector D2 with time separation τ is $R_a = R_2 R_1 t_r P_0(\tau)$, where $P_0(\tau)$ is the probability that no particle is emitted within τ. Therefore, after a total counting time t, the number of accidental coincidences will be

$$N_a = R_a t = R_2 R_1 t_r t P_0(\tau) = \overline{N}_1 \overline{N}_2 \frac{t_r}{t} P_0(\tau) . \tag{2.155}$$

Field-emission of electrons is for the most part a Poissson process whereby the probability that n particles are emitted within a certain time interval τ is $P_n = e^{-\overline{n}} \overline{n}^n / n!$, in which $\overline{n} = R|\tau|$ is the mean number of particles emitted in that interval. It therefore follows that the probability that *no* particle is emitted during the interval – or, equivalently, the probability of receiving a second particle τ seconds after receipt of a first – is $P_0 = e^{-\overline{n}}$. Thus, the number of accidental coincidences to be expected is

$$N_a(\tau) = \overline{N}_1 \overline{N}_2 \frac{t_r}{t} e^{-R|\tau|} . \tag{2.156}$$

Because electrons are fermions, the probability that a second electron will be emitted into a coherence area within an interval τ of the first, is less than that predicted by Poisson statistics. Ideally that probability is 0 for $\tau = 0$ and approaches unity as $|\tau|$ greatly exceeds the coherence time of the source. The exact dependence of the probability on τ is given by the second-order correlation function, which, for a field-emission source, takes the form of (2.111),

$g^{(2)}(\xi, \tau) = 1 - F(\xi, \tau)$, with the coherence function $F(\xi, \tau) = |g^{(1)}(\xi, \tau)|^2$ given by (2.112) under the currently assumed experimental conditions of a 100% polarized electron beam and two detectors situated an equal distance from the source within a coherence length of one another ($\xi = 0$). The actual number of coincidences would then be

$$N_c(\tau) = \overline{N}_1 \overline{N}_2 \frac{t_r}{t} e^{-R|\tau|} \left[1 - F(\tau) \right] , \qquad (2.157)$$

if the detectors (and associated circuitry) could respond instantaneously, which of course is not possible. Thus the coincidence count must be averaged over the resolution time of the detector

$$\langle N_c(\tau) \rangle = \int_{\tau - t_r/2}^{\tau + t_r/2} N_c(\tau') F_D(\tau - \tau') d\tau' , \qquad (2.158)$$

where we will adopt a detector window function

$$F_D(t) = \frac{1}{t_r} \begin{cases} 1 & |t| \leq t_r , \\ 0 & |t| > t_r , \end{cases} \qquad (2.159)$$

of the same form as the spatial aperture function previously used in Chap. 1 [see (1.57) and (1.58)]. For an unpolarized electron beam, the coherence function in (2.157) must be multiplied by $1/2$.

The temporal coherence function $F(\tau)$ of a field-emission source modeled by the Fowler–Nordheim theory is calculable from the probability density (2.151)

$$F(\tau) = \left| \frac{\dfrac{1}{d} \displaystyle\int_0^\mu e^{-(\varepsilon - \mu)/d} e^{-i\varepsilon\tau/\hbar} d\varepsilon}{\dfrac{1}{d} \displaystyle\int_0^\mu e^{-(\varepsilon - \mu)/d} d\varepsilon} \right|^2 = \frac{1 + e^{-2\mu/d} - 2e^{-\mu/d} \cos(\mu\tau/\hbar)}{1 + (d\tau/\hbar)^2} ,$$

$$(2.160)$$

or

$$F(\tau) \xrightarrow[\mu/d \gg 1]{} \frac{t_c^2}{t_c^2 + \tau^2} , \qquad (2.161)$$

where definition of the coherence time $t_c = d/\hbar$ was employed in the reduction (2.161) for a beam of sharply defined energy. One can readily perform the integral (2.158) analytically by treating the exponential factor $e^{-R|\tau|}$ in (2.157) as a constant and removing it from the integrand. This will have little effect on the value of the integral because the exponential factor is very close to unity for τ close to 0 (where electron correlation is important) and approaches 0 at large delays (where electrons are essentially uncorrelated). The averaged coincident count is then

$$\langle N_c(\tau) \rangle = \overline{N}_1 \overline{N}_2 \frac{t_r}{t} e^{-R|\tau|} \left[1 - \langle F(\tau) \rangle \right] , \qquad (2.162)$$

where convolution of the coherence function with the detector response function leads to

$$\langle F(\tau) \rangle = \frac{t_c}{t_r} \left(\tan^{-1} \frac{\tau + t_r/2}{t_c} - \tan^{-1} \frac{\tau - t_r/2}{t_c} \right) . \qquad (2.163)$$

For zero delay and a short response time compared to the coherence time, $\langle F(\tau) \rangle \sim 1$; for a long response time, which is the condition applicable to all experiments to date with free electron beams, $\langle F(\tau) \rangle \sim \pi t_c/t_r \equiv t_0/t_r$. Then $\langle F(\tau) \rangle$ vanishes in the limit $\tau \to \infty$ irrespective of the response and coherence times.

Figure 2.20 shows plots of the logarithm of the coincidence count (2.162) as a function of τ for a decreasing sequence of response times. The plot in Fig. 2.20a portrays the purely Poissonian statistics of uncorrelated particles, the straight lines of equal positive or negative slope corresponding to whether the initiating particle was received by detector D1, whereupon $\tau > 0$, or by D2, for which $\tau < 0$. The drop in coincidence counts around $\tau = 0$ in plots Figs. 2.20b through d is greater the closer the response time approaches the coherence time. Values of this ratio were chosen to accentuate the fermion anticorrelation beyond what is currently realizable in the laboratory.

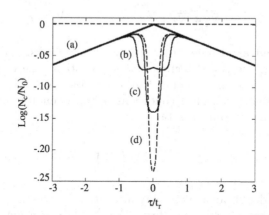

Fig. 2.20. Theoretical variation of electron coincidence counts as a function of delay time τ for different ratios of apparatus response time to coherence time t_r/t_c: (a) ∞, (b) 20, (c) 10, (d) 6. The linear rise and decline shown in (a) marks an incoherent source whose particle emission is governed by Poisson statistics. The decrease in coincidences around $\tau = 0$ in (b), (c), (d) signifies electron antibunching. For purposes of comparison, the number of coincidences is divided by $N_0 = N^2 t_r/t$, where N is the number of particles at each detector in time $t = 1$ hour from an incident coherent particle current of $\sim 10^{-8}$ A

The signal of interest, which is the difference between the total number of coincidences and the accidental number of coincidences observed in a counting time t, is

$$S(\tau) = \langle N_c(\tau) \rangle - \langle N_a(\tau) \rangle = \overline{N}_1 \overline{N}_2 \frac{t_r}{t} e^{-R|\tau|} \langle F(\tau) \rangle . \tag{2.164}$$

At $\tau = 0$, the signal assumes its largest value, which our model is

$$S(0) = \overline{N}_1 \overline{N}_2 \frac{t_r}{t} \langle F(0) \rangle = \overline{N}_1 \overline{N}_2 \frac{2t_c}{t} \tan^{-1} \frac{t_r}{2t_c} . \tag{2.165}$$

Whether this difference count is statistically significant or not depends on how it compares to the fluctuations in the number of accidental coincidences, which we quantify by the root-mean-square value

$$\Delta N_a = \sqrt{\overline{N_a^2} - \overline{N_a}^2} \sim \sqrt{\overline{N}_1 \overline{N}_2 (\overline{N}_1 + \overline{N}_2)} \frac{t_r}{t} . \tag{2.166}$$

The second (approximate) relation in (2.166) was derived by treating the fluctuations at the two detectors as uncorrelated Poisson processes with mean counts much larger than unity. A comparison of the signal in (2.165) to the 'noise' in (2.166) thus leads to

$$\frac{S}{N} = \sqrt{\frac{\overline{N}_1 \overline{N}_2}{\overline{N}_1 + \overline{N}_2}} \frac{t_r}{t} \langle F(0) \rangle \longrightarrow \sqrt{\frac{Rt}{2} \frac{t_0}{t_r}} , \tag{2.167}$$

where the second expression in (2.167) results from assuming symmetric counting rates at the two detectors and a resolution time long in comparison to the coherence time. From (2.167) it follows that the total counting time required (under the ideal conditions assumed for this illustration) to achieve a given signal-to-noise ratio is

$$t = \frac{2(S/N)^2}{R} \left(\frac{t_r}{t_0} \right)^2 . \tag{2.168}$$

I leave to Appendix 2D an analogous treatment of the experimental procedure to observe the cross-correlation in fluctuations of electron counts at two detectors, which would be a fermionic counterpart to the HBT optical experiment manifesting a correlation between photons in two coherent beams of light [39].

In the 1980s, I began to examine the prospects of observing the antibunching of electrons and other manifestations (discussed in the following chapter) of quantum interference resulting from the antisymmetry of multi-fermion wave functions. The feasibility of detecting the fermionic contribution in autocorrelation, cross-correlation, and coincidence experiments with electron beams was marginal, but not hopeless, as would have been (and still is) the case with beams of neutrons and other elementary or composite fermions. Few laboratories were equipped to perform such experiments, and, to my knowledge, the attempt to observe fermionic correlations with electron beams was pursued principally by two groups in friendly rivalry, one led by Franz Hasselbach at the University of Tübingen, and the other comprising Michael Scheinfein, John Spence, and me at the NSF High-Resolution Electron Microscope Facility at Arizona State University. Regrettably, the second group had to abandon the experiments when anticipated funding was not forthcoming.

With progress in electronic microfabrication techniques, electron antibunching was eventually observed in the correlation of currents in small solid-state devices [81] (which I discussed in a previous book, *A Universe of Atoms* [82]). In such devices, where electrons fill nearly all levels up to the Fermi level, the degeneracy approaches unity, and manifestations of fermion quantum statistics are more readily demonstrable. Nevertheless, there are advantages to the use of free beams in many types of experiments. It is a tribute to the persistence of the Tübingen group that in 2002, more than a decade after I visited Dr. Hasselbach to discuss the initiation of such experiments, he and his associates successfully observed the anticorrelations of free electrons by means of coincidence measurements with a low-voltage field-emission beam [83].

The electron-optical apparatus of the Tübingen experiment resembled a miniature version of the HBT stellar interferometer. A 900 eV beam of electrons with effective dispersion of 0.13 eV was focused by quadrupole lenses so as to produce an elliptically shaped beam of coherent electrons. With the major axis of the ellipse oriented so that it passed through the centers of the two detectors (in a configuration like that of Fig. 2.19), the entire surface of each detector was coherently illuminated. However, with the ellipse rotated 90°, the detectors were only partially coherently illuminated. A statistically significant difference (3 standard deviations) between the number of coincidences obtained under the two conditions manifested the antibunching of electrons.

With a coherent particle current $R = 4.7 \times 10^9$ electrons/sec and a coherence time of $t_0 = 3.25 \times 10^{-14}$ s, the beam degeneracy was approximately $\delta = Rt_0 = 1.5 \times 10^{-4}$. The theoretically expected fractional reduction in coincidences within the time-resolution window $t_r = 26 \times 10^{-12}$ s of the apparatus can be estimated from (2.156) and (2.165) to be

$$\frac{N_a - N_c}{N_a} \sim \frac{t_0}{t_r} = \frac{Rt_0}{Rt_r} = \frac{\delta}{N_r} = 1.2 \times 10^{-3} \,,$$

in accord with the experimental result.

Although the effect seen is still weak, the experiment represents an important first step towards realizing the experimental prototypes of the following chapter that manifest new quantum interference effects involving correlations of particles.

Appendix 2A   Consequences of Spectral Width on Photon Correlations

In real light sources, the photon spectrum is not infinitely sharp, but is broadened by various physical mechanisms like spontaneous decay, collisional interactions, and the presence of a nonuniform electric or magnetic field. An examination of the effect of spectral broadening on the correlation function, (2.14), reproduced below

$$P_{12}(\theta, \theta_2) = \cos^2(\theta_1 - \theta_2) \cos^2\left[\frac{1}{2}\omega_{\alpha\beta}\left(\Delta t - \frac{\Delta z}{c}\right)\right] , \qquad (2A.1)$$

is instructive, for it reveals to what extent spectral broadening may degrade the perfect correlation produced by monochromatic waves. The simple case to be analyzed here is that in which the frequencies of the correlated photons are governed by a Gaussian spectral profile

$$f(\omega) = \frac{1}{\sqrt{2\pi}\sigma} e^{-(\omega-\omega_0)^2/2\sigma^2} , \qquad (2A.2)$$

with central angular frequency ω_0 and width σ. The spectral average of the correlation function (2A.1) is simplified by the fact that it is the frequency difference $\omega_{\alpha\beta} = \omega_\alpha - \omega_\beta$ that appears in the argument of the cosine. Thus, rather than performing separate integrations over the frequency distribution of each photon of the pair, we can use the properties of a Gaussian distribution to determine from (2A.2) the frequency distribution of the difference of two normal random variables. In the case of a random variable $\overline{\omega} = \omega_\alpha - \omega_\beta$, with both terms in the difference governed by the distribution (2A.2), the distribution of the difference frequency takes the form[9]

$$f(\overline{\omega}) = \frac{1}{\sqrt{2\pi}(\sqrt{2}\sigma)} e^{-\overline{\omega}^2/4\sigma^2} . \qquad (2A.3)$$

In other words, $\overline{\omega}$ itself is a normal random variable of zero mean and standard deviation $\sqrt{2}\sigma$.

Averaging the correlation function (2A.1) by means of the spectral distribution (2A.3) leads to

$$\langle P_{12}(\theta_1, \theta_2)\rangle_\omega = \frac{1}{2}\cos^2(\theta_1 - \theta_2)\left[1 + e^{-2\sigma^2(\Delta t - \Delta z/c)^2}\right] , \qquad (2A.4)$$

---

[9] A normal random variable X of mean μ and standard deviation σ, represented symbolically by $N(\mu, \sigma^2)$, is uniquely defined by its moment-generating function (mgf) $g_X(t) = \langle e^{Xt}\rangle = e^{\mu t + \sigma^2 t^2/2}$, in which the angular brackets signify the operation $\langle h\rangle = \int h(\omega)f(\omega)d\omega$. The mgf of the difference of two random variables $Y = X_1 - X_2$ is the product $g_Y(t) = g_{X_1}(t)g_{X_2}(-t)$. If X_1 and X_2 are independent random variables distributed according to $N(\mu, \sigma^2)$, then $g_Y(t) = e^{2\sigma^2 t^2/2}$ and $Y = N(0, 2\sigma^2)$. For a comprehensive discussion of moment-generating functions and the statistics of functions of normal random variables, see [84].

where the subscript ω explicitly denotes an average with respect to frequency. Recall that the parenthetical expression in the exponent of (2A.4) is the difference in the retarded time intervals of detection, i.e., $\tau = \tau_1 - \tau_2$, in which $\tau_i = t_i - z_i/c$, $i = 1, 2$. If correlated photon pairs are produced at the rate $1/T$, then the time-average of the correlation function (2A.4) yields

$$\langle P_{12}(\theta_1, \theta_2)\rangle_{\omega,\tau} = \frac{1}{T}\int_0^T \langle P_{12}(\theta_1, \theta_2)\rangle_\omega d\tau$$

$$\overset{\sigma T \gg 1}{\longrightarrow} \frac{1}{2}\cos^2(\theta_1 - \theta_2)\left(1 + \frac{\sqrt{\pi}}{2\sqrt{2}\sigma T}\right). \qquad (2A.5)$$

Thus, if the product of the spectral width and production time interval is sufficiently large that the second term in the bracketed expression of (2A.5) can be neglected, the pair correlation function reduces to one-half that for a monochromatic beam of entangled photons. The result $(1/2)\cos^2(\theta_1 - \theta_2)$ is what one would have obtained initially by *not* symmetrizing the state vector in (2.6); i.e., not taking account of the invariance of the state vector under particle exchange.

Appendix 2B   Chemical Potential at $T = 0$ K

At $T = 0$ K, the chemical potential μ is equal to the Fermi energy, the highest level filled by a system of N fermions of mass m in a volume V. Since the Pauli exclusion principle limits the number of fermions in each distinct quantum state to 0 or 1, the number of fermions filling all levels to μ is equal to the product of the number of states per cell of phase space (degeneracy factor $g = 2S + 1 = 2$ for spin-1/2 particles) by the number of filled cells, or

$$N = g\int\frac{d^3x d^3p}{h^3} = \frac{2(4\pi)V}{h^3}\int_0^{p_F} p^2 dp = \frac{8\pi p_F^3}{3h^3}V, \qquad (2B.1)$$

where p_F is the Fermi momentum. Thus, the Fermi momentum and number density of particles $n = N/V$ are related by

$$p_F = \left(\frac{3h^3 n}{8\pi}\right)^{1/3} = (3\pi^2\hbar^3 n)^{1/3}. \qquad (2B.2)$$

The nonrelativistic expression for chemical potential is then

$$\mu = \frac{p_F^2}{2m} = \frac{(3\pi^2\hbar^3 n)^{2/3}}{2m}. \qquad (2B.3)$$

Tungsten is a chemical element with molar mass 184 g, specific gravity 19.25, and (for purposes of this illustration) 2 outer-shell electrons. The number density of conduction electrons in tungsten is then approximately

$$n = (2 \text{ e/atom}) \times (N_{\text{Av}} \text{ atoms/mol}) \times (1 \text{ mol}/0.184 \text{ kg}) \times (1.93 \times 10^4 \text{ kg/m}^3)$$

$$= 1.26 \times 10^{29} \text{ m}^{-3} ,$$

leading to $\mu = 1.46 \times 10^{-18}$ J $= 7.7$ eV, where $N_{\text{Av}} \sim 6.02 \times 10^{23}$ is Avogadro's number.

The exact relativistic expression for chemical potential will play an important part in the physics of neutron stars and black holes which I will discuss in the last chapter.

Appendix 2C   Probability Density of a Sum of Random Variables

Consider the random variable $Z = X + Y$, which is a sum of two independent random variables with respective probability densities $p_X(x)$ and $p_Y(y)$. As is frequently the convention in statistical analyses, upper-case letters will designate random variables and corresponding lower-case letters will designate the realizations or values that the variable can have. We assume that X and Y can range over the entire set of real numbers. If a random variable, let us say X, is restricted to a given range (e.g., just positive numbers), then the probability density $p_X(x)$ can be defined to be 0 outside the specified range.

By definition, the cumulative probability that Z is less than or equal to a certain value z is then

$$P(Z \leq z) = \int_{-\infty}^{z} p_Z(z')\mathrm{d}z' = P(X + Y \leq z)$$

$$= \int_{-\infty}^{\infty} p_X(x)\mathrm{d}x \int_{-\infty}^{z-x} p_Y(y)\mathrm{d}y . \tag{2C.1}$$

From the first equality in (2C.1), it follows that the probability density $p_Z(z)$ is obtained by taking the first derivative of the cumulative probability with respect to z,

$$p_Z(z) = \frac{\mathrm{d}P(Z \leq z)}{\mathrm{d}z} = \int_{-\infty}^{\infty} p_X(x)P_Y(z - x)\mathrm{d}x . \tag{2C.2}$$

The second equality in (2C.2) is a special case of Leibniz's rule for differentiating integrals

$$\frac{d}{dx}\int_{a(x)}^{b(x)} f(x,y)dy = \frac{db(x)}{dx}f(x,b) - \frac{da(x)}{dx}f(x,a) + \int_{a(x)}^{b(x)} \frac{\partial f(x,y)}{\partial x}dy \ .$$

Appendix 2D   Correlated Fluctuations of Electrons at Two Detectors

Consider a field-emission beam of 100% spin-polarized electrons (coherence time t_c) split so as to illuminate equally and coherently (at the rate R particles/sec) each of two detectors (D1 and D2) with the same response time t_r. In contrast to the previously treated case of coincidence counting, the procedure now is for the counting circuitry associated with each detector to register all particles that arrive within an interval t_r, to multiply the two counts together, and to repeat this operation over a large number of sequential intervals. The experimental outcome of the acquired statistics is the cross correlation in fluctuations $\overline{\Delta N_1 \Delta N_2}$, where the mean-square fluctuation in counts at each detector takes the form

$$\overline{(\Delta N_i)^2} = \overline{N}_i \big(1 - \overline{N}_i \langle F \rangle \big) \approx \overline{N}_i \left(1 - \overline{N}_i \frac{t_0}{t_r}\right) \qquad (i = 1,2) \ , \qquad (2D.1)$$

in which the mean counts achieved in time t are $\overline{N}_1 = \overline{N}_2 = Rt$, and the average of the coherence function over the response time ($t_r \gg t_c$) is $\langle F \rangle \sim \pi t_c / t_r \equiv t_0 / t_r$.

To evaluate the cross-correlation in fluctuations we employ an insightful method of reasoning applied by Purcell [44] to the original HBT intensity interferometry experiment. Imagine connecting the two detectors together to form a single detector that has registered a total of $N = N_1 + N_2$ counts. Then the variance in this total count would be

$$\overline{(\Delta N)^2} = \overline{(\Delta N_1 + \Delta N_2)^2} = \overline{(\Delta N_1)^2} + \overline{(\Delta N_2)^2} + 2\overline{\Delta N_1 \Delta N_2} \ , \qquad (2D.2)$$

as well as have the form of (2D.1)

$$\overline{(\Delta N)^2} = \overline{N} \left(1 - \overline{N}\langle F \rangle\right) \ . \qquad (2D.3)$$

Substitution of (2D.1) into (2D.2) and comparison with (2D.3) gives immediately

$$\overline{\Delta N_1 \Delta N_2} = -\overline{N}_1 \overline{N}_2 \langle F \rangle \ . \qquad (2D.4)$$

For the signal (2D.4) to be statistically significant, it must be detectable above the noise, which we quantify by the product of the root-mean-square fluctuations at each detector

$$\sqrt{\overline{(\Delta N_1)^2}\,\overline{(\Delta N_2)^2}} = \sqrt{\overline{N}_1\overline{N}_2}\,, \tag{2D.5}$$

estimated to good approximation by Poisson statistics. The resulting signal-to-noise ratio is then

$$\frac{S}{N} = \frac{|\overline{\Delta N_1\Delta N_2}|}{\sqrt{\overline{N}_1\overline{N}_2}} = \sqrt{\overline{N}_1\overline{N}_2}\langle F\rangle \approx \frac{Rtt_0}{t_{\rm r}}\,. \tag{2D.6}$$

The observing time t to achieve a desired S/N

$$t = \frac{S/N}{R}\frac{t_{\rm r}}{t_0} \tag{2D.7}$$

is proportional to S/N and the ratio of resolving time to correlation time, in contrast to the corresponding expression (2.168) for a coincidence-counting experiment in which the observing time increases with the square of these quantities.

One further point of interest is to consider an initially unpolarized electron beam that is divided into two components of opposite 100% spin polarization. In that case, (2D.3) for the combined detector outputs must be modified by inserting a factor of 1/2 in the second term

$$\overline{(\Delta N)^2} = \overline{N}\left(1 - \frac{1}{2}\overline{N}\langle F\rangle\right)\,, \tag{2D.8}$$

whereas the fluctuations of the polarized electrons at the individual detectors expressed by (2D.1) remains unchanged. A repetition of the same mode of reasoning that led to (2D.4) now leads to $\overline{\Delta N_1\Delta N_2} = 0$, showing that electrons of opposite spin are uncorrelated.

3

Correlations and Entanglements II: Interferometry of Correlated Particles

3.1 Interferometry of Correlated Particles

The quantum processes discussed in the first chapter involved exclusively the self-interference of single particles passing through apertures or around obstacles or either around or through an external field, depending on whether the field in question was the magnetic field or vector potential. These variations on the theme of Young's two-slit experiment are all examples of the first-order coherence of quantum particles. The Hanbury Brown–Twiss experiment, discussed in the previous chapter, was the first to manifest the second-order coherence of light and provides a prototype for innovative ways to probe the phenomena arising from multiparticle quantum states.

In the following sections I will consider examples of the interferometry of correlated charged particles that combine various features of the basic Young's two-slit experiment, the Aharonov–Bohm (AB) effect, the Einstein–Podolsky–Rosen (EPR) paradox, and the Hanbury Brown–Twiss (HBT) experiments. These novel processes manifest simultaneously three distinct kinds of quantum interference: (1) interference, dependent upon optical path length difference, resulting from the wave-like propagation of particles (or, more accurately, the wave-like description of particle ensembles); (2) interference, dependent upon confined magnetic flux, resulting from particle charge and spatial topology; and (3) interference, dependent upon quantum statistics, resulting from particle indistinguishability under exchange. Although the self-interference of single particle amplitudes has been called (by Feynman, especially) a mystery that cannot be explained but only described mathematically, physicists have lived with it long enough to develop an intuition for what is likely to occur in given circumstances. However, the study of correlated particle interference in vacuum or in the presence of potential fields is still sufficiently unexplored theoretically and untried experimentally that results can be strange even by the familiar standards of quantum mechanics.

3.2 The Aharonov–Bohm (AB) Effect with Entangled Electrons

It has been noted before in Chap. 1 that the AB effect is an intrinsically nonlocal physical phenomenon in which the spatial distribution of particles diffracting around the solenoid or toroid is modified by a magnetic field through which the particles never pass. The nonlocal nature of this effect takes on even stranger dimensions when produced in a context reminiscent of the EPR paradox [85].

In all AB experiments performed to date, the interference pattern is effectively built up by transit of one particle at a time through the apparatus as with a Young's two-slit experiment with weak light source. Consider, however, an experimental configuration like that of Fig. 3.1 with two well separated solenoids and a source that produces pairs of charged particles simultaneously. After production, the two particles separate, each propagating around one of the regions of confined magnetic flux to be received afterward at one of two local detectors. The charged particles, which we will suppose to be electrons, are identical and cannot be distinguished, but are required to be correlated in linear momentum – that is, if one particle propagates toward a particular mirror to the right of the source (interferometer 1), the other particle propagates toward a particular mirror to the left of the source (interferometer 2).

To be correlated in this manner, the two electrons had to have once been part of a common system. For example, perhaps the source contains a supply of the exotic atomic species $\mu^+ e^- e^-$ (analogous to the H^- ion). Upon decay of the muon into a positron, neutrino, and antineutrino, the electrons, which are

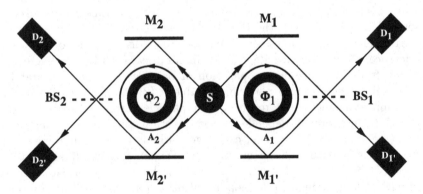

Fig. 3.1. Hybrid Aharonov–Bohm and Einstein–Podolsky–Rosen (AB–EPR) experiment. Electrons issue from source S in pairs with one electron entering interferometer 1 and the other electron, whose linear momentum is correlated with the first, entering interferometer 2. The AB effect is manifested, not in the counts at individual detectors, but only in the correlation of counts between one detector at interferometer 1 and another detector at interferometer 2

in a $^1S_0$ ground state, would fly off in opposite directions in order to conserve the total linear momentum of the system. Thus, if one electron propagated toward mirror M_1, the other would propagate toward mirror $M_{2'}$, and likewise for the directions leading to mirrors $M_{1'}$ and M_2. There is nothing intrinsically quantum mechanical about a system at rest splitting into two constituents with the foregoing velocity correlation. It is worth noting, however, that this particular correlation is not required, and that the essential physics would be unchanged if a different momentum correlation were presumed.

We are interested in the following questions: What is the probability of receiving an electron at a left-side detector D_2 or $D_{2'}$ given that an electron was received at one of the right-side detectors, e.g., D_1? And what is the probability of receiving an electron at D_1 *irrespective* of the detector to which the other electron may have gone?

One may be tempted to answer these questions according to ideas and imagery drawn from classical physics that the two electrons – pictured perhaps as mutually receding wave packets of some specified coherence length – propagate independently (except for the velocity correlation posed by momentum conservation) and cease to belong to a common system once the packet overlap has become negligibly small. Then, since the experimental conditions are such that each observer cannot tell by which pathway (source $\longrightarrow$ mirror $\longrightarrow$ detector) a locally detected electron has gone around a solenoid, quantum interference should occur. The electrons, therefore, should give rise independently to AB effects at each end of the double interferometer with a flux-dependent probability of a form similar to that of (1.31)

$$P_j(\Phi_i) = \alpha_j + \beta_j \cos\left(\delta_j + 2\pi\frac{\Phi_i}{\Phi_0}\right) ,$$ (3.1)

where $i = 1, 2$, respectively, denotes the right-side or left-side solenoid, and $j = 1, 1', 2, 2'$ designates one of the detectors; $\Phi_0 = e/hc$ is the fluxon. Moreover, the joint probability of particle detection – as for any independent events, classical or quantum – would simply be the product of the corresponding probabilities (3.1), e.g.,

$$P_{j,k}(\Phi_1, \Phi_2) = P_j(\Phi_1)P_k(\Phi_2) .$$ (3.2)

Reasonable as it may appear, these deductions are not correct.

Let $A(M_1, M_2; D_1, D_2)$ be the amplitude for propagation of one particle via mirror M_1 to detector D_1 and the other particle via mirror M_2 to detector D_2 with similarly represented amplitudes for the other pathways. Further, let r_1, t_1 be the respective reflection and transmission amplitudes for a particle incident on beam splitter BS_1 from above and r_1', t_1' the corresponding amplitudes for incidence from below. The analogous amplitudes for beam splitter BS_2 will be subscripted with 2. It may seem redundant to ascribe left-side and right-side reflection and transmission amplitudes to a device (the beam splitter) that is apparently symmetric under mirror inversion. Although it is indeed the case that the *probability* of a particle being transmitted or reflected at a lossless beam splitter does not depend on whether the particle is incident at one side or the other, the corresponding *amplitudes* for these two processes can in fact differ by a relative phase. The origin of this phase may be traced to the mathematical requirement that the transfer matrix (the elements of which are the reflection and transmission amplitudes) connecting input and output states must be unitary. As a consequence of this unitarity, one can derive the relations [86]

$$r_i r_i' - t_i t_i' = |r_i|^2 + |t_i|^2 = 1 , \qquad r_i^* t_i' + t_i^* r_i' = 0 , \tag{3.3}$$

with $(i = 1, 2)$, from which it follows that

$$r_i' = r_i^* , \qquad t_i' = -t_i^* . \tag{3.4}$$

On the basis of relations (3.3) and (2.7) one can take the reflection amplitudes in the present case to be purely real $(r_i = r_i' \equiv R_i)$ and the transmission amplitudes to be purely imaginary $(t_i = t_i' \equiv iT_i)$.

The eight possible amplitudes for joint receipt of particles at the detector pairs (D_1, D_2) or (D_1, D_2') are then expressible as

$$A(M_1, M_2; D_1, D_2) = R_1 R_2 \exp\left[i(\alpha_1 + \alpha_2)\right] , \tag{3.5}$$

$$A(M_{1'}, M_{2'}; D_1, D_2) = -T_1 T_2 \exp\left[i(\alpha_{1'} + \alpha_{2'})\right] , \tag{3.6}$$

$$A(M_1, M_{2'}; D_1, D_2) = iR_1 T_2 \exp\left[i(\alpha_1 + \alpha_{2'})\right] , \tag{3.7}$$

$$A(M_{1'}, M_2; D_1, D_2) = iT_1 R_2 \exp\left[i(\alpha_{1'} + \alpha_2)\right] , \tag{3.8}$$

$$A(M_1, M_2; D_1, D_{2'}) = iR_1 T_2 \exp\left[i(\alpha_1 + \alpha_2)\right] , \tag{3.9}$$

$$A(M_{1'}, M_{2'}; D_1, D_{2'}) = iT_1 R_2 \exp\left[i(\alpha_{1'} + \alpha_{2'})\right] , \tag{3.10}$$

$$A(M_1, M_{2'}; D_1, D_{2'}) = R_1 R_2 \exp\left[i(\alpha_1 + \alpha_{2'})\right] , \tag{3.11}$$

$$A(M_{1'}, M_2; D_1, D_{2'}) = -T_1 T_2 \exp\left[i(\alpha_{1'} + \alpha_2)\right] , \tag{3.12}$$

where the relative phases incurred by passage of a particle (with charge e) to one side or the other of the solenoids with vector potential fields $\boldsymbol{A}_1$ and $\boldsymbol{A}_2$ take the form of relations (1.21), (1.22)

$$\alpha_1 = \frac{e}{\hbar c} \int\limits_{\text{path } S \to M_i \to BS_k} \boldsymbol{A}_k \cdot d\boldsymbol{s}_i , \tag{3.13}$$

with $k = 1$ for $i = 1, 1'$ and $k = 2$ for $i = 2, 2'$. In order not to obscure the principal physics of interest with more complicated notation than is necessary, I have assumed that:

- the geometrical path lengths of the four specified pathways are equal,
- the mirrors contribute no differential phase shifts,
- the detectors have 100% efficiency.

If the momentum correlation of the two particles corresponds to that required by the conservation of linear momentum, i.e., back-to-back separation, then the amplitude for joint detection of one particle at D_1 and the other particle at D_2 (or $D_{2'}$) is given by the linear superpositions

$$A(D_1, D_2) = \frac{1}{\sqrt{2}}\left[A(M_1, M_{2'}; D_1, D_2) + A(M_{1'}, M_2; D_1, D_2)\right], \qquad (3.14)$$

$$A(D_1, D_{2'}) = \frac{1}{\sqrt{2}}\left[A(M_1, M_{2'}; D_1, D_{2'}) + A(M_{1'}, M_2; D_1, D_{2'})\right]. \qquad (3.15)$$

The resulting probabilities of joint detection are then

$$P(D_1, D_2) = |A(D_1, D_2)|^2 \qquad (3.16)$$
$$= \frac{1}{2}\left[(R_1 T_2)^2 + (T_1 R_2)^2 + 2R_1 R_2 T_1 T_2 \cos\frac{2\pi(\Phi_1 - \Phi_2)}{\Phi_0}\right],$$

$$P(D_1, D_{2'}) = |A(D_1, D_{2'})|^2 \qquad (3.17)$$
$$= \frac{1}{2}\left[(R_1 R_2)^2 + (T_1 T_2)^2 - 2R_1 R_2 T_1 T_2 \cos\frac{2\pi(\Phi_1 - \Phi_2)}{\Phi_0}\right],$$

where, by application of Stokes' law,

$$\Phi_1 = \alpha_1 - \alpha_{1'}, \qquad \Phi_2 = \alpha_2 - \alpha_{2'} \qquad (3.18)$$

are the magnetic fluxes through interferometers 1 and 2, respectively. As indicated by the sense of circulation of the vector potential fields in Fig. 3.1, the flux Φ_1 is positive for a magnetic field directed into the paper, and the flux Φ_2 of the other solenoid is positive for a magnetic field directed out of the paper.

The joint detection probabilities of relations (3.16), (3.17) manifest a quantum interference that depends on the difference of the magnetic fluxes in solenoids 1 and 2. In other words, from the perspective of one of the observers – let us say the one by detector D_1 on the right side – the number of electrons counted is influenced not only by the nearby solenoid, but *also* by the distant solenoid around which the electrons received at D_1 could not have propagated. If the flux through solenoid 1 is null, i.e., $\Phi_1 = 0$, then the AB effect inferred by the observer at D_1 is attributable *entirely* to the flux Φ_2 in the remote solenoid.

The situation is in fact stranger still. Suppose the arrivals of the electrons emitted into the interferometer on the left side are not observed. Then, from (3.16), (3.17), the probability of detecting an electron at D_1 *irrespective* of the path taken by the corresponding electron of the pair is

$$P(D_1) = P(D_1, D_2) + P(D_1, D_{2'}) = \frac{1}{2}, \qquad (3.19)$$

which is a constant and displays no quantum interference effect at all.

Phrased somewhat differently, an observer counting electrons at D_1 and oblivious to the existence of other observers at D_2 and $D_{2'}$ – who may be arbitrarily far away – will see no AB effect. Only when these particle counts are correlated with those of a distant observer is the AB effect manifested. Thus, the inferences drawn by the observer on the right are strongly influenced by the measurements made (or not made) by a remote observer on the left. And yet in either case, as seen from the perspective of the right-side observer, it is the 'same' beam of electrons diffracting to one side or the other of his solenoid.

Had the alternative correlation of electrons propagating via mirror pair (M_1, M_2) or $(M_{1'}, M_{2'})$ been chosen, the resulting joint detection probabilities would depend on different combinations of reflection and transmission amplitudes, but the disappearance of the AB effect, as expressed in (3.19), would not have changed. The essential condition is that the electron motion be correlated. If, however, the jointly produced electrons are *uncorrelated*, in which case each pair of pathways from source to mirrors has equal probability, then the resulting joint detection probability, as expected, is simply the product of the single-particle detection probabilities for each detector

$$P(D_1, D_2) = \frac{1}{4}\left(1 + 2\sqrt{R_1 T_1}\sin\frac{2\pi\Phi_1}{\Phi_0}\right)\left(1 + 2\sqrt{R_2 T_2}\sin\frac{2\pi\Phi_2}{\Phi_0}\right) , \quad (3.20)$$

$$P(D_1, D_{2'}) = \frac{1}{4}\left(1 + 2\sqrt{R_1 T_1}\sin\frac{2\pi\Phi_1}{\Phi_0}\right)\left(1 - 2\sqrt{R_2 T_2}\sin\frac{2\pi\Phi_2}{\Phi_0}\right) . \quad (3.21)$$

In this case the probability of detecting an electron at D_1 irrespective of the fate of the companion electron now becomes

$$P(D_1) = \frac{1}{4}\left(1 + 2\sqrt{R_1 T_1}\sin\frac{2\pi\Phi_1}{\Phi_0}\right) . \quad (3.22)$$

Equation (3.22) is a function of the magnetic flux of solenoid 1 only, as would be expected for the ordinary single-solenoid AB effect with uncorrelated single-particle wave packets. The appearance of a sine in (3.22) instead of a cosine as in (3.1) merely reflects the 90° phase shift between the amplitudes for reflection and transmission at a beam splitter.

In the foregoing example, the intrinsic spin of the correlated particles has played no role. Consider an alternative experimental configuration, shown in Fig. 3.2, in which the correlated paths of the particles are directly determined by particle spin. This configuration has the interesting feature of distinguishing fermions from bosons.

We suppose, as before, that the source produces two-particle singlet wave packets, and that the first set of beam splitters $BS_{1'}$ and $BS_{2'}$ reflect spin-up particles and transmit spin-down particles with 100% probability. This results in correlated paths between the source and detectors involving mirror pair (M_1 and $M_{2'}$) for a spin-up particle propagating to the right and spin-down particle propagating to the left, and mirror pair ($M_{1'}$ and M_2) for the

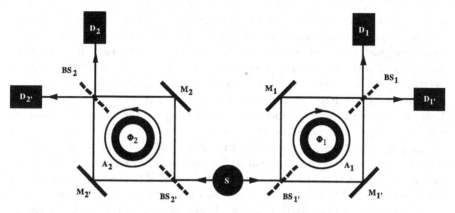

Fig. 3.2. Alternative configuration of the AB–EPR experiment sensitive to particle spin. The correlation in particle counts between one detector at interferometer 1 and a second detector at interferometer 2 can reveal whether the particles are fermions or bosons

opposite situation. The spin-statistics connection, however, gives rise to a relative phase of 0 ($e^{i0} = +1$) or π ($e^{i\pi} = -1$) between the amplitudes for the direct and exchange processes depending on whether the particles are bosons or fermions, respectively. The beam splitters BS_1 and BS_2 are not sensitive to spin and have the reflection and transmission amplitudes R_1, R_2, iT_1, iT_2 as designated previously.

By an analysis largely the same as that of the experimental configuration of Fig. 3.1 which is insensitive to spin, one may demonstrate that the probabilities for joint arrival of particles at the detector pairs (D_1, D_2) and (D_1, $D_{2'}$) are

$$P(D_1, D_2) = \frac{1}{2}\left[(R_1 T_2)^2 + (T_1 R_2)^2 \pm 2R_1 R_2 T_1 T_2 \cos\frac{2\pi(\Phi_1 - \Phi_2)}{\Phi_0}\right],$$
$$(3.23)$$

$$P(D_1, D_2') = \frac{1}{2}\left[(R_1 R_2)^2 + (T_1 T_2)^2 \mp 2R_1 R_2 T_1 T_2 \cos\frac{2\pi(\Phi_1 - \Phi_2)}{\Phi_0}\right],$$
$$(3.24)$$

where upper and lower signs of the interference terms, respectively, refer to bosons and fermions. As before, the single-particle detection probability shows no quantum interference, i.e., $P(D_1) = 1/2$. Also, if the particles are again uncorrelated in their motions – i.e., if one particle can take any available pathway irrespective of the pathway taken by the other particle – then the detectors on one side of the interferometer would register a quantum interference effect independent of the detectors on the other side, and, of course, the joint detection probability would factor just as in relation (3.2).

When looked at with expectations based on classical physics, there is a sort of double irony in the outcomes of these proposed AB–EPR experiments. First, as with the ordinary AB effect, the *local* magnetic field *through* which electrons do not pass can influence the electron spatial distribution. And second, the *distant* magnetic field *around* which electrons do not pass can also influence their spatial distribution. In other words, the act of *not* observing where electrons go on one side of the interferometer apparently destroys the quantum interference of the detected electrons on the other side of the interferometer. There is no restriction in principle on how far apart the separated pairs of electrons can be at the time of detection.

Although these results may seem to defy common sense – which they do because common sense is rarely tutored by quantum mechanical experiences – they are a direct consequence of the entanglement of the electron wave function as represented, for example, by expressions (3.14), (3.15). In general, the wave function of a multiparticle system is entangled if it cannot be factored into a product of single-particle wave functions. From the standpoint of quantum mechanics, the entangled system of particles remains a *single* system, despite subsequent separation of its components, until the entanglement is destroyed by a measurement, thermal contact with the environment, or some other external interaction.

3.3 Hanbury Brown–Twiss Correlations of Entangled Electrons

In the first chapter, the AB effect was introduced by means of a Young's two-slit experiment with a flux-bearing solenoid inserted between the slits. Let us now reconsider this experimental configuration modified, as shown in Fig. 3.3, by the use of a source which generates pairs of electrons and the addition of a second detector at the viewing screen [87]. Of interest is not only the probability of electron arrival at each detector singly, but also, as in the examples of the previous section, the joint probability of electron arrival at two detectors. For this purpose, the outputs of the two detectors in Fig. 3.3 are schematically linked to a correlator characteristic of the HBT experiment. In particular, we would like to ascertain the effect of the confined magnetic flux on the probability of coincident electron detection.

If the AB and AB–EPR effects, as well as a general 'feel' for quantum mechanics are any guide, one may think that the joint detection probability will be some harmonic function of the confined magnetic flux. But quantum mechanical intuition can be just as misleading as classical mechanical common sense. If the AB effect is a subtle one, then the two-slit AB–HBT experiment, which involves the quantum states of identical particles, is even more so and illustrates strikingly the potential pitfalls of adopting too literally the visual imagery of wave packets.

We assume again for simplicity – although the assumption can be readily relaxed in a more general analysis – that the electrons produced by the source are spin-polarized so they can interfere, and that one electron is to issue from

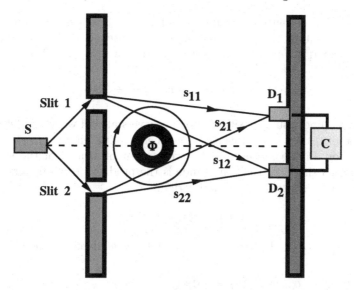

Fig. 3.3. Hybrid AB–HBT experiment. Pairs of electrons issuing from source S diffract around the solenoid and are rece4ived at two detectors whose outputs are correlated. The electron antibunching revealed by the correlated counts in unaffected by the magnetic flux, although the count rate of singly emitted electrons manifests the AB effect

each slit and be received at each detector. By design of the apparatus, the passage of two particles through one slit and none through the other can be made negligibly small. Also, the arrival of two particles at one detector and none at the other would give rise to a null coincidence count, and therefore contribute nothing to the signal.

Unlike the momentum correlation required of the (spin-independent) AB–EPR effect, the entanglement of the electrons in the present case arises in an entirely different way. From a classical perspective, one can imagine two alternative emission–detection processes:

- the arrival at detectors D_1 and D_2 of electrons that issued respectively from slits 1 and 2,
- the converse situation with the particles exchanged.

Because it is impossible to distinguish these two alternatives experimentally (without an intrusive observation that would disturb the system and destroy the quantum interference), the correct quantum mechanical procedure requires that the wave function characterizing the arrival of the electrons at the two detectors be antisymmetrized under particle exchange as follows:

$$\Psi(1,2) = \frac{1}{\sqrt{2}}\Big[\phi_1(D_1)\phi_2(D_2) - \phi_1(D_2)\phi_2(D_1)\Big] \tag{3.25}$$

where $\phi_i(\mathrm{D}_j)$ represents the amplitude of the electron that has propagated from the ith slit to the jth detector. In the absence of the solenoid, this amplitude could be appropriately represented by a wave packet of the form

$$\phi_i(\mathrm{D}_j) = \int g(k)\exp(iks_{ij})\mathrm{d}k \equiv \varphi(s_{ij}) , \qquad (3.26)$$

comprising a linear superposition of plane wave states of wave number k and amplitude $g(k)$ that have traversed a geometric path length s_{ij} between slit and detector. In the presence of the vector potential field of the solenoid, however, each single-particle amplitude is the product of two factors, one of geometric and the other of magnetic origin

$$\phi_i(\mathrm{D}_j) = \varphi(s_{ij})\exp(i\alpha_{ij}) , \qquad (3.27)$$

where, as in (3.13) and (3.18), the phase shifts engendered by the field of the solenoid

$$\alpha_{ij} = \frac{e}{\hbar c}\left(\int \boldsymbol{A} \cdot \mathrm{d}\boldsymbol{s}\right)_{\text{source}\to\text{slit }i\to\text{detector }j} \qquad (3.28)$$

are again related to the confined magnetic flux $\varPhi$ by Stokes' law

$$\alpha_{11} - \alpha_{21} = \alpha_{12} - \alpha_{22} = \frac{e}{\hbar c}\varPhi = \frac{2\pi\varPhi}{\varPhi_0} . \qquad (3.29)$$

The joint probability that one electron is received at each detector (to which the coincidence count rate is proportional) is

$$P(\mathrm{D}_1, \mathrm{D}_2) = |\varPsi(\mathrm{D}_1, \mathrm{D}_2)|^2 , \qquad (3.30)$$

which, upon substituting relations (3.26)–(3.28) into (3.25), leads to an expression of the form

$$P(\mathrm{D}_1, \mathrm{D}_2) = \frac{1}{2}\Big\{|\varphi(s_{11})\varphi(s_{22})|^2 + |\varphi(s_{12})\varphi(s_{21})|^2$$

$$- 2\mathrm{Re}\big[\varphi(s_{11})\varphi(s_{22})\varphi(s_{12})^*\varphi(s_{21})^*\big]\Big\} . \qquad (3.31)$$

There is indeed a quantum interference term, but all dependence on the magnetic flux has vanished! The magnetic phase shifts for the direct and exchange processes have mutually cancelled.

Suppose one of the detectors (e.g., D_2) is turned off so that only electrons at the other detector (D_1) are counted. What the inactive detector now receives is seemingly irrelevant, and it may therefore appear reasonable that the total electron amplitude at the active detector is the linear superposition of electron 'waves' from slits 1 and 2, i.e.,

$$\varPsi(\mathrm{D}_1) \sim \varphi(s_{11})\exp(i\alpha_{11}) + \varphi(s_{21})\exp(i\alpha_{21}) . \qquad (3.32)$$

It would then follow that the (appropriately normalized) single-particle detection probability

$$P(D_1) = |\Psi(D_1)|^2 \tag{3.33}$$

$$= \frac{1}{2}\left\{ |\varphi(s_{11})|^2 + |\varphi(s_{21})|^2 + 2\mathrm{Re}\left[\varphi(s_{11})\varphi(s_{21})^* \exp(i2\pi\varPhi/\varPhi_0)\right] \right\}$$

clearly depends on the confined magnetic flux. One must be careful, however. Can it truly be the case that the AB effect occurs if one looks for electrons at *one* location, and does not occur if one looks for electrons at *two* locations? If so, the present example leads to an extraordinarily puzzling consequence which I illustrate concretely by resorting to a monochromatic plane wave description of the electrons. The above expressions for single and joint detection probabilities then reduce to the following relations for electrons of wave number k:

$$P(D_1) = \frac{1}{2}\left\{ 1 + \cos\left[k(s_{11} - s_{21})\frac{2\pi\varPhi}{\varPhi_0} \right] \right\}, \tag{3.34}$$

$$P(D_2) = \frac{1}{2}\left\{ 1 + \cos\left[k(s_{12} - s_{22})\frac{2\pi\varPhi}{\varPhi_0} \right] \right\}, \tag{3.35}$$

$$P(D_1, D_2) = \frac{1}{2}\left\{ 1 - \cos\left[k(s_{11} - s_{21} + s_{22} - s_{12})\right] \right\}. \tag{3.36}$$

By arranging experimental conditions so that the geometrical phases are

$$k(s_{11} - s_{21}) = k(s_{22} - s_{12}) = -\frac{\pi}{2},$$

that is, symmetrical location of D_1 and D_2 above and below the forward beam direction, and adjusting the magnetic flux so that $\varPhi/\varPhi_0 = 1/4$, one finds that $P(D_1, D_2) = 1$, $P(D_1) = 1$, and $P(D_2) = 0$. How can it be that there is a 100% coincidence count rate if the individual count rate at one of the detectors is zero?

It is the inconsistent treatment of correlated and uncorrelated electron states that lies at the origin of the foregoing paradoxical results. The probabilities $P(D_1)$ and $P(D_2)$ are derived from *single-particle* relations (3.32), (3.33) and therefore depict the case of *uncorrelated* electron propagation through the two slits. Although it may seem reasonable, when only one of the two detectors is registering particles, to imagine the electrons as arriving independently at the detector in single-particle wave packets, this is not correct. The joint probability $P(D_1, D_2)$ for uncorrelated particles is simply the product $P(D_1)P(D_2)$ and is, as expected, a function of the magnetic flux.

For correlated electron pairs, however, the probability that one particle arrives at a particular detector (D_1) irrespective of the subsequent fate of the unobserved companion electron is determined by integrating the joint probability over the full coordinate range of the second detector. Thus, for the special case of plane wave states, one can show that relation (3.36) leads to

$$P(D_1) = \int_{-\infty}^{+\infty} P(D_1, D_2)\mathrm{d}D_2 \sim \frac{1}{2}\left[1 - J_0(kd)\cos(kd\sin\theta) \right], \tag{3.37}$$

where $J_0(kd)$ is the zeroth order Bessel function, d is the slit separation, and θ is the angle (with respect to the forward direction) of detector D_1 seen from the midpoint between the slits. (As in the case of ordinary Fraunhofer diffraction, the approximate far-field relation $s_{21} - s_{11} \sim d \sin \theta$ has been assumed.) Relation (3.37) does not depend on magnetic flux and cannot vanish for nonvanishing $P(D_1, D_2)$. Whether or not an actual coincidence experiment of the type proposed manifests a flux dependence therefore depends on the quantum composition of the electron beam, i.e., on the nature of the particle correlations. This point will be discussed further in the next section.

Although emphasis has so far been placed on the effect (or non-effect) of the isolated magnetic field, the physical implication of the interference term displayed in (3.31) or (3.36) is also of interest, for it is exhibits the electron avoidance behavior referred to as antibunching in the previous chapter. The joint probability of coincident electron detection vanishes when D_1 is coincident with D_2, i.e., $s_{11} - s_{21} = s_{12} - s_{22}$. In contrast to the bunching of photons manifested in the HBT experiments, indistinguishable electrons tend not to arrive in pairs.

3.4 Correlated Particles
in a Mach–Zender Interferometer

The Young's two slit configuration represents what in optics is termed a wave-front splitting interferometer. Portions of a primary incident wave front give rise either directly or indirectly to coherent sources of secondary waves. An alternative type of interferometer, of which the Mach–Zehnder type schematically shown in Fig. 3.4 is an example, makes use of amplitude splitting, i.e., the division of an incident beam into two components which travel different paths before recombining and interfering. Although the wave-front splitting configuration of the previous section resulted in a quantum cancellation effect for the two-particle AB–HBT experiment, this need not be the case for other interferometer configurations. In this section we will examine the interferometry of correlated fermions in a Mach–Zehnder interferometer paying particular attention to the nature of possible fermion correlations and the effects of external potential fields on fermion fluctuations [88].

It is well known theoretically and verified experimentally that photons (which are massless bosons) can manifest a variety of clustering behaviors depending upon the nature of the source. For example, it has already been mentioned that the fluctuations of chaotic light such as black-body radiation manifest a positive cross-correlation, or bunching. However, laser light above threshold, characterized by so-called coherent states,[1] ideally give rise to no fluctuations. The coherent state – which corresponds to the state of a classical

---

[1] The coherent states $|\alpha\rangle$ of a single-mode harmonic oscillator (from which model the optical states are derived) take the following form

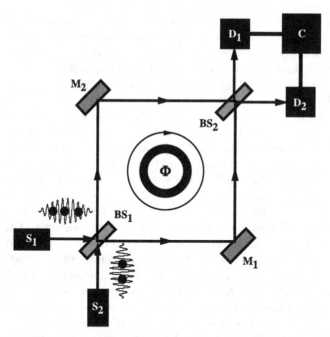

Fig. 3.4. AB–HBT experiment employing a four-port Mach–Zehnder interferometer. Electrons from one or two sources enter the interferometer and propagate around a region of confined magnetic flux. For chaotic states, the cross-correlation in counts at the two detectors manifests a magnetic flux-dependent electron antibunching. SU(2) correlated fermion states lead to other kinds of fluctuation behavior

harmonic oscillator as closely as quantum mechanics allows – has the same coherence properties as a classical stable wave of well-defined amplitude and phase. Finally, under appropriate circumstances the fluorescent emission from single two-level atoms manifests an antibunching effect as characterized by a second-order correlation function less than unity. The photon antibunching arises from the fact that the atom can radiate only from its excited state. Thus, having emitted one photon, it can not emit a second until it has been re-excited.

What about electrons? Is it the case, as has long been thought, that the antisymmetrization of multielectron wave functions required by the Pauli principle *always* leads to electron *anticorrelation*, i.e., fluctuations in which electrons tend to avoid arriving in pairs? Surprisingly (perhaps), the answer is 'no' [90].

$$|\alpha\rangle = \exp\left(-\frac{|\alpha|^2}{2}\right) \sum_{n=0}^{\infty} \frac{\alpha^n}{\sqrt{n!}}|n\rangle \,,$$

in a basis of energy (or excitation number) states $|n\rangle$. For the properties of coherent states and a detailed exposition of photon statistics see [89].

In the general case, particles distributed in some specifiable way over linear momentum and spin states can enter the interferometer of Fig. 3.4 through either or both of the two entrance ports. The entering beam is divided at beam splitter BS_1, reflects from mirrors M_1 and M_2 to propagate around a tube of magnetic flux confined to the center of the interferometer. At beam splitter BS_2 it is divided again and directed into detectors D_1 and D_2 whose outputs are correlated as in the HBT experiment. It is assumed that the mirrors are 100% reflecting and that no particle loss occurs at the beam splitters whose reflectance and transmittance amplitudes are respectively (r_1, t_1) and (r_2, t_2) for beams incident at the left surface and (r_1', t_1') and (r_2', t_2') for beams incident at the right surface. The relations between these amplitudes have been given by (3.3), (3.4).

As in the case of photon interference with multiphoton states where discrete attributes of light play an important role, the analysis of the Mach–Zehnder fermion interferometer is also facilitated by means of a field theoretic (or second-quantized) description. In this approach, similar to that employed in the correlation of spontaneous emission from coherently excited atoms in the previous chapter, the fermion fields – which superficially resemble the wave functions of familiar first-quantized quantum mechanics – are not functions, but operators independent of the state of the system. All information concerning the latter is represented by the density or statistical operator ρ. The fermion field operators at the two input and output ports of the interferometer of Fig. 3.4 can be expressed in a plane-wave basis as follows (with $i = 1, 2$)

$$\Psi_i^{(\text{in})}(x, t) = \sum_{k,s} b_i(k, s) \exp\left[i(kx - \omega t)\right] , \qquad (3.38)$$

$$\Psi_i^{(\text{out})}(x, t) = \sum_{k,s} d_i(k, s) \exp\left[i(kx - \omega t)\right] , \qquad (3.39)$$

where the annihilation $[b_i(k, s)$ and $d_i(k, s)]$ and creation $[b_i(k, s)^\dagger$ and $d_i(k, s)^\dagger]$ operators satisfy the standard fermion anticommutation relations of the form

$$\{b_i(k, s), b_j(k', s')\} = \{b_i(k, s)^\dagger, b_j(k', s')^\dagger\} = 0 ,$$
$$\{b_i(k, s), b_j(k', s')^\dagger\} = \delta_{ij}\delta_{ss'}\delta(k - k') . \qquad (3.40)$$

By taking account of the changes in phase and amplitude resulting from free propagation between entrance and exit ports and interaction at each beam splitter one can establish to within a global phase factor the following relationship between input and output annihilation operators

$$\begin{pmatrix} d_1 \\ d_2 \end{pmatrix} = \begin{pmatrix} t_1 t_2' e^{-i\theta/2} + r_1 r_2 e^{i\theta/2} & r_1' t_2' e^{-i\theta/2} + t_1' r_2 e^{i\theta/2} \\ t_1 r_2' e^{-i\theta/2} + r_1 t_2 e^{i\theta/2} & r_1' r_2' e^{-i\theta/2} + t_1' t_2 e^{i\theta/2} \end{pmatrix} \begin{pmatrix} b_1 \\ b_2 \end{pmatrix} , \qquad (3.41)$$

where θ is the total phase difference incurred between the left (via M_2) and right (via M_1) beam components. Upon implementation of the unitarity relations (3.3) and (3.4), equation (3.41) leads to

$$d_1(k, s) = \left(|r_1 r_2| e^{i\theta/2} + |t_1 t_2| e^{-i\theta/2} \right) b_1(k, s)$$
$$+ i \left(|r_1 t_2| e^{-i\theta/2} - |t_1 r_2| e^{i\theta/2} \right) b_2(k, s) , \qquad (3.42)$$

$$d_2(k, s) = \left(-i|t_1 r_2| e^{-i\theta/2} - |r_1 t_2| e^{i\theta/2} \right) b_1(k, s)$$
$$+ \left(|r_1 r_2| e^{-i\theta/2} + |t_1 t_2| e^{i\theta/2} \right) b_2(k, s) . \qquad (3.43)$$

It will be assumed in what follows that the beam is quasi-monochromatic, so that the momentum spread is much less than the mean particle momentum ($\Delta k \ll k_0$), in which case θ is independent of the momentum distribution to a good approximation. If the relative phase is produced entirely by the AB effect, then $\theta = 2\pi(\Phi/\Phi_0)$ depends only on an external parameter (the magnetic flux) and is rigorously independent of the geometric path length and momentum distribution of the particles. Since we are concerned principally with the quantum effects of fields and statistics, we adopt at the outset a square interferometer geometry to eliminate the relative phase arising from unequal geometric path lengths. This simplification can be readily relaxed whenever necessary, and the resulting relative phase easily determined.

From the fermion creation and annihilation operators one can construct (as in the standard quantum mechanical analysis of a harmonic oscillator) Hermitian operators corresponding to the number of particles entering the input ports ($i = 1, 2$)

$$N_i = \sum_{k,s} b_i(k, s)^\dagger b_i(k, s) , \qquad (3.44)$$

and the number of particles leaving the output ports to be received respectively at detectors D_1 and D_2

$$N(D_i) = \sum_{k,s} d_i(k, s)^\dagger d_i(k, s) . \qquad (3.45)$$

Substitution of relations (3.42) and (3.43) into (3.45) for the special, but useful, case of 50–50 beam splitters ($|r_i| = |t_i| = 1/2$) allows one to relate the input and output particle number operators as follows:

$$N(D_1) = \frac{1}{2} \left[N_1 \cos^2 \frac{\theta}{2} + N_2 \sin^2 \frac{\theta}{2} + M \cos \frac{\theta}{2} \sin \frac{\theta}{2} \right] , \qquad (3.46)$$

$$N(D_2) = \frac{1}{2} \left[N_1 \sin^2 \frac{\theta}{2} + N_2 \cos^2 \frac{\theta}{2} - M \cos \frac{\theta}{2} \sin \frac{\theta}{2} \right] . \qquad (3.47)$$

In the preceding relations

$$M = \sum_{k,s} \left[b_1(k,s)^\dagger b_2(k,s) + b_2(k,s)^\dagger b_1(k,s) \right] \qquad (3.48)$$

is an operator that removes a particle from the beam at one input port and adds it to the beam of the other input port. Note that

$$N(\mathrm{D}_1) + N(\mathrm{D}_2) = N_1 + N_2 \equiv N , \qquad (3.49)$$

as is required by conservation of particle number. For charged particles, this is tantamount to the conservation of electric charge.

It is worth noting that if the second beam splitter is opaque ($|r_2| = 1, |t_2| = 0$) the configuration is equivalent to an electron version of the original HBT split-beam photon-counting experiment. The resulting operator expressions are similar to those of (3.46) and (3.47) but without the trignonometric factors, since the configuration does not constitute an interferometer, and there is consequently no phase angle.

The characteristics of any input beam are specified by the mathematical form of the density operator ρ. Theoretical expressions corresponding to the expectation values of dynamical observables are determined by taking the trace of the appropriate operator with ρ. Thus, the mean number of counts received by each detector in a given sampling time (the time during which a known number of particles enters the interferometer) is

$$\overline{N(\mathrm{D}_i)} = \mathrm{Tr}\left[\rho N(\mathrm{D}_i)\right] , \qquad (3.50)$$

the number of coincident counts at the two detectors is

$$\overline{N(\mathrm{D}_1)N(\mathrm{D}_2)} = \mathrm{Tr}\left[\rho N(\mathrm{D}_1)N(\mathrm{D}_2)\right] , \qquad (3.51)$$

and the cross-correlation in fluctuations in counts at the two detectors is

$$C(\mathrm{D}_1, \mathrm{D}_2) = \mathrm{Tr}\left\{ \rho \left[N(\mathrm{D}_1) - \overline{N(\mathrm{D}_1)} \right] \left[N(\mathrm{D}_2) - \overline{N(\mathrm{D}_2)} \right] \right\} \qquad (3.52)$$

$$= \overline{N(\mathrm{D}_1)N(\mathrm{D}_2)} - \overline{N(\mathrm{D}_1)}\,\overline{N(\mathrm{D}_2)} .$$

Consider the simple case of a beam of electrons of reasonably well-defined linear momentum and energy entering the interferometer through only one of the two entrance ports, e.g., port 1. In a basis $|n_1\{k,s\}; n_2\{k,s\}\rangle$ of eigenstates of the input particle number operators, relation (3.44), the density operator of the proposed system would consist of states of the form $|n_1\{k,s\}; 0\rangle$, where the designation $n_i\{k,s\}$ is the total number of particles entering port i ($i = 1, 2$) with a given spectrum of linear momentum and spin eigenvalues. (Note that the eigenvalues of N_i are independent of how the particles are distributed over momentum and spin states.) In the case characteristic of thermal and field-emission electron sources where the number of particles in the input beam fluctuates about a mean value N_1 with dispersion $(\Delta N_1)^2$, the density operator is constructed from a mixture of states of different particle number

$$\rho_{\text{chaotic}} = \sum_{\{k,s,n\}} \rho(n_1\{k,s\})|n_1\{k,s\};0\rangle\langle n_1\{k,s\};0| \, . \tag{3.53}$$

The source is designated chaotic in analogy to chaotic optical sources for which the density operator is diagonal in a photon number basis.

Substitution of the density matrix (3.53) into the expressions (3.50)–(3.52) leads to mean particle counts at each detector

$$\overline{N(\mathrm{D_1})} = \frac{\overline{N_1}}{2}(1 + \cos\theta) \, , \tag{3.54}$$

$$\overline{N(\mathrm{D_2})} = \frac{\overline{N_1}}{2}(1 - \cos\theta) \, , \tag{3.55}$$

the joint count rate

$$\overline{N(\mathrm{D_1})N(\mathrm{D_2})} = \frac{1}{4}\overline{N_1(N_1 - 1)}\sin^2\theta \, , \tag{3.56}$$

and the cross-correlation in particle fluctuations

$$C(\mathrm{D_1}, \mathrm{D_2}) = \frac{1}{4}\left[(\Delta N_1)^2 - \overline{N_1}\right]\sin^2\theta \, , \tag{3.57}$$

all of which clearly show an influence of the magnetic flux through the phase angle. Note that the coincidence count rate (3.56) vanishes, as it must, for an input of $\overline{N_1} = N_1 = 1$, since the one particle must be received at only one of the detectors, and thus there can be no coincidence. This result contrasts with that deduced from a classical wave analysis in which a split incident beam of arbitrarily weak amplitude could still illuminate both detectors.

For a beam of randomly emitted classical particles the fluctuation in particle number is governed by Poisson statistics for which the variance is equal to the mean, and the cross-correlation (3.57) vanishes. However, for a chaotic electron source, the antisymmetrization of the state vectors leads to a variance that is smaller than the mean incident particle number by an amount proportional to the beam degeneracy δ, as we have seen in the previous chapter. In the experimentally realistic case of a counting interval t_r (the detector response time) long in comparison to the beam coherence time t_c, the variance takes the form [91]

$$(\Delta N_1)^2 \approx \overline{N_1}\left[1 - \overline{N_1}\sigma F(0)\frac{t_c}{t_r}\right] \, , \tag{3.58}$$

where σ is a polarization factor (unity for total spin polarization and one-half for an unpolarized beam), and $F(0)$ is a spatial coherence function (unity for a beam cross-section equal to a coherence area A_c). Expression (3.58) is the fermion analogue of (2.26) for thermal light. The beam degeneracy, determined practically from

$$\delta = \overline{N}_1 \frac{t_c}{t_r} \, , \tag{3.59}$$

is interpretable as the mean number of particles per cell of phase space. This number can never exceed unity for fermions, since there can be at most one fermion per quantum state. Substitution of (3.58) and (3.59) into relation (3.57) leads to a negative cross-correlation

$$C(D_1, D_2) \approx -\frac{1}{4} \overline{N}_1 \sigma F(0) \delta \sin^2 \theta \, , \tag{3.60}$$

indicative again of electron antibunching.

Although a vector potential field is ostensibly responsible for the quantum phase θ in the foregoing analysis, one can envision different experimental configurations for which the phase is attributable to other external potentials. One interesting example is that of gravity [92]. The force of gravity is the dominant influence in shaping the macroscopic universe, but it is intrinsically so weak that its effects on the elementary constituents of matter are for the most part negligible.[2] There are very few physical systems – in particular, systems directly accessible to laboratory investigation – whose dynamical behavior requires a quantum mechanical description with simultaneous inclusion of a gravitational interaction.

One example, however, is the Colella–Overhauser–Werner (COW) experiment [93] which provided the first experimental demonstration of the effect of gravity in a circumstance unique to quantum mechanics, i.e., through a gravitational influence on the phase of a wave function leading to quantum interference. The COW experiment employed a neutron beam split and recombined in a single-crystal neutron interferometer, a configuration effectively equivalent to that shown in Fig. 3.4 but with the solenoid removed and the plane of the interferometer oriented vertically so that the upper path (between mirror M_2 and beam splitter BS_2) is actually a distance z above the lower path (between BS_1 and M_1). Then, for a nonrelativistic particle of mass m and mean momentum $\hbar k_0$ there is a gravitationally induced phase shift

$$\theta = \frac{m^2 g z d}{\hbar k_0} \, , \tag{3.61}$$

---

[2] One significant exception is the influence of gravity on constituents of matter, primarily neutrons and possibly quarks, during the terminal collapse of massive stars that have exhausted their nuclear fuel. I discuss the role played by fermion and boson statistics in the end states of such stars in the last chapter of this book.

where g is the local acceleration of gravity and d is the horizontal path length between a mirror and beam splitter at the same height. Equations (3.54) and (3.55) illustrate the effect of the gravitationally induced quantum interference on the particle distribution at the output ports. Similarly, (3.57) or (3.60) manifests the effect of gravity on particle correlations. Although the gravitational field is present throughout the interferometer and the particles are therefore subject to the force of gravity, (in contrast to the AB effect where the magnetic field is confined to the interior of the solenoid and the particles are not subject to a magnetic force), it is *not* the gravitational force but the gravitational potential that plays a direct role here. The relative phase expressed in (3.61), which is proportional to the area zd of the interferometer, may be thought to accrue only over the horizontal segments of the motion where the classical effect of the gravitational force on the split beam is zero, since no work is done.

The utilization of both input ports of a fermion interferometer permits, at least in principle, the construction of fermion ensembles manifesting a variety of new and surprising statistical properties. One finds that the clustering behavior of fermions need not be limited by Fermi–Dirac statistics to antibunching, but, like light, can depend as well on the specific state composition of the ensemble. To be sure, each multiparticle basis state that contributes to the ensemble description must, itself, be compatible with the antisymmetrization restrictions of quantum statistics. Nevertheless, there is room for considerable diversity among possible fermion ensembles.

Of particular interest are fermion beams described by basis states that are labeled by the total number of particles entering the two ports and by the difference in number of particles between the two input ports. For such states one does not in general know the number of particles entering each port individually. It will be seen that these states span representations of an SU(2) algebra constructed from the fermion creation and annihilation operators. In other words, the mathematical description is in many ways similar to that of the familiar treatment of orbital or spin angular momentum. There is a vector operator $\boldsymbol{J}$ with components J_i $(i = x, y, z)$ from which can be constructed raising and lowering operators J_+ and J_- that effectively transfer particles between input ports, and a Casimir operator $J^2 = \boldsymbol{J} \cdot \boldsymbol{J}$ which, under appropriate circumstances, gives the total number of particles entering the interferometer through the two ports. [A Casimir operator is a nonlinear invariant operator that commutes with all other members of the algebra; in general, there are $n - 1$ of them for the special unitary group SU(n).] The basis states $|j, m\rangle$, which will be termed correlated two-port states, are labeled by the quantum numbers specifying the eigenvalues of J^2 $[= j(j + 1)]$ and j_z $(= m)$, in contrast to the uncorrelated two-port states $|n_1; n_2\rangle$ which are eigenvectors of the input number operators N_1 and N_2. To aid in avoiding confusion between these two different representations, the quantum numbers of uncorrelated two-port basis states will be separated by a semicolon while those of correlated two-port basis states will be separated by a comma.

Let us now look at the development of the correlated two-port states in detail [90, 94]. From the fermion annihilation and creation operators one can construct the following bilinear superpositions

$$J_x = \frac{1}{2} \sum_{k,s} \left[b_1(k,s)^\dagger b_2(k,s) + b_2(k,s)^\dagger b_1(k,s) \right] , \tag{3.62}$$

$$J_y = \frac{1}{2i} \sum_{k,s} \left[b_1(k,s)^\dagger b_2(k,s) - b_2(k,s)^\dagger b_1(k,s) \right] , \tag{3.63}$$

$$J_z = \frac{1}{2} \sum_{k,s} \left[b_1(k,s)^\dagger b_1(k,s) - b_2(k,s)^\dagger b_2(k,s) \right] = \frac{1}{2}(N_1 - N_2) , \tag{3.64}$$

and the Casimir invariant

$$J^2 = \frac{N}{2} \left(\frac{N}{2} + 1 \right) - W , \tag{3.65}$$

where N is the total particle number operator in (3.49), and

$$W = \sum_{k,s;k',s'} \left[b_1(k,s)^\dagger b_2(k',s')^\dagger b_2(k,s) b_1(k',s') \right. \tag{3.66}$$

$$\left. + b_1(k,s)^\dagger b_2(k',s')^\dagger b_2(k',s') b_1(k,s) \right]$$

is an operator that transfers two particles between input ports. The vector operator J commutes with both N and W, although N and W do not commute with each other.

When the relative phase θ incurred by passage of a wave packet through the interferometer is independent of the particle momentum and spin (as is the case for the AB effect) – or very nearly so in the case of a quasi-monochromatic beam – the particle number operators for both input and output ports are expressible in terms of the above SU(2) operators in a relatively simple way

$$N_1 = \frac{1}{2}N + J_z , \tag{3.67}$$

$$N_2 = \frac{1}{2}N - J_z , \tag{3.68}$$

$$N(D_1) = \frac{1}{2}N + J_z \cos\theta + J_x \sin\theta , \tag{3.69}$$

$$N(D_2) = \frac{1}{2}N - J_z \cos\theta - J_x \sin\theta . \tag{3.70}$$

It is also useful to denote the operator $N(D)$ which gives the difference in particle counts at the output ports

$$N(D) = N(D_1) - N(D_2) = 2(J_z \cos \theta + J_x \sin \theta) . \tag{3.71}$$

Recall from relation (3.64) that $2J_z$ gives the particle number difference at the input ports. In anticipation of our interest in the cross-correlation of particle counts at the two output ports, we record as well the product

$$N(D_1)N(D_2) = \frac{1}{2}\left[N(D_1)^2 + N(D_2)^2 - N(D)^2\right] , \tag{3.72}$$

which follows straightforwardly from squaring $N(D)$ in the first equality of (3.66).

By means of the raising and lowering operators

$$J_+ = J_x + iJ_y = \sum_{k,s} b_1(k,s)^\dagger b_2(k,s) , \tag{3.73}$$

$$J_- = J_x - iJ_y = \sum_{k,s} b_2(k,s)^\dagger b_1(k,s) , \tag{3.74}$$

one can construct all members of a family of states $|j,m\rangle$ of given j by knowing one of the states. For example, start with the state that corresponds to all N particles entering the interferometer through port 1, i.e., $|N;0\rangle$ in the uncorrelated two-port representation. This state is an eigenstate of W with eigenvalue 0, and consequently from relations (3.64) and (3.65) an eigenstate of J^2 and J_z with quantum numbers $j = m = N/2$ in the correlated two-port representation. By sequential application of J_- to the correlated two-port state $|N/2, N/2\rangle$, one can generate the full spectrum of states $|j,m\rangle$ in the manner below

$$J_-^{N/2-m}\left|\frac{N}{2}, \frac{N}{2}\right\rangle = \left[\frac{N!(N/2-m)!}{(N/2+m)!}\right]^{1/2}\left|\frac{N}{2}, m\right\rangle . \tag{3.75}$$

As an explicit illustration of the connection between the two representations, consider a two-particle input beam ($N = 2$) where the particles can have momentum–spin values (k_α, s_α) or (k_β, s_β). To avoid encumbering our notation with unnecessary symbols, the set of eigenvalues (k_γ, s_γ) will be represented simply by the label γ. Thus, the state corresponding to two particles entering port 1 and no particles entering port 2 can be expressed by the state vectors

$$|1,1\rangle = |2;0\rangle = |\{\alpha, \beta\};0\rangle , \tag{3.76}$$

where the first vector is a $|j,m\rangle$ state and the second vector is a $|n_1;n_2\rangle$ state specified in greater detail by the third vector which is implicitly antisymmetric under exchange of particle labels α, β. Application of J_- to (3.76) leads to the state

$$|1, 0\rangle = |1; 1\rangle = -\frac{1}{\sqrt{2}}(|\alpha; \beta\rangle - |\beta; \alpha\rangle) , \qquad (3.77)$$

in which one particle enters each port, but we do not know through which port a particle of particular spin and linear momentum goes. There is a 50% probability for each particle to enter through each port. Finally, application of J_- a second time results in the state

$$|1, -1\rangle = |0; 2\rangle = -|0; \{\alpha, \beta\}\rangle , \qquad (3.78)$$

in which both particles enter port 2 and none enters port 1.

Since the correlated two-port states can not in general be factored into state vectors of the two ports individually, they exhibit a novel form of entanglement beyond that deriving strictly from the spin–statistics connection. It is also important to note – and easily verified in the special case of the above example – that a system characterized by a density operator diagonal in a representation of SU(2) correlated states is generally *not* diagonal in a basis of momentum-spin states. Thus, the fluctuation behavior characteristic of the correlated states need not necessarily be similar to that of the chaotic states examined previously.

Let us first examine the quantum statistical behavior of an ensemble of entering fermions characterized by a density operator diagonal in a basis of SU(2) correlated states. From (3.69)–(3.72), the cross-correlation of outputs at D_1 and D_2 can be expressed in the form

$$C(D_1, D_2) = \frac{1}{4}(\Delta N)^2 - (\Delta m)^2 \cos^2 \theta - \frac{1}{2}\left(\frac{1}{4}\overline{N^2} + \frac{1}{2}\overline{N} - m^2\right) \sin^2 \theta , \quad (3.79)$$

where $(\Delta N)^2$ is the variance about the mean total particle number entering the interferometer, and $(\Delta m)^2$ is the variance about $\overline{m}$, which is one half the difference in mean numbers of particles $\overline{N}_1$ and $\overline{N}_2$ entering the two ports. Equation (3.79) can also be written in the form

$$C(D_1, D_2) = C_{12} \cos^2 \theta + \frac{1}{4}\left[(\Delta N)^2 - \overline{N} - 2\overline{N_1 N_2}\right] \sin^2 \theta , \qquad (3.80)$$

where

$$C_{12} = \overline{N_1 N_2} - \overline{N}_1 \overline{N}_2 \qquad (3.81)$$

is the cross-correlation in beam fluctuations at the input ports.

As a check of consistency, one can verify that if all the particles enter the interferometer through one of the ports – let us say port 1 (in which case $N = 2m$) – then upon substitution of $(\Delta N)^2 = 4(\Delta m)^2 = (\Delta N_1)^2$, the cross-correlation reduces to relation (3.57) as expected.

Suppose, however, the input beam is in a pure SU(2) correlated state of precisely known particle number N and particle difference number m. Then the variances in N and m vanish, and the values $N_1 = N/2 + m$ and $N_2 = N/2 - m$ are also sharp. The cross-correlation in output counts reduces in that case to

$$C(\mathrm{D}_1, \mathrm{D}_2) = -\frac{1}{2} \left[N_1 N_2 + \frac{1}{2}(N_1 + N_2) \right] \sin^2 \theta , \qquad (3.82)$$

which is again intrinsically negative except at selected phase angles for which it vanishes.

In the more general case of an input beam comprising an appropriate mixture of SU(2) correlated states there is a range of phase angles, given by

$$\frac{4C_{12}}{2\overline{N_1 N_2} + \overline{N} - (\Delta N)^2} > \tan^2 \theta , \qquad (3.83)$$

for which the cross-correlation can be positive. For example, at $\theta = 0$, the condition $(\Delta N)^2 > 4(\Delta m)^2$ results in a positive cross-correlation. This condition is most directly met by an ensemble of distributed total particle number, $(\Delta N)^2 > 0$, but sharp particle number difference, $(\Delta m)^2 = 0$, at the input ports. An input composed of the states $|N/2, m\rangle$ and $|N/2 + 2, m\rangle$ with $|m| \le N/2$ leads to $C(\mathrm{D}_1, \mathrm{D}_2) = C_{12} = 1/4$.

Consider next an ensemble described by a mixture of states of the form

$$|S(N)\rangle = \frac{1}{\sqrt{2}} \left(\left| \frac{1}{2}N, \frac{1}{2} \right\rangle + \left| \frac{1}{2}N, -\frac{1}{2} \right\rangle \right) , \qquad (3.84)$$

where N is an odd integer and $(\Delta N)^2 > 1$. The analogy with angular momentum suggests that expression (3.84) is a type of singlet state, hence the designation $|S(N)\rangle$. Since the cross-correlation

$$C(\mathrm{D}_1, \mathrm{D}_2) = \frac{1}{4} \left[(\Delta N)^2 - \cos^2 \theta \right] + \frac{1}{8} \left[(\Delta N)^2 - \frac{1}{2}(\overline{N} + 3)(\overline{N} - 1) \right] \sin^2 \theta$$
$$(3.85)$$

is positive for $(\Delta N)^2 > 0$ (which is characteristic of the proposed ensemble) and $\theta = 0$, the system provides another example of fermionic bunching.

It is unfortunate that there is no way at present (to my knowledge) to produce these SU(2) correlated states in the laboratory, because not only do they manifest a complex statistical behavior of theoretical interest, but they can be of practical use in particle interferometry. A desirable objective is to enhance the sensitivity of an interferometer to the relative phase angle θ. In general, one would expect that the dispersion (square root of the variance) of θ, which can be calculated from expectation values of $N(D)$

$$\Delta \theta = \frac{\sqrt{[\Delta N(\mathrm{D})]^2}}{|\partial \overline{N}(\mathrm{D})/\partial \theta|} , \qquad (3.86)$$

to diminish as some function of the number of particles, i.e., the larger the number of particles that pass through the interferometer, the more sharply is the relative phase angle θ determined. Ordinarily, this inverse dependence goes as the square root of the number of particles. For example, in the case of a beam described by a correlated state $|j, m\rangle$, (3.86) reduces to

$$\Delta\theta = \left\{ \frac{1}{2} \left[\frac{\frac{1}{2}N\left(\frac{1}{2}N+1\right)}{m^2} - 1 \right] + \frac{(\Delta m)^2}{m^2} \cot^2\theta \right\}^{1/2} . \tag{3.87}$$

Consider the special, but widely applicable, example of a single-port input of N particles described by the state $|N/2, N/2\rangle$. From relation (3.87) it is seen that the *minimum* dispersion of θ (for an uncorrelated input) is

$$(\Delta\theta)^{\text{unc}}_{\text{min}} = \sqrt{\frac{N}{N^2}} \longrightarrow \frac{1}{\sqrt{N}}, \tag{3.88}$$

where the limiting expression is rigorously valid for sharp N. For the two-port singlet states of relation (3.84), however, the dispersion in θ is given by

$$\Delta\theta = \left[\frac{4}{(N+1)^2} + \frac{(N+3)(N-1)}{(N+1)^2} \tan^2\theta \right]^{1/2}, \tag{3.89}$$

from which it follows that the expression comparable to (3.88)

$$(\Delta\theta)^{S(N)}_{\text{min}} = \frac{2}{N+1} \tag{3.90}$$

is inversely proportion to the number of particles (rather than the square root). Thus, the use of two-port correlated states can enhance the sensitivity of a fermion interferometer.

4

Quantum Boosts and Quantum Beats

4.1 Superposing Pathways in Time

We have seen in preceding chapters that the potential for quantum interference exists whenever a particle can propagate from its source to the detector by alternative spatial pathways under experimental conditions such that the exact pathway taken can not be known. The archetypal example is the Young's two-slit experiment in which the particle, when probed, passes through one slit or the other. Unprobed, the resulting particle distribution is explicable only in terms of probability amplitudes for propagation through both slits. There is a direct temporal analogue to the two-slit experiment in which the linearly superposed amplitudes represent – not alternative spatial pathways – but the evolution of alternative indistinguishable events in time.

Of the many ways in which the structure of atoms, molecules, and other bound-state quantum systems differs from that of macroscopic classical systems one of the most striking is the discreteness or quantization of energy. The internal energy of a classical planetary system is an accident of its formation and likely to differ from one system to another even though the two systems may be composed of identical masses. By contrast, every ground-state hydrogen atom has the same energy irrespective of its formation; likewise for identical atoms in corresponding excited states.

Although the idea of electrons populating quantized energy eigenstates is a familiar one, it is nevertheless necessary to be careful lest uncritical usage provide a misleading picture of the atom and its interactions. The Russian spectroscopist E.B. Aleksandrov expressed this point well when he wrote [95]:

> In connection with the remarkable progress made in the interpretation of atomic spectra, the concepts of energy levels and their populations have become so firmly entrenched in atomic physics and spectroscopy that they became gradually independent concepts, losing the meaning attributed to them by quantum mechanics. Yet the statement commonly made in spectroscopy, that the atom is at a given (excited) level, is incorrect in the overwhelming majority of cases.

In such circumstances the energy of an atom is actually indeterminate, the atomic state being represented by a linear superposition (with appropriate amplitudes) of all possible stationary states that can be reached by the particular type of excitation employed. One optical consequence of this coherent superposition is that the rate of spontaneous emission (i.e., the fluorescence intensity) from an ensemble of atoms excited in this manner can oscillate in time while diminishing exponentially, such as illustrated in Fig. 4.1 for the case of a four-level atom with two closely spaced excited states. For these 'quantum beats' to be observable, the emission events among the various atoms must be in some sense synchronized; otherwise the superposition of oscillating intensities widely out of phase would display no net modulation. Had the atoms decayed from well-defined energy states, the temporal variation in fluorescence would have been strictly an exponential decay. Thus, there is a significant conceptual and experimental distinction between a quantum ensemble designated a *mixture* of states, wherein each constituent is endowed with definite, although statistically distributed energy values, and another ensemble designated a *superposition* of states, with each constituent in a linear superposition of energy eigenstates encompassing the same energy values and populations as the first ensemble [96].

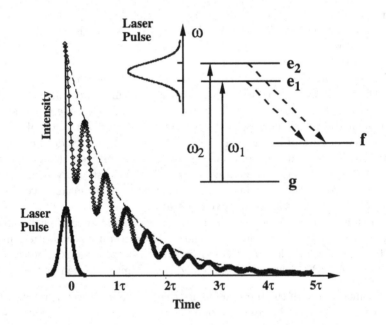

Fig. 4.1. Coherent excitation of two close-lying excited states by a pulsed laser giving rise to oscillations (quantum beats) in the spontaneous emission as a function of time. A light pulse narrow in the time domain has a broad frequency spectrum which, shown in the *upper insert*, contains Fourier components capable of inducing transitions from the ground state to both excited states

We have discussed these distinctions to a limited extent in Chap. 2 in the context of the information carried by first- and second-order correlation functions. In the present chapter we will examine in more detail the features – some rather surprising – of the quantum interference in time of the radiation from an ensemble of coherently excited atoms.

Production of an atom (or molecule) in a linear superposition of excited energy states ordinarily requires an impulsive excitation ('quantum boosts'), i.e., a process of sufficiently short duration that its Fourier spectrum contains frequency components corresponding to the energy intervals between ground and excited states. Thus, for the example shown in the insert of Fig. 4.1, the spectral width of the excitation must satisfy the relation

$$\Delta\omega \geq \omega_0 \equiv \frac{E_2 - E_1}{\hbar} = \omega_2 - \omega_1 , \qquad (4.1)$$

where ω_0 is the Bohr angular frequency of the excited level. Although the theoretical feasibility of such light oscillations was discussed in papers shortly following the creation of quantum mechanics [97], the actual experimental implications were not conceptually appreciated or experimentally realized until some thirty years later [98] principally by those involved with optical pumping. (Optical pumping refers to the use of light to populate a set of energy levels with a distribution different from that of a normal Boltzmann distribution at the temperature of the experiment.) There are a variety of ways of achieving the impulsive excitation required to generate a superposition state and the ensuing modulated fluorescence, as, for example, by light pulses [99], pulsed electron impact [100], and electron capture collisions with a thin carbon foil target [101].

Like the quantum interference phenomena described in earlier sections, the phenomenon of quantum beats is intrinsic to each atom and not a cooperative interaction between atoms. In other words, the spontaneous emission from single atoms is *not* modulated, but registers at the detector as one quantum of light at a time; the pattern of beats (measured at one location in real time or, equivalently, at different spatial locations along an accelerated atomic beam) can nevertheless be built up by the decay of many such single atoms. This is again the old 'mystery' of quantum interference translated to the time domain: How can independently excited, randomly decaying, noninteracting atoms produce a pattern of photon arrivals that oscillates in time? Note that the synchronization required for the beats to survive ensemble averaging does not imply that emitting atoms communicate with or influence one another. Rather, an apt classical analogy, if there be any, would be that of a large number of independent clocks all separately wound and set to the same time by the clockmaker.

There is no visualisable classical mechanism for the interference (built up one photon at a time through repeated excitations), but quantum mechanics allows one to analyze the phenomenon mathematically. Consider, again, the system illustrated in Fig. 4.1 in which it is assumed for simplicity that the two excited states have the same lifetime $\tau = 1/\Gamma$ where Γ is the decay rate. Let a_{gi} be the amplitude for a transition from the ground state g to excited state e_i ($i = 1, 2$), and b_{if} be the corresponding amplitude for radiative decay from e_i to the lower state f (which could be the ground state). Then, the net amplitude for excitation to state e_i and subsequent decay after a time interval t takes the form

$$A_{gif}(t) = a_{gi}e^{-iE_it/\hbar}e^{-\Gamma t/2}b_{if} = a_{gif}e^{-iE_it/\hbar}e^{-\Gamma t/2} , \qquad (4.2)$$

where $a_{gif} = a_{gi}b_{if}$. We have made the heuristic assumption – to be examined further in the next section – that the excitation occurs effectively instantaneously. If the energy dispersion of the excitation mechanism is sufficiently broad, and quantum selection rules do not forbid the indicated transitions, there are two indistinguishable pathways in time by which the atom can pass from the initial level g to the final level f; the corresponding total amplitude for the process is

$$A_{gf}(t) = \left(a_{g1f}e^{-iE_1t/\hbar} + a_{g2f}e^{-iE_2t/\hbar}\right)e^{-\Gamma t/2} . \qquad (4.3)$$

Hence, the probability for the transition to occur at time t with emission of one quantum of light – although from which state the observer does not know – is

$$P_{gf}(t) = |A_{gf}(t)|^2 = \left[|a_{g1f}|^2 + |a_{g2f}|^2 + 2|a_{g1f}a_{g2f}|\cos(\omega_0 t - \phi)\right]e^{-\Gamma t} , \qquad (4.4)$$

where each amplitude can be a complex number $a_{gif} = |a_{gif}|e^{i\phi_{gif}}$ and $\phi = \phi_{g2f} - \phi_{g1f}$. If, by some means – for example, by placing an optical narrowband filter before the detector – the observer can select only photons of energy E_1 or E_2 and thereby determine the temporal pathway by which the system evolved, the quantum beats would disappear and the system would simply decay exponentially in time.

As a spectroscopic method, the observation of quantum beats has a number of significant advantages compared with alternative procedures that probe the atom with resonant oscillatory fields. For one thing, the presence of an external oscillating field, as will be shown in the next chapter, can affect the level separation and decay of the states being examined. With quantum beat spectroscopy the atoms 'ring out' their level structure without being probed. Since the strong impulsive fields that produce the excited states can be made to vanish substantially by the time the atoms are likely to decay, one can investigate a sample of interest in a resonance cell (or 'bottle') – rather than employ the more complicated technology of an accelerated beam – and still be able to separate the processes of excitation and detection. Secondly, even though the atoms in a gas or vapor may be moving randomly about the interior of a resonance cell, rather than moving with well-defined linear momentum along a beam, the optical *signal*, i.e., the low-frequency modulation, as opposed to the high-frequency optical carrier wave, is largely free of Doppler broadening. Since the beat frequency is proportional to the difference in energy of two levels of the *same* atom, the Doppler shifts of the photons *potentially emissable* from each level nearly cancel. The italicized words again emphasize the fact that under the given circumstances only one quantum or light, not two, actually emerges from the spontaneous decay of a single atom. The beat is intrinsic to each atom, but made manifest only by the decay of a large number of similarly prepared atoms.

4.2 Laser-Generated Quantum Beats

Of the many ways to excite an atom impulsively into a linear superposition of states, one of the most practically useful and conceptually interesting is the application of pulsed laser light [102, 103]. The use of light in general – as opposed to impulsive excitation by some form of particle bombardment, for example – is most amenable to theoretical analysis, since the interaction of matter with light is completely accounted for by quantum electrodynamics, the most thoroughly understood of all physical interactions. (By contrast, coherent atomic excitation by particle bombardment is not as simply analyzable or well understood.) It was, in fact, by means of pulsed light from shuttered spectral (i.e., nearly monochromatic) lamps that the phenomenon of field-free quantum beats was first observed. These sources, however, were weak in the sense that the probability of atomic excitation was low; at any moment the likelihood was greatest to find the illuminated atom in its ground state.

The development of tunable pulsed lasers has made possible the high intensity, short pulse duration, and broad spectral width that are advantageous in the study of some intriguing, if not outright exotic, physical systems such as Rydberg atoms [104], i.e., atoms so highly excited that they begin to resemble in many ways (but not all [105]) minute classical planetary systems. The wide range over which such lasers can be tuned allows one to excite a large number of UV, visible, and IR transitions. The high power makes it possible to saturate even weakly allowed transitions. And when, because of parity restrictions, a single laser can not induce transitions between two states of interest, two or more lasers used sequentially can effect the desired result by a stepwise excitation. These three advantages, besides those already cited common to quantum beat spectroscopy irrespective of the method of excitation, permit an experimenter to select almost any Rydberg state of interest or to study with facility an entire series of states.

It is the high intensity of pulsed lasers, however, that enriches (or complicates – depending on one's point of view) the interaction between the atoms and light. The weak-pumping approximation implicit in the simple analysis of the previous section is, strictly speaking, a first-order perturbation calculation in which the atomic system interacts at most once with the exciting light. In other words, the atom is presumed to absorb a single photon during the passage of the pulse and to spontaneously emit a photon once after the pulse has passed. Thus, this mode of treatment is also known as the linear absorption approximation. While valid for classical light sources, the linear absorption approximation is no longer a priori justified when the light source is a high-power pulsed laser. For one thing, while illuminated by the light pulse, a particular atom can be driven back and forth a number of times between the ground and excited states by the processes of photon absorption followed by stimulated emission. In addition, if the spectral profile of the exciting light is not symmetric, or if it is not centered precisely on the transition to be effected, the light could displace the atomic energy levels from their vacuum values. These processes, absent under conditions of weak pumping, can modify the amplitude and phase of the ensuing quantum beats – sometimes in an extraordinary way as will be discussed shortly.

The theory of light-induced quantum interference that will be discussed in this section is valid under the relatively nonrestrictive limitation to *broadband* pulse excitation. This condition, realized in most quantum-beat experiments, permits the theory to be formulated within the framework of the classical optical pumping cycle developed by Barrat and Cohen-Tannoudji [106] for weak light sources and subsequently generalized by others for continuous lasers [107].

Let us generalize somewhat the simple atomic structure assumed in Sect. 4.1 by considering an atom with three groupings (or manifolds) of states:

- ground g,
- excited e,
- final f,

where each manifold can contain more than one state. In a quantum beat experiment envisioned here the atoms are excited from the g to e manifold by means of a light pulse of polarization ε and duration T and subsequently decay optically at rate Γ to the lower manifold f. (Different excited states could have different lifetimes, but this would unnecessarily complicate the conceptual ideas to be elucidated here.) After passage of the pulse, fluorescence of a particular polarization ε_d is observed. In order that the preparation of the excited states be well separated in time from the detection of the spontaneous emission, it is necessary that T be short in comparison to the lifetime $\tau = \Gamma^{-1}$.

The existence of a multiplicity of states (at least two) in the e manifold is essential to the production of quantum beats. While substructure in the lower g and f manifolds is not essential, such structure can affect the phase and amplitude of the beats. Similarly, the energy resolution of the detector – and therefore the number of final states involved in the decay process – may also influence the beat signal. It is of interest to note that a semi-classical theory of radiative phenomena (termed the neoclassical theory [108]) proposed in the 1960s predicted quantum beats from single atoms with one excited state and a multiplicity of lower states. This is forbidden by quantum electrodynamics (QED) which – excluding processes involving the weak nuclear interactions (discussed in Chap. 6) – remains unchallenged within its domain of validity by any reliably reproducible experiment. QED does, however, permit quantum beats arising from lower-state splittings in the case of *cooperative* interactions between two or more identical atoms [109].

In contrast to the heuristic linear-approximation analysis of Sect. 4.1, which determines the net *amplitude* for transition from the g to the f manifolds, the optical pumping equations determine the atomic *density matrix* $\rho(t)$, which enters directly into the calculation of the mean value of an observable (represented by operator O) through the trace relation

$$\langle O \rangle = \mathrm{Tr}(\rho O) . \tag{4.5}$$

The elements of the density matrix are ensemble-averaged bilinear combinations of transition amplitudes.

In determining the time evolution of a quantum system subject to an external interaction, it is often convenient to eliminate at the outset from the equations of motion the time-dependence arising from the internal interactions governed by the field-free Hamiltonian H_0, since this evolution is already known and generally involves the highest frequencies. To do this, one transforms the equations of motion into the interaction representation. The transformed density matrix

$$\tilde{\rho}(t) = e^{iH_0t/\hbar}\rho(t)e^{-iH_0t/\hbar} \tag{4.6}$$

is independent of t before and after the passage of the light pulse when the external interaction vanishes. Let $\rho_- = \tilde{\rho}(-\infty)$ and $\rho_+ = \tilde{\rho}(+\infty)$ characterize the atom at the temporal limits $t \rightarrow -\infty$ and $t \rightarrow +\infty$, respectively. The effect of the laser pulse is then entirely known if one can determine the time evolution of ρ_- into ρ_+.

The time evolution of the atomic density matrix can be determined from a set of optical pumping equations generalized to include pulsed laser excitation [96]. Although the derivation of these equations will not be given here, the structure of the equations and the assumptions underlying the derivation will be discussed. First, one decomposes the atomic density matrix into a submatrix of excited states ρ_e and of ground states ρ_g by means of projection operators P_e and P_g in the following standard way

$$\rho_e = P_e \rho P_e , \qquad \rho_g = P_g \rho P_g , \tag{4.7}$$

where each projection operator has the form

$$P_\mu = \sum_i |\mu_i\rangle\langle\mu_i| , \tag{4.8}$$

with the summation extending over the states of the appropriate manifold ($\mu = e$ or g). This leads to the following system of equations coupling the elements of ρ_e and ρ_g

$$\frac{d\rho_e}{dt} = -\frac{i}{\hbar}[H_0, \rho_e] - \frac{1}{2}\{\Gamma_e, \rho_e\} + \frac{1}{T_p(t)} P_e \boldsymbol{\varepsilon} \cdot \boldsymbol{DP}_g \rho_g P_g \boldsymbol{\varepsilon}^* \cdot \boldsymbol{DP}_e \tag{4.9}$$

$$- \frac{1}{2T_p(t)}\{P_e \boldsymbol{\varepsilon} \cdot \boldsymbol{DP}_g \boldsymbol{\varepsilon}^* \cdot \boldsymbol{DP}_e, \rho_e\} - i\Delta E(t)\big[P_e \boldsymbol{\varepsilon} \cdot \boldsymbol{DP}_g \boldsymbol{\varepsilon}^* \cdot \boldsymbol{DP}_e, \rho_e\big] \ ,$$

$$\frac{d\rho_g}{dt} = -\frac{i}{\hbar}[H_0, \rho_g] + \frac{1}{T_p(t)} P_g \boldsymbol{\varepsilon}^* \cdot \boldsymbol{DP}_e \rho_e P_e \boldsymbol{\varepsilon} \cdot \boldsymbol{DP}_g \tag{4.10}$$

$$- \frac{1}{2T_p(t)}\{P_g \boldsymbol{\varepsilon}^* \cdot \boldsymbol{DP}_e \boldsymbol{\varepsilon} \cdot \boldsymbol{DP}_g, \rho_g\} + i\Delta E(t)\big[P_g \boldsymbol{\varepsilon}^* \cdot \boldsymbol{DP}_e \boldsymbol{\varepsilon} \cdot \boldsymbol{DP}_g, \rho_g\big] \ .$$

Although seemingly complicated, the various terms of the above equations (where $[x, y]$ represents the commutator and $\{x, y\}$ the anticommutator of two operators x and y) are amenable to simple interpretation. The first term of (4.9) and (4.10) involving the commutator with H_0 represents the free evolution in time of the two sets of states. The other terms depending only on the elements of the ground state matrix ρ_g represent the effects of optical absorption. This process in (4.9) populates the excited states at a rate proportional to the reciprocal of the so-called pumping time $T_p(t)$ which is, itself, inversely related to the strength of the exciting pulse in the following way

$$\frac{1}{T_p(t)} = \frac{\pi}{\hbar^2} F(t, \omega_{eg}) |(D_r)_{eg}|^2 \ . \tag{4.11}$$

Here $F(t, \omega)$ is the spectral density of the electric field $\boldsymbol{E}$ of the laser pulse, i.e., the Fourier transform (at angular frequency ω) of the electric field autocorrelation function

$$F(t, \omega) = \int \langle E(t) E^*(t - t') \rangle e^{i\omega t'} dt' \ . \tag{4.12}$$

The transform in relation (4.11) is evaluated at the Bohr frequency ω_{eg} characterizing the energy interval between the ground and excited states. The factor

$$(D_r)_{eg} = \langle e | D_r | g \rangle \tag{4.13}$$

is the radial matrix element of the electric dipole operator $\boldsymbol{\mu}_E = D_r \boldsymbol{D}$ here written as the product of a scalar radial part D_r and vectorial angular component $\boldsymbol{D}$. Absorption also serves in (4.10) to depopulate the ground states at a rate proportional to $1/2T_{\mathrm{p}}(t)$ and to displace the different substates of g by an amount $\Delta E(t)$ (expressed as a frequency shift)

$$\Delta E(t) = \frac{1}{\hbar}|(D_{\mathrm{r}})_{eg}|^2 \mathcal{P} \int\limits_{-\infty}^{\infty} \frac{F(t,\omega)}{\omega - \omega_{eg}}\mathrm{d}\omega \, . \tag{4.14}$$

The symbol P in (4.14) represents the Cauchy principal value of the integral.

Correspondingly, except for the second term of (4.9) which is responsible for state decay by spontaneous emission, all other terms in the optical pumping equations depending on the elements of the excited state density matrix ρ_e represent the effects of stimulated emission. This process is symmetrical to that of absorption; it serves to populate the ground states in (4.10) and to depopulate and displace the excited states in (4.9). The process of spontaneous emission in (4.9) is accounted for by an anticommutation operation with a phenomenological decay operator Γ_e which takes the form (in an energy representation) of a diagonal matrix whose elements are the decay rates (inverse lifetimes) of the excited states. This procedure is justifiable by rigorous application of quantum electrodynamics [110] and is applicable even if the atomic system is subjected to external fields [111]. To a good approximation the theoretical effect of decay is to multiply the atomic density matrix ρ_+ in the absence of spontaneous emission by the factor $\mathrm{e}^{-\Gamma_e t}$. (I discuss the interaction of atoms with optical and radiofrequency electromagnetic fields, both classical and quantized, in more detail in my book *Probing The Atom* [56].)

In the derivation of (4.9) and (4.10), it has been assumed that the correlation time t_{c} of the light pulse, i.e., the inverse of the spectral width Δ, is much shorter than the pulse duration T, i.e.,

$$t_{\mathrm{c}} = \frac{1}{\Delta} \ll T \, . \tag{4.15}$$

The correlation time is a measure of the time interval over which the phase of the electric field is well-defined. The further assumption that

$$t_{\mathrm{c}} \ll T_{\mathrm{p}} \tag{4.16}$$

signifies that the phase of the incident pulse undergoes many random fluctuations over the period required to pump an atom out of its ground state. Equation (4.16) places an upper limit on the pulse intensity – since the stronger the pulse, the shorter the pumping time – beyond which the optical pumping equations may no longer be valid. The above two conditions are not too restrictive, however, and are generally satisfied in standard quantum beat experiments. For example, a dye laser of a few hundred watts peak power yields temporal parameters on the order of:

pulse width	pumping time	coherence time
$T \sim 10^{-9}$ s $>$	$T_{\mathrm{p}}(t) \sim 10^{-10}$ s $>$	$t_{\mathrm{c}} \sim 10^{-11}$ s

for an allowed pumping transition.

The assumption has also been made in deriving (4.9), (4.10) that the Bohr frequencies in the manifolds e and g are small in comparison to Δ. This simplifies the analysis by leading to a unique pumping time and level shift for all the substates of a given manifold, but is not essential to the validity of the basic theoretical approach. The added complication of $T_p(t)$ and ΔE varying with the states of a manifold can be incorporated into the theory whenever necessary.

It is to be noted that no density matrix elements of the form ρ_{eg} or ρ_{ge}, which characterize optical coherence terms, appear in the optical pumping equations. These contributions – interpretable according to classical imagery as a macroscopic electric dipole moment precessing at optical frequencies – vanish when averaged over many correlation times of the field. In the broadband approximation, as the above theoretical approach is known, all relevant time parameters are *long* in comparison to the correlation time.

Solving the equations of motion and implementing the appropriate initial conditions lead to the matrix ρ_+ and, from (4.6), to the expression

$$\rho_e(t) = P_e \rho(t) P_e = P_e e^{-iH_0 t/\hbar} \rho_+ e^{iH_0 t/\hbar} P_e e^{-\Gamma_e t} \qquad (4.17)$$

for the total atomic density matrix projected onto the excited states. The optical signal, observed in the spontaneous emission after the pulse excitation is concluded, is obtained from (4.5)

$$I(\varepsilon_d) = K \mathrm{Tr}\left[\rho_e(t) O_{\det}^\dagger(\varepsilon_d)\right] e^{-\Gamma_e t}, \qquad (4.18)$$

where the detection operator $O_{\det}(\varepsilon_d)$ describes the optical transition (with polarization ε_d) between the e and f manifolds

$$O_{\det}(\varepsilon_d) = P_e \varepsilon_d \cdot \boldsymbol{D} P_f \varepsilon_d^* \cdot \boldsymbol{D} P_e . \qquad (4.19)$$

Note that only the angular component of the electric dipole operator appears in (4.19); the radial part has already been incorporated in the definition of the pumping time T_p. The constant K depends on geometrical factors such as the solid angle of acceptance of the detector and distance of the detector from the decaying atoms. As a scaling factor, it does not affect the form of the quantum beat signal and will henceforth be disregarded.

From the mathematical form of (4.18) one clearly sees that superposed on the exponential decay are modulations of the fluorescence at the various Bohr frequencies of the excited manifold. The amplitude of each Fourier component of the beat signal depends on the elements of $O_{\det}(\varepsilon_d)$ and ρ_+. To illustrate the use and physical content of the optical pumping equations, with an eye on the specific effects of nonlinearity in the interaction between the atoms and the light pulse, we will examine in the next section the quantum beat signals produced by a three-level system subjected to different excitation conditions.

4.3 Nonlinear Effects in a Three-Level Atom

Consider an atom with nondegenerate ground state g and two excited states e_1 and e_2, both of which decay at the rate Γ; the Bohr frequency is $\omega_0 = \omega_2 - \omega_1$. This system is almost the same as that treated in Sect. 4.1, except that states g and f are here taken to be identical. The processes of excitation and decay, summarized in Fig. 4.2, are characterized by the respective matrix elements

$$a_j \equiv \langle e_j | \boldsymbol{\varepsilon} \cdot \boldsymbol{D} | g \rangle \qquad (j = 1, 2) \, , \tag{4.20}$$

$$b_j \equiv \langle e_j | \boldsymbol{\varepsilon}_{\mathrm{d}} \cdot \boldsymbol{D} | g \rangle \qquad (j = 1, 2) \, , \tag{4.21}$$

which, for the sake of simplicity, are assumed to be real. The matrix representation of the detection operator then takes the form

$$O_{\mathrm{det}} = \begin{pmatrix} b_1^2 & b_1 b_2 \\ b_1 b_2 & b_2^2 \end{pmatrix} . \tag{4.22}$$

There are four independent elements to the density matrix of the atomic system which are designated as follows

$$x \equiv \langle g | \rho | g \rangle \, ,$$

$$y_j \equiv \langle e_j | \rho | e_j \rangle \qquad (j = 1, 2) \, , \tag{4.23}$$

$$z \equiv \langle e_2 | \rho | e_1 \rangle = \langle e_1 | \rho | e_2 \rangle^* \, ,$$

where the initial conditions before the laser excitation

$$x_- = 1 \, , \qquad y_{1-} = y_{2-} = z_- = 0 \tag{4.24}$$

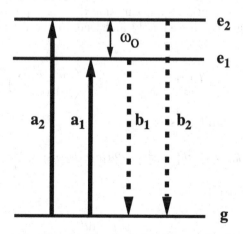

Fig. 4.2. Nondegenerate three-level system showing amplitudes for excitation and decay processes

characterize a ground-state atom. The corresponding density matrix elements describing the state of the atom after passage of the pulse will be subscripted with a plus sign. The quantum beat signal (4.18) then takes the form of (4.4)

$$ I = \left[b_1^2 y_{1+} + b_2^2 y_{2+} + b_1 b_2 (z_+ + z_+^*) \cos(\omega_0 t) \right] e^{-\Gamma t} . \tag{4.25} $$

To calculate the signal intensity I explicitly one must solve the optical pumping equations

$$ \frac{dx}{dt} = \frac{1}{T_p(t)} \left[a_1^2 y_1 + a_2^2 y_2 + a_1 a_2 (z + z^*) - (a_1^2 + a_2^2) x \right] , \tag{4.26} $$

$$ \frac{dy_j}{dt} = \frac{1}{T_p(t)} \left[a_j^2 (x - y_j) - \frac{a_1 a_2}{2} (z + z^*) \right] \pm i a_1 a_2 \Delta E(t)(z^* - z) , \tag{4.27} $$

with $\pm$ referring to $j = 1,\ 2$ respectively, and

$$ \frac{dz}{dt} = \frac{1}{T_p(t)} \left[a_1 a_2 \left(x - \frac{y_1 + y_2}{2} \right) - \frac{a_1^2 + a_2^2}{2} z \right] $$
$$ - i\omega_0 z - i\Delta E(t) \left[a_1 a_2 (y_1 - y_2) + (a_2^2 - a_1^2) z \right] \tag{4.28} $$

to determine the time evolution of the density matrix elements y_1, y_2, z. [The element x does not appear in (4.25).]

Under the conditions of weak pumping ($T/T_p \ll 1$) and short pulses ($\omega_0 T \ll 1$) one can neglect in the right hand side of the above equations all terms involving the elements of ρ_e (i.e., y_1, y_2, and z) and assume $x = 1$. This is the linear absorption approximation discussed previously. The equations can be integrated immediately and lead to the density matrix (after pulse passage)

$$ (\rho_e)_+ = \begin{pmatrix} y_{2+} & z_+ \\ z_+^* & y_{1+} \end{pmatrix} = k_0(\infty) \begin{pmatrix} a_2^2 & a_1 a_2 \\ a_1 a_2 & a_1^2 \end{pmatrix} , \tag{4.29} $$

with resulting quantum beat signal

$$ I_0 = k_0(\infty) \left[a_1^2 b_1^2 + a_2^2 b_2^2 + 2 a_1 a_2 b_1 b_2 \cos(\omega_0 t) \right] e^{-\Gamma t} , \tag{4.30} $$

where the preparation factor

$$ k_0(\infty) = \int_{-\infty}^{\infty} \frac{dt'}{T_p(t')} \tag{4.31} $$

is a measure of the efficiency of excited state preparation. The beats occur at frequency $\omega_0/2\pi$ with a modulation depth η_0 (equivalent to the visibility of fringes in a Young's two-slit experiment) given by

$$\eta_0 = 2\frac{a_1a_2b_1b_2}{a_1^2b_1^2 + a_2^2b_2^2} . \tag{4.32}$$

Under conditions where $a_1 = a_2$ and $b_1 = b_2$ the modulation depth can reach 100% This occurs, for example, in the case of Zeeman quantum beats following a $J = 0$ to $J = 1$ transition where both the excitation and decay radiation are polarized perpendicular to the external magnetic field, and only $m_J = \pm 1$ substates are excited.

Now let us continue to assume that the pulse duration T is short compared to the quantum beat period $\omega_0/2\pi$, and that the effects of light shifts are negligible

$$\omega_0 T \ll 1 , \qquad (\Delta E)T \ll 1 , \tag{4.33}$$

but that the optical pumping need no longer be weak. One can therefore drop from (4.27) and (4.28) those terms in which ω_0 and ΔE appear. Surprisingly, integration of the optical pumping equations gives rise to an excited state density matrix and quantum beat signal of the same form as in (4.29) and (4.30), but with $k_0(\infty)$ replaced by the time-dependent preparation factor

$$k(t) = \frac{1}{2(a_1^2 + a_2^2)}\left\{1 - \exp\left[-\int_{-\infty}^{t}\frac{2(a_1^2 + a_2^2)dt'}{T_{\mathrm{p}}(t')}\right]\right\} . \tag{4.34}$$

If the pumping is weak, relation (4.34) reduces to (4.31) upon a first-order expansion of the exponential. If the pumping is sufficiently strong, the exponential term becomes negligible, and $k(\infty)$ reduces to a constant and is independent of the pumping time $T_{\mathrm{p}}(t)$. Regardless of the pumping strength, however, the signals I and I_0 are proportional. In other words, the strength of pumping, as usually expressed by the so-called saturation parameter

$$S \equiv \frac{T}{T_{\mathrm{p}}} , \tag{4.35}$$

has no influence on the modulation depth under the experimental condition of short pulse duration.

The physical significance of this result can be understood as follows. The evolution of the system may be thought of as a sequence of absorption and stimulated emission processes occurring on average every T_{p} seconds. In the weak-pumping approximation (with $S \ll 1$) one takes account only of the first absorption process which, starting from the ground state g, creates the excited-state populations y_1 and y_2 and the coherence z proportional to a_1^2, a_2^2, and a_1a_2, respectively. For an intense pulse (with $S \gg 1$) many such sequences of absorption and stimulated emission can occur during the passage of one pulse. Nevertheless, stimulated emission destroys the excited state populations and coherence in the same proportions as absorption creates them with the consequence that the modulation depth is unaltered. Equation (4.25) shows that the modulation depth can change only if the relative magnitude of the excited state coherence and populations change.

If the system has a more complex level structure than that of a three-level atom, the form of the quantum beat signal can change with increasing saturation. Consider, for example, an atomic system with a pair of ground states, each one coupled by the laser pulse (and by spontaneous emission) to a distinct pair of excited states. Analysis of such a system leads to an optical signal that is the sum of two contributions of the form of (4.30) with different preparation factor, excitation and decay matrix elements, and beat frequency for the two uncoupled sets of three states. In this case the ratio of the preparation factors and the modulation depth of each frequency component will depend on the saturation parameter T/T_p although, in general, this dependence is not very marked.

The effects of saturation show up in a much more interesting and surprising way when one considers the example of a long pulse excitation, i.e., a pulse length no longer negligible in comparison to the beat period ($\omega_0 T \gg 1$). Intuitively, one might anticipate that the contributions to the signal from each (differentially) small time interval during passage of the pulse would interfere destructively when spread over a total interval T that exceeds the period of beats to be observed. The modulated component of the optical signal should then diminish and eventually vanish as T is lengthened beyond $2\pi/\omega_0$. In the case of *weak* optical pumping, this expectation is indeed correct as will now be demonstrated.

Retaining ω_0 in (4.28) for the coherence z – but neglecting the excited state populations and level shifts – leads to

$$z_+ = a_1 a_2 \int_{-\infty}^{\infty} \frac{\exp(-\omega_0 t')\mathrm{d}t'}{T_p(t')} . \qquad (4.36)$$

In the weak-pumping limit the populations y_{1+} and y_{2+} are still independent of ω_0 and given in the density matrix (4.29). The quantum beat signal then takes the form

$$I_0 = \left[k_0(\infty)(a_1^2 b_1^2 + a_2^2 b_2^2) + 2k_{\omega_0}(\infty)a_1 a_2 b_1 b_2 \cos(\omega_0 t) \right] e^{-\Gamma t} , \qquad (4.37)$$

where

$$k_{\omega_0}(\infty) = \int_{-\infty}^{\infty} \frac{e^{-i\omega_0 t'}\,\mathrm{d}t'}{T_p(t')} \qquad (4.38)$$

is the Fourier transform of the pulse profile at the beat frequency ω_0. For the time-independent or dc component of the beat signal, the preparation factor

$k_0(\infty)$ is the Fourier transform at frequency 0. Compared to the corresponding case of a short pulse width, the modulation depth is now a function of the beat frequency

$$\eta = \eta_0 \frac{k_{\omega_0}(\infty)}{k_0(\infty)} \ . \tag{4.39}$$

From the integral in relation (4.38) it is seen that the modulation depth – and hence the visibility of the beats – vanishes when the pulse duration is sufficiently long to permit variation of the phase $\omega_0 t'$ over at least 2π radians.

Intuition fails, however, in the case of both long pulse duration and strong pumping. If the linear approximation is no longer valid and the pulse duration no longer negligible, the entire set of coupled optical pumping equations must be solved as they stand. We will disregard for the moment, however, the light shifts and set $\Delta E = 0$. This is always possible if one restricts attention to a symmetric excitation profile centered on the frequency of the optical transition. Nevertheless, this simplification does not permit the equations to be solved in a simple, physically interpretable analytic form, and it is necessary to resort to numerical analysis by computer.

Let us adopt a Gaussian pulse profile

$$\frac{1}{T_{\mathrm{p}}(t)} = \zeta \exp\left[-\frac{t^2 \ln 2}{(T/2)^2}\right] \ , \tag{4.40}$$

with an amplitude ζ which determines the saturation parameter

$$S = \int_{-\infty}^{\infty} \frac{dt}{T_{\mathrm{p}}(t)} \ . \tag{4.41}$$

In Fig. 4.3 are shown computer simulations of the quantum beat signal for two pumping strengths – one low ($S = 0.1$) and one high ($S = 4000$). For purposes of illustration, the excitation and detection matrix elements have all been chosen to be unity, and the condition of long pulse duration is expressed by the assignment $\omega_0 T = 5$. Since it is the modulation depth, and not the exponential decay, that is of significance here, the beat profiles are shown for decay rate $\Gamma = 0$.

Although the modulation depth is seen to be small, as expected, in the lower trace where the pumping is weak, the striking feature of Fig. 4.3 is that strong pumping gives rise to large beat visibility in the upper trace even though the pulse duration is long compared to the beat period. Why are the beats not washed out as in the case of weak pumping?

This effect, which is known as the saturation regeneration of quantum beats, can be understood qualitatively in terms of a random-walk model. It is useful to recall first that a coupled two-level quantum system (with periodic transitions between the two levels) may be likened to a classical electric dipole

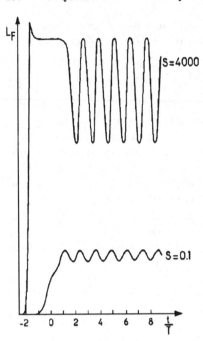

Fig. 4.3. Restitution of quantum beats for excitations of sufficiently high intensity (as gauged by the saturation parameter S) and pulse length long with respect to the beat period. Under the conditions of weak pumping (small S), a long pulse length leads, as shown, to low beat visibility. The theoretical parameters for the above plots are: $a_1 = a_2 = b_1 = b_2 = 1$, $\omega_0 T = 5$, $\Gamma = 0$. (Adapted from Silverman et al. [102])

moment undergoing precession. (This point will be examined in greater detail in the next chapter when we consider the effects of resonant radiofrequency fields on excited states.) It is the off-diagonal elements – the coherence term z – in the density matrix that corresponds to the classical precession; the conjugate pair z and z^* represent precessional motion in opposite senses. We have previously described the time evolution of the atomic system as a succession of absorption and stimulated emission processes each occurring on average every T_p seconds. When the atom evolves freely in the excited manifold, the coherence z is said to precess at the angular frequency ω_0. However, this precession is eventually interrupted by a stimulated emission coupling z to the ground-state population x. (Since the state lifetime $\tau \gg T$, the effect of spontaneous emission can be neglected over the duration of the pulse.) When a subsequent absorption process occurs, it renews the excited states (both populations and coherence), but the precession of the coherence term can take place either in the original sense (x coupled to z) or in the opposite sense (x coupled to z^*).

During passage of the pulse the number of elementary absorption and emission processes is of the order of T/T_p. Over the interval of time ($\sim T_p$) that an atom temporarily remains in the excited manifold, the phase of the coherence term – i.e., the precession angle of the analogous classical dipole – is of the order $\omega_0 T_p$. It is assumed in this heuristic argument that the laser pulse is sufficiently intense, and the pumping time correspondingly short, that $\omega_0 T_p \ll 1$. Although the precession can occur sometimes in one direction, sometimes in the other, so that the mean precession angle is zero, the *dispersion* in phase is not zero, but is given by the root mean square of the individual phase variations

$$(\Delta\phi)^2 = (\omega_0 T_{\mathrm{p}})^2 \frac{T}{T_{\mathrm{p}}} = \omega_0^2 T_{\mathrm{p}} T \quad \Longrightarrow \quad \Delta\phi = \omega_0 \sqrt{T_{\mathrm{p}} T} \,. \tag{4.42}$$

In order that the pulse create a substantial coherence in the excited state, the dispersion in phase must be less than about 1 radian ($\Delta\phi \leq 1$) which, from (4.42) is equivalent to requiring

$$S = \frac{T}{T_{\mathrm{p}}} \gg (\omega_0 T)^2 \,. \tag{4.43}$$

Thus, one learns from (4.43) that even if the pulse duration is long ($\omega_0 T \gg 1$), as long as the saturation parameter – or, equivalently, the pumping strength – is sufficiently high, it should still be possible to observe quantum beats in the fluorescence signal.

An alternative way of considering this phenomenon is that the effect of saturation is to slow down the coherence precession rate – in the analogous way that virtual absorption and emission of light propagating through a transparent material lead to a phase velocity below c (or refractive index greater than unity) – and thereby prevent destructive interference from wiping out the beats during passage of the pulse. Once the pulsed excitation is completed, the coherence term resumes its normal precession rate, and the modulation of the fluorescence occurs at the frequencies characterizing the energy level structure of the field-free atom. Although, to my knowledge, the predicted saturation regeneration of quantum beats has not yet been observed in spontaneous emission from excited states because of the high value of S required ($S > 100$), the slackening of the precession rate, as a result of nonlinear interactions with the exciting light, has been reported in other types of optical pumping experiments.

One such example is the Hanle effect [112]. The Hanle effect refers to the variation in fluorescent intensity of specified polarization when an external static magnetic field is varied about its null value at which point the radiating Zeeman substates become degenerate. A second example, also employing a static magnetic field, is the occurrence of quantum beats in the radiofrequency domain from optically-induced Zeeman coherences within an atomic *ground* state, a process referred to as free-induction decay [113]. This process is particularly interesting and warrants a closer examination.

The basic idea can be simply illustrated with a three-state atom comprising two degenerate ground states (1 and 2) and one excited state (3) as shown in the left frame of Fig. 4.4. The atom is irradiated by a continuous-wave (cw) laser which drives transitions 1–3 and 2–3, but not 1–2. There is no initial coherence between states 1 and 2, but a steady-state (i.e., time-independent)

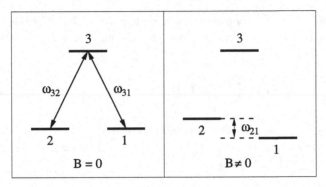

Fig. 4.4. Schematic level diagram for observation of ground-level Zeeman quantum beats (free-induction decay) in a three-state atom. In the absence of a magnetic field (*left*), the ground level is doubly degenerate; an incident cw laser drives transitions between the two ground states and the excited state, thereby creating a ground-state coherence ρ_{12}. When the laser irradiation is terminated and a magnetic field switched on (*right*), the ground-level degeneracy is broken and the coherence evolves in time at the Bohr frequency ω_{21}, which can be observed as an oscillatory decay of the system magnetization

coherence develops as a consequence of the laser irradiation. Upon extinction of the laser light and sudden initiation of a static field (e.g., magnetic or electric) of appropriate symmetry (right frame), the atomic coherence begins evolving in time at the Bohr frequency $\omega_{21} = (E_2 - E_1)/\hbar$ and can be detected by observing the oscillatory decay of a macroscopic magnetization or polarization of the sample.

The outcome stated above may seem surprising at first, if not in outright violation of quantum electrodynamics (QED) which predicts that radiative decay into substates of a ground level cannot lead to ground-state quantum beats. (Mathematically, one must sum the QED probabilities for transitions from states of an excited manifold to a given ground state.) There is, in fact, no inconsistency, as may be seen from the following heuristic argument, which outlines what, in effect, is the reasoning behind a Raman scattering process. In the presence of the cw laser radiation, the ground eigenstates of the system are no longer the eigenstates of the field-free Hamiltonian, but linear superpositions that may be written as

$$\psi_1 = a_{11}|1\rangle + a_{13}|3\rangle , \qquad \psi_2 = a_{22}|2\rangle + a_{23}|3\rangle ,$$

where amplitudes a_{11} and a_{22} are close to unity and amplitudes a_{13} and a_{23} are much smaller in the case of weak optical pumping. The ground-state component of the total system density matrix, $\rho_g = \overline{\Psi_g \Psi_g^\dagger}$, in which

$$\Psi_g = \begin{pmatrix} \psi_1 \\ \psi_2 \end{pmatrix}$$

and the overbar signifies an average with respect to distributed quantum phases, then takes the form

$$\rho_g = \begin{pmatrix} |a_{11}|^2 & \overline{a_{13}a_{32}^*} \\ \overline{a_{13}^* a_{32}} & |a_{22}|^2 \end{pmatrix},$$

from which it is clear that a nonvanishing ground-state coherence $(\rho_g)_{12} = \overline{a_{13}a_{32}^*}$ has been established.

The elements of the density matrix can be calculated, although by a mathematically more expedient method than that suggested by the preceding heuristic argument, and leads to a steady-state coherence

$$(\rho_g)_{12} = \frac{\frac{1}{4}V_{13}V_{23}}{(\omega - \omega_0)^2 + (V_{13}^2 + V_{23}^2)},$$

where V_{ij} is the electric dipole matrix element coupling states i and j, ω is the pump laser angular frequency, and ω_0 is the resonance frequency of the optical transition between state 3 and either of the two degenerate ground states. In this illustrative example, the natural lifetime of the excited state and the pumping time of the ground states have been disregarded. Since the solution of this problem employs concepts and techniques introduced in the next chapter on the interactions of atoms with radiofrequency fields, I defer discussion of it to Appendix 5B.

In the cited experiment [113], Zeeman coherences were induced in the $3s^2 S_{1/2}$ ground state of atomic sodium by a long laser pulse close to the D_1 resonance line. A static magnetic field was switched on and the induced ground-state coherence was subsequently observed (in the absence of the pump beam) by measuring the magnetization along the axis of a separate cw probe beam by means of the difference of the dispersion of right and left circularly polarized components. Figure 4.5 shows the quantum beat signal (free-induction decay) after termination of the pump pulse of duration much longer than the period of the beats.

Up to this point in the analysis of laser-induced quantum beats, there has been no (or very small) light-induced displacement of energy levels. We consider next situations in which light shifts could have a potentially significant effect on the quantum beat signal. To isolate this effect from other consequences of saturation connected with pulse length, assume at first that the free precession of the atom during passage of the pulse is negligibly small ($\omega_0 T \ll 1$). It may seem plausible that, if the light shifts are large and different for each substate of the excited manifold, the dephasing of the atomic coherence during passage of the pulse would lead to disappearance of the quantum beats. This reasoning is deceptive, however. One can show rigorously that, regardless of the magnitude of the displacement-and even if $(\Delta E)T \gg 1$, the level displacement has no effect on the signal under the currently assumed experimental conditions. This result follows from integrating the optical pump-

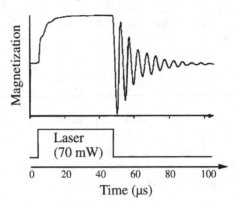

Fig. 4.5. Free-induction decay (ground-state Zeeman quantum beats) arising from a 70 mW square-wave pulse of cw laser radiation of atomic sodium, demonstrating the regeneration of quantum beats from a long pulse (pulse width $\gg$ beat period) of sufficiently large intensity. (Adapted from Rosatzin et al. [113])

ing equations (4.27) and (4.28) with $\omega_0 = 0$. The solutions y_1, y_2, z obtained for $t \to \infty$ are all independent of ΔE.

An explanation of the above puzzling result may be found in the symmetry of the original density matrix equations (4.9) and (4.10). It will be seen that the effective Hamiltonian characterizing light shifts in the excited states

$$\Delta E \, P_e \varepsilon \cdot DP_g \varepsilon^* \cdot DP_e$$

has exactly the same structure as the product of operators

$$\frac{1}{T_\mathrm{p}} P_e \varepsilon \cdot DP_g \rho_g P_g \varepsilon^* \cdot DP_e$$

governing the preparation of the excited manifold *provided the ground state is isotropic* (i.e, ρ_g is proportional to P_g). In other words, absorption of a photon puts the atom into an eigenstate of the effective Hamiltonian governing level displacement, and therefore cannot create an atomic coherence among these eigenstates evolving in time at frequencies of the order of ΔE during passage of the pulse even if $(\Delta E)T$ is large. The saturation effects tied to light shifts in this case are rigorously null.

If, however, the phase $\omega_0 T \gg 1$, the foregoing reasoning is no longer valid. It is then necessary to consider contributions of both the atomic Hamiltonian and the effective light-shift Hamiltonian to the evolution of the excited state during passage of the pulse. The quantization axis for energy eigenstates depends in that case upon the relative magnitudes of ω_0 and ΔE and, in addition, varies in time. Furthermore, if the atomic system is more complex than the one studied here and is characterized by a nondiagonal ground-state density matrix, then the pumping operator above need no longer have the same symmetry as the effective light-shift Hamiltonian. In that case, optical excitation could create coherence terms during the passage of the pulse even for $\omega_0 = 0$.

In general, the cases for which light shifts play a role in the determination of the quantum beat signal are complicated situations in which other effects of saturation are involved as well. The manifestation of these effects demands particular experimental conditions (such as light pulses with strong nonresonant spectral components) not likely to be realized in an actual quantum beat experiment.

4.4 Quantum Beats in External Fields

Although the introductory remarks of this chapter specifically pointed to the avoidance of time-varying external fields as one of the advantages of quantum beat spectroscopy, there are nevertheless circumstances under which the use of external fields can be conceptually and practically helpful. Indeed, as will be illustrated in the following chapter, the interaction of a coherently prepared atom with time-varying fields can lead to some very interesting physics. In this section, however, we are concerned primarily with theoretically interesting effects of magnetic fields on laser-induced quantum beats.

To determine the time evolution of an atomic system one must add to the field-free Hamiltonian H_0 the Hamiltonian of the appropriate external interaction H_{ext} whose specific form differs according to whether an electric or magnetic field is involved. In the case of coupling to a static magnetic field $\boldsymbol{B}_0$, the Zeeman Hamiltonian (in the absence of hyperfine structure) is given by

$$H_{\text{ext}} = -\mu_{\text{B}} \boldsymbol{B}_0 \cdot (\boldsymbol{L} + 2\boldsymbol{S}) , \qquad (4.44)$$

where μ_{B} is the Bohr magneton

$$\mu_{\text{B}} = \frac{|e|\hbar}{2mc} = 9.274 \times 10^{-21} \text{ erg G}^{-1} = 9.274 \times 10^{-24} \text{ J T}^{-1} , \qquad (4.45)$$

and $\boldsymbol{L}$ and $\boldsymbol{S}$ are respectively the orbital and spin angular momentum operators (in units of $\hbar$). For coupling of a fine-structure state $|nLJm_J\rangle$, where $\boldsymbol{J} = \boldsymbol{L} + \boldsymbol{S}$, to a static electric field $\boldsymbol{E}_0$ the effective Stark Hamiltonian can be written in the form

$$H_{\text{ext}} = \sum_{(n'L'J'm')} \frac{\boldsymbol{E}_0 \cdot \boldsymbol{\mu}_{\text{E}} |n'L'J'm'\rangle \langle n'L'J'm'| \boldsymbol{E}_0 \cdot \boldsymbol{\mu}_{\text{E}}}{E_{nLJ} - E_{n'L'J'}} . \qquad (4.46)$$

The energy level structure, which determines the possible quantum beat frequencies, is then obtained by diagonalizing $H_0 + H_{\text{ext}}$.

In general, the actual spectral composition of the quantum beats following excitation by one or more lasers depends on the polarization of the light and the relative strengths of the parameters governing the internal (e.g., fine structure) and external interactions. Even for hydrogenic systems like atomic Rydberg states, the calculation of the quantum beat profiles is by no means a trivial matter, and I will leave the details to the original scientific papers [102, 103]. It is instructive, however, to illustrate the effects of an external magnetic field in the case of an atom whose internal Hamiltonian includes a spin-orbit coupling term

$$H_{\mathrm{so}} = \hbar A \boldsymbol{L} \cdot \boldsymbol{S} , \tag{4.47}$$

where A is a measure of the strength of this fine-structure interaction (in units of angular frequency since $\boldsymbol{L}$ and $\boldsymbol{S}$ are in units of $\hbar$).

Figure 4.6 shows the variation in energy of the nD sublevels of atomic sodium as a function of magnetic field strength. Note that the ordering of the $J = 3/2$ and $J = 5/2$ levels is *opposite* that of normal ordering – i.e., the ordering of atomic hydrogen fine structure – as a result of complex interactions with the core electrons.[1] Since D and S states have the same parity,

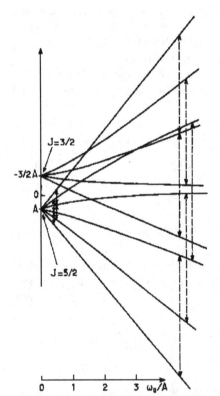

Fig. 4.6. Variation in energy with magnetic field of the nD sublevels of the sodium atom. The fine structure ordering is reversed from that of hydrogen. *Arrows* indicate the expected transitions leading to quantum beats in the region of low and high fields. (Adapted from Silverman et al. [102, 103])

---

[1] The origin of the anomalous fine structure ordering is discussed in my books *And Yet It Moves: Strange Systems and Subtle Questions in Physics* [114] and *A Universe of Atoms, An Atom in the Universe* [82]. For a detailed relativistic treatment, see [115].

a direct transition from the $3S$ ground level cannot be effected by single laser excitation. Two laser pulses, however, properly timed and of appropriate frequency, can induce sequential $3S$–$3P$ and $3P$–nD transitions thereby exciting the atom into a linear superposition of $D_{3/2}$ and $D_{5/2}$ states [116]. The relative orientation of the two excitation polarizations influences strongly the visibility of ensuing quantum beats; examination of a number of special cases suggests that greater beat contrast occurs for crossed polarizations.

In Figs. 4.7, 4.8, and 4.9 are illustrated the quantum beat transients calculated numerically for the respective conditions of zero, weak, and strong magnetic fields. The level of field strength is defined by the relative magnitude of the fine structure interaction parameter A and the cyclotron frequency

$$\omega_c = \frac{\mu_B B_0}{\hbar} \, , \tag{4.48}$$

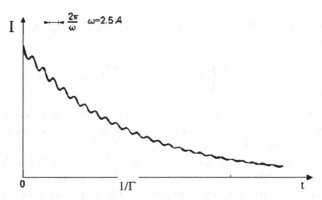

Fig. 4.7. Theoretical zero-field quantum beat signal. (Adapted from Silverman et al. [102, 103])

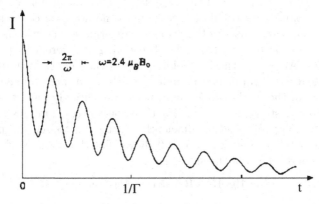

Fig. 4.8. Theoretical weak-field ($\omega_0 \ll A$) quantum beat signal. B_0 is the applied magnetic field, and μ_B is the Bohr magneton. (Adapted from Silverman et al. [102, 103])

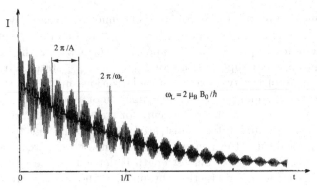

Fig. 4.9. Theoretical strong-field ($\omega_0 \gg A$) quantum beat signal. ω_L is the Larmor frequency, and A is the fine-structure interaction parameter. (Adapted from Silverman et al. [102, 103])

both of which enter the expression for the eigenvalues (in units of angular frequency)

$$\omega_{L\pm1/2,m} = -\frac{A}{4} + m\omega_c \pm \frac{1}{2}\left\{(Am + \omega_c)^2 + A^2\left[L(L+1) - \left(m^2 - \frac{1}{4}\right)\right]\right\}^{1/2} \tag{4.49}$$

of the basis vectors $|Jm\rangle'$ of the Hamiltonian $H_0 + H_{\text{ext}}$. A configuration was chosen such that the two light polarizations are parallel to each other and perpendicular to the external magnetic field.

The theoretical zero-field quantum beat signal (Fig. 4.7) displays a beat frequency $\omega_0 = 5A/2$, corresponding to the fine-structure level separation, with shallow modulation depth as expected. The application of a weak magnetic field ($\omega_0 \ll A$), however, enhances the beat visibility as shown in Fig. 4.8. The signal is seen to be modulated primarily at the frequency $2g_J\mu_B B_0$ where $g_J = 1.2$ is the Landé factor of the $D_{5/2}$ level. The calculation shows that in this case the signal is particularly sensitive to the $\Delta m = 2$ coherence in the $D_{5/2}$ level. Contributions of other coherence terms either within the $D_{3/2}$ level or between the $D_{3/2}$ and $D_{5/2}$ levels evolving at different frequencies are much weaker. Arrows in the low-field region of the energy level diagram of Fig. 4.6 indicate the couplings contributing to the quantum beat signal. Thus, measurement of the weak-field Zeeman beats can directly yield the Landé g factor for excited states, a point of practical spectroscopic interest.

When the magnetic field is sufficiently strong ($\omega_0 \gg A$), the pattern of beats is completely changed, as Fig. 4.9 shows. For the example of sodium D states the analytic form of the signal is

$$I = 198.15 + 105.55 \cos(At) \cos(\omega_L t) , \tag{4.50}$$

where the two frequencies appearing in the signal are indicated by arrows on the high-field region of the energy level diagram of Fig. 4.6. The signal now consists of a carrier wave at the Larmor frequency $\omega_L = 2\omega_c$ modulated strongly at the frequency A of the fine-structure interaction. What is the origin of this beats-within-beats structure?

A simple interpretation can be given in terms of the classical vector model of the atom [117]. In a high magnetic field the angular momenta L and S are decoupled and each can precess freely about the magnetic field. This explains the appearance of the Larmor frequency ω_L in place of $2g_J\omega_c$. One must also take account, however, of the diagonal part $(Am_L m_S)$ of the fine structure interaction (4.47), which acts as a perturbation adding to the applied field a small internal magnetic field $B' = \hbar A m_S/\mu_B$. This internal magnetic field may be oriented parallel or antiparallel to the applied field according to the sign of $m_S = \pm 1/2$, the projection of spin along the quantization axis (i.e., the direction of the external magnetic field). There are therefore two Larmor frequencies $\omega_{L\pm} = \omega_L \pm A$, and correspondingly the sum of two cosinusoidal terms at these frequencies gives the frequency dependence of relation (4.50) as

$$\cos\left[(\omega_L + A)t\right] + \cos\left[(\omega_L - A)t\right] = 2\cos(At)\cos(\omega_L t) .$$

The foregoing high-field quantum beam experiment, which to my knowledge has yet to be performed, suggests an alternative to zero-field experiments for measuring the fine structure interaction constant. As seen in the foregoing example, high-field quantum beats show a marked enhancement in the signal-to-noise ratio.

4.5 Correlated Beats from Entangled States

An entangled state, so designated by Schrödinger [118] in 1935, refers to a quantum state of a multiparticle system that cannot be expressed as a product of single-particle states. Such states give rise to correlations between separated particles which are unaccountable within the framework of classical physics and sometimes bizarre even by the standards of quantum physics if one's expectations are based on the study of single-particle systems or systems of uncorrelated particles. Entangled states are therefore of considerable interest to those concerned with the foundations of quantum physics. Schrödinger, himself, regarded the property of entanglement as the foremost characteristic property of quantum mechanics.

We have already encountered quantum entanglement in the discussion of the EPR paradox and the correlations manifested in the AB–EPR and AB–HBT experimental configurations of previous chapters. It will be recalled that the EPR paradox was advanced as an argument that quantum mechanics could not be a complete theory and most likely had to be supplemented by additional (and possibly unknowable) variables. Since the appearance in 1935 of the seminal EPR paper, a number of experiments have been performed with entangled photon states for the purpose of revealing the possible existence of such local hidden variables by testing the correlations expressed in a set of inequalities derived by J.S. Bell [119]. An essential feature common to these EPR-type experiments is the correlation in space, time, or polarization of two photons emitted during radiative decay of an excited atomic state. In these experiments the two-photon entangled states were produced by cascade transitions from single atoms excited incoherently by optical or electronic transitions or by the nonlinear optical process of parametric down-conversion [120].

In this section we examine a remarkable example of nonlocal correlations manifested by quantum beats in the radiative decay of entangled states of two identical, but widely separated, excited atoms [121]. What makes the example of particular interest is that this quantum interference effect occurs even though each individual atom is *not* prepared in a linear superposition of excited states.

To appreciate the extraordinary nature of the effect recall that the spontaneous emission from incoherently populated atomic states decays exponentially in time. A percussional excitation sharply defined in time – produced, for example, by pulsed laser – and sufficiently uncertain in energy can put an atom into a linear superposition of close-lying excited states. Only then, it would seem, should the subsequent atomic fluorescence be modulated at the Bohr frequencies corresponding to the energy intervals of the superposed states. If the spectral width of the exciting pulse is much smaller than the Bohr frequencies, the atom will be prepared in a sharp energy eigenstate, and no quantum beats will be induced.

In striking contrast to the conditions characteristic of standard quantum beat spectroscopy, the quantum interference phenomenon to be described now can be produced by spectrally *narrow* but correlated photons. The quantum beats then appear – not in the fluorescence from individual atoms – but in the *joint* detection of photon pairs, one light quantum arising from each of two separated atoms.

Figure 4.10 shows a schematic diagram of a possible experimental configuration. A light source – created, for example, by decay of S-state sys-

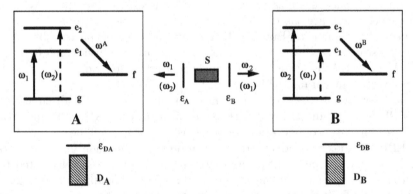

Fig. 4.10. Configuration of a long-distance or EPR-type quantum-beat experiment. Pairs of photons emerge in opposite directions from source S, exciting separated atoms A and B into state e_1 or e_2, but *not* a superposition of the two. Photodetectors D_A and D_B individually reveal no quantum beats although beats appear at frequency $\omega_2 - \omega_1$ in their correlated output. (Adapted from Silverman [121])

tems – produces photons in pairs, one photon of frequency ω_1 and the other of frequency ω_2, that propagate in opposite directions through polarizers ε_A and ε_B to excite two arbitrarily separated atoms. Which direction a photon of particular frequency takes is entirely random and unpredictable, although if it is known that a photon of frequency ω_1 has propagated to the left, then a photon of frequency ω_2 must necessarily have propagated to the right. Depending on whether the frequency is ω_1 or ω_2, a photon can induce a transition from the atomic ground state g to the excited state e_1 *or* e_2 respectively. Neither photon, however, prepares an atom in a linear superposition of states e_1 *and* e_2. Finally, the fluorescent photons created by radiative decay of the atoms from states e_1, e_2 to some final state f pass through polarizers ε_{D_A} and ε_{D_B} and are monitored without energy selection at detectors D_A and D_B.

Time-ordered diagrams representing the two modes of excitation and decay are illustrated in Fig. 4.11. Since the two modes give rise to indistinguishable final states of the whole system, their amplitudes are to be added, and the resulting probability for the process will display a quantum interference term. However, the *single-atom* density matrix, which reveals all that can be known by measurements on one of the two atoms (A or B), is equivalent (as will be demonstrated shortly) to that of an atom with *incoherently* populated excited states. Thus, no beats will appear at the output of one detector alone. The quantum beats occur only in the joint probability of receiving photons at the two detectors.

Pursuant to the foregoing description, the state of the two-atom system immediately following excitation by two correlated photons is representable by the entangled superposition of state vectors

$$|\phi(0)\rangle = a_{12}|e_1^A; e_2^B\rangle + a_{21}|e_2^A; e_1^B\rangle , \qquad (4.51)$$

where the label e_j^K designates excited state j (1 or 2) of atom K (A or B). To within a constant factor unimportant for our present purposes the coefficients of the superposition can be written as

$$a_{ij} \sim \langle g|\boldsymbol{\mu}_A{\cdot}\boldsymbol{\varepsilon}_A|e_i\rangle\langle g|\boldsymbol{\mu}_B{\cdot}\boldsymbol{\varepsilon}_B|e_j\rangle , \qquad (4.52)$$

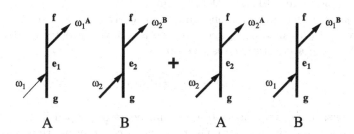

Fig. 4.11. Time-ordered diagrams for the excitation and decay processes giving rise to long-distance quantum beats. *Vertical lines* represent the evolution of atoms A and B from ground state g to final state f via intermediate excited states e_1 or e_2. *Oblique lines* represent absorption (*arrows in*) and emission (*arrows out*) of photons

where $\boldsymbol{\mu}_K$ is the electric dipole moment of atom K. The density operator of the total system is then

$$\rho(0) = |\phi(0)\rangle\langle\phi(0)| \, , \tag{4.53}$$

from which it readily follows by tracing over the states of atom B that the single-atom density operator of atom A takes the form

$$\rho_A(0) = \text{Tr}_B\left[\rho(0)\right] = |a_{12}|^2|e_1^A\rangle\langle e_1^A| + |a_{21}|^2|e_2^A\rangle\langle e_2^A| \, . \tag{4.54}$$

By tracing (4.53) over the states of atom A, one arrives at a similar expression for the single-atom density operator of atom B. Since the matrix represented by expression (4.54) is diagonal in an energy representation – i.e., displays only populations and no coherence terms – measurements made on A or B alone would not distinguish systems represented by state vector (4.51) from two ensembles of atoms with incoherently populated excited states.

After the initial excitation, the states of the two atoms evolve freely and independently under their respective Hamiltonians. Thus, the state of atom A at time t_A and of atom B at time t_B (where the sharp excitation defines the time origin) is represented by the vector

$$|\phi(t_A, t_B)\rangle \tag{4.55}$$

$$= a_{12} \exp\left[-i\left(\omega_1^A - \frac{i}{2}\Gamma_1^A\right)t_A\right] \exp\left[-i\left(\omega_2^B - \frac{i}{2}\Gamma_2^B\right)t_B\right] |e_1^A; e_2^B\rangle$$

$$+ a_{21} \exp\left[-i\left(\omega_2^A - \frac{i}{2}\Gamma_2^A\right)t_A\right] \exp\left[-i\left(\omega_1^B - \frac{i}{2}\Gamma_1^B\right)t_B\right] |e_2^A; e_1^B\rangle \, ,$$

where ω_i^K is the angular frequency corresponding to excited state i of atom K, and Γ_i^K is the associated spontaneous emission decay rate. Although the two atoms A and B are of identical kind, the Bohr frequencies and decay rates are specifically labeled by A or B since the atoms may be moving with different velocities with respect to the stationary detectors of the laboratory frame and therefore subject to Doppler shifts.

The joint probability that one photon is received at detector D_A within an interval Δt_A about time t_A and a second is received at detector D_B within an interval Δt_B about time t_B is given by

$$P(t_A, t_B) = I(t_A, t_B)\Delta t_A \Delta t_B \, . \tag{4.56}$$

By generalization of the theory of quantum beats developed in the previous sections, one can determine $I(t_A, t_B)$ to within a constant instrumental factor by calculating the mean value of the detection operator

$$O_{\text{det}} = \left[(\boldsymbol{\varepsilon}_{D_A}\cdot\boldsymbol{\mu}_A)(\boldsymbol{\varepsilon}_{D_B}\cdot\boldsymbol{\mu}_B)\right]|f_A f_B\rangle\langle f_A f_B|\left[(\boldsymbol{\varepsilon}_{D_A}\cdot\boldsymbol{\mu}_A)(\boldsymbol{\varepsilon}_{D_B}\cdot\boldsymbol{\mu}_B)\right]^\dagger \tag{4.57}$$

as follows

$$I(t_A, t_B) = \text{Tr}\Big[|\phi(t_A', t_B')\rangle\langle\phi(t_A', t_B')|O_{\det}\Big] , \tag{4.58}$$

where

$$t_K' = t_K - \frac{r_K}{c} \tag{4.59}$$

is the retarded emission time of atom K (A or B) a distance r_K from detector D_K.

Evaluation of the trace in (4.58) leads to the joint intensity function

$$I(t_A, t_B) = |a_{12}b_{12}|^2 \exp\Big[-(\Gamma_1^A t_A' + \Gamma_2^B t_B')\Big] \tag{4.60}$$

$$+ |a_{21}b_{21}|^2 \exp\Big[-(\Gamma_2^A t_A' + \Gamma_1^B t_B')\Big]$$

$$+ 2\exp\Big\{-\frac{1}{2}\big[(\Gamma_1^A + \Gamma_2^A)t_A' + (\Gamma_1^B + \Gamma_2^B)t_B'\big]\Big\}$$

$$\times \text{Re}\,\Big\{a_{12}a_{21}^* b_{12}b_{21}^* \exp\Big[-i(\omega_{21}^A t_A' - \omega_{21}^B t_B')\Big]\Big\} ,$$

in which

$$b_{ij} \sim \langle f|\boldsymbol{\mu}_A\cdot\boldsymbol{\varepsilon}_{D_A}|e_i\rangle\langle f|\boldsymbol{\mu}_B\cdot\boldsymbol{\varepsilon}_{D_B}|e_j\rangle \tag{4.61}$$

are the spontaneous emission matrix elements, and

$$\omega_{21}^K = \omega_2^K - \omega_1^K \tag{4.62}$$

is the beat frequency of the fluorescent emission from atom K in the laboratory frame. This frequency is related to the corresponding frequency ω_{21} in the atomic rest frame by the Doppler effect

$$\omega_{21}^K = \omega_{21}\left(1 + \boldsymbol{k}_K\cdot\frac{\boldsymbol{v}_K}{c}\right) , \tag{4.63}$$

where $\boldsymbol{k}_K$ is a unit vector from detector K to atom K and $\boldsymbol{v}_K$ is the velocity of atom K.

Since it is a Doppler-shifted *beat* frequency, rather than optical frequency, that comprises the signal, the quantum interference effect described here is largely insensitive to atomic motion. For example, for a thermal distribution of atoms at room temperature ($T \sim 300$ K), the atomic velocities span an approximate range $0 < v/c < 10^{-3}$, and the Doppler effect leads to a spread of $\sim 10^{11}$ s^{-1} about an optical frequency of $\omega_2 \sim \omega_1 \sim 10^{14}$ s^{-1}. This may well be much larger than the quantum beat frequency ω_{21} (derived, for example, from atomic fine structure) which could be typically of the order of 10^8 s^{-1}. or lower. The Doppler spread of the beats, however, would span the much smaller range $0 < \omega_{21} < 10^5$ s^{-1}, and therefore have negligible effect on dephasing of the observed signal. Similarly, for an atom with sharply defined energy eigenvalues, $\omega_i \gg \omega_{21} \gg \Gamma_i$ ($i = 1, 2$), one can ignore the effect of atomic motion on the excited state decay rates.

Assume for purposes of illustration that the excitation and decay matrix elements expressed by relations (4.52) and (4.61) are real-valued numbers, and let us disregard the weak effects of atomic motion. Equation (4.60) then reduces to the following simpler expression for the jointly detected two-photon fluorescent signal

$$I(t_A, t_B) = (a_{12}b_{12})^2 \exp\left[-\left(\Gamma_1 t'_A + \Gamma_2 t'_B \right) \right] \tag{4.64}$$

$$+(a_{21}b_{21})^2 \exp\left[-\left(\Gamma_2 t'_A + \Gamma_1 t'_B \right) \right]$$

$$+2a_{12}a_{21}b_{12}b_{21} \exp\left[-\frac{1}{2}(\Gamma_1 + \Gamma_2)(t'_A + t'_B) \right]$$

$$\times \cos\left\{ \omega_{21}\left[(t_B - t_A) - \frac{r_B - r_A}{c} \right] \right\} .$$

Note in particular the argument of the cosine factor of the interference term in which the retarded times have been replaced by the actual detection times by means of defining relation (4.59). A significant feature of the above expression is that a large dispersion in the difference in retardation times can be tolerated without the quantum interference term being averaged away. Although in the foregoing discussion reference is made to 'atom A' or 'atom B', what is really implied, of course, are two distinct atomic ensembles within each of which many identical atoms are in motion and instantaneously located at different distances from the closest light detector. Thus, the spatial intervals r_A and r_B are distributed quantities. For the quantum beats to persist, the term $\omega_{21}|r_B - r_A|/c$ must remain less than about one radian as the optical path lengths from atoms A and B to all points on the detecting surfaces of D_A and D_B, respectively, vary. For a beat frequency ω_{21} on the order of 10^8 s^{-1}, one must have $|r_B - r_A| < c/\omega_{21} \sim 300$ cm, which is experimentally easy to satisfy.

The possibility of quantum beats in the fluorescence of a multi-atom system has long been known [109] for the standard procedure of single-photon excitation of a localized system of atoms. In that case nearly contiguous atoms constitute a single quantum system, and quantum beats arise from the interference of decay pathways involving one or another of the associated atoms. These beats are highly sensitive to the dispersion in retardation times, as they are in general to the Doppler effect, and would be totally dephased for mean atom separations that exceed an optical wavelength ($\sim 10^{-5}$ cm). Moreover, such an interference can occur in elastic scattering only – i.e., the states g and f must be the same, otherwise the emitting atom can in principal be identified.

In striking contrast, the 'long-distance' quantum beat phenomenon arising from entangled states can occur in inelastic ($f \neq g$) as well as elastic scattering, and is largely insensitive to atomic motion or location.

5

Sympathetic Vibrations:
The Atom in Resonant Fields

5.1 Beams, Bottles, and Resonance

Science, according to the articulate writer and Nobel Laureate in medicine, Peter Medawar, is the "art of the soluble" [122], and as there is virtually no system in nature that is *exactly* soluble, the application to the real world of physics – the quintessential science – is to a great extent the art of modeling. To recount the great successes of physics is in large measure to unfold a historical record of aptly chosen, albeit hypothetical, models of reality such as frictionless free-fall, point electrical charges, and ideal gases. Among these immensely useful abstractions is the two-level quantum system.

Quantum systems (depending on the nature of the potential energy function) ordinarily have many, perhaps an infinite, number of eigenstates, but can be regarded as having only two to the extent that one may neglect all couplings except those between a given pair of interest. Then the quantitative description of such a system can be cast into the same aesthetic mathematical form irrespective of whether the two states are the spin states of an electron, the hyperfine states of a hydrogen atom, the inversion modes of an ammonia molecule, or the macroscopic states of the two superconducting regions comprising a Josephson junction. With an anticipated application to atomic physics in mind, we shall take the two-level system of this chapter to be an atom and consider its interaction with a classical linearly polarized monochromatic oscillating field. We have then the 'purest' form of wave inducing transitions in the most elementary bound-state matter – and yet no exact analytical solution of this deceptively simple system has yet been found. How one can handle this situation will be addressed shortly.

The importance of the two-level atom to physics, however, can hardly be overestimated. Investigations of a two-level atom in an oscillating field go back at least to the mid-1930s, to the magnetic resonance experiments of I.I. Rabi whose interest in determining the signs of nuclear magnetic moments led him to derive the fundamental Rabi 'flopping formula' [123] which has since served as a basis for nearly all successive magnetic resonance experiments. In this paper Rabi calculated the probability that an atom subject to a 'gyrating' (or rotating) magnetic field of specified frequency (primarily in the radiofrequency domain) undergoes a transition from one to the other of its quantum states – or, viewed in terms of classical imagery, that the atomic magnetic moment undergoes a reorientation. The significance of this paper is succinctly captured by Rabi's biographer, J.S. Rigden [124]: "Today, fifty years later, this paper is cited by laser physicists who use Rabi's 'flopping formula', derived in the 1937 paper, thus showing how a great paper can be applicable far beyond the immediate intentions of its author."

One particularly significant issue of Rabi's resonance method was the subsequent development of the atomic beam *electric* resonance technique pioneered by Lamb and Retherford [125] to make high-precision tests of relativistic atomic structure and quantum electrodynamics. Profiting from the development of radiofrequency and microwave sources needed for radar during the Second World War, the Lamb–Retherford experiment marked the first major innovation in the study of excited-state atomic structure beyond traditional optical spectroscopy. The advantage of an atomic beam, as already mentioned, was that atoms could be studied in a region separate from the source of their production which was generally filled with rapidly fluctuating electromagnetic fields (as in the vicinity of an electrical discharge or an electron beam). Use of radiofrequency (rf) techniques permitted the direct coupling of excited states within the same electronic manifold with greatly reduced Doppler broadening.

In the Lamb–Retherford experiment, a beam of excited hydrogen atoms was produced by thermal dissociation of H_2 in a tungsten oven and then bombarded by a transverse beam of 10.8 eV electrons (whose energy corresponds to the transition $n = 1$ to $n = 2$). Except for the $2^2S_{1/2}$ metastable state, which has a relatively long lifetime of 1/7 second due primarily to a two-photon transition to the $1^2S_{1/2}$ ground level (since single photon transitions are forbidden by angular momentum conservation), all the other excited states were rapidly extinguished. A virtually pure beam of $2^2S_{1/2}$ atoms then passed through a microwave interaction region where $2^2S_{1/2}$–$2^2P_{1/2}$ electric dipole transitions were driven. This diminished the metastable population which ultimately reached the detector where they were observed by monitoring the electrons ejected by impact of the surviving atoms upon a tungsten plate.

The resonance curve or line shape, from which the $2^2S_{1/2}$–$2^2P_{1/2}$ level separation could be determined, is ideally a measure of the variation of transition probability with the frequency of the applied oscillating field. From a practical standpoint, however, it was not feasible at the time for the experimenters to sweep a microwave or rf oscillator without simultaneously encountering marked changes in the output power. Had they conducted the experiment under these conditions, the line shapes would have been distorted and therefore useless for rendering accurate information on atomic structure. Lamb circumvented this problem by keeping the oscillator frequency constant and varying, instead, the strength of a transverse, homogeneous magnetic field through which the atomic beam passed. The magnetic field served at least two purposes. First, by deflecting charged particles away from the detector, it reduced the noise background. Principally, however, by altering the separation of the magnetic substates within each fine-structure level (the Zeeman effect), it afforded a means of selecting specific pairs of substates to investigate and of 'sweeping out' the resonance line shape by scanning, not the applied frequency of the oscillator, but the resonance frequency of the atom.

The Lamb–Retherford experiment demonstrated conclusively the long-suspected failure of the Dirac relativistic theory of the hydrogen atom to predict correctly the energy separation of states of the same n, J quantum numbers. In addition, many fascinating points of atomic physics were brought to light. However, as a general method for investigating excited atomic states, this approach left much to be desired. For one, it was restricted to the study of $2S$ metastables, since only atoms thus prepared could survive traveling a macroscopic distance before decaying. The lifetime of a $2P$ state, for example, is approximately 2 ns; emerging from a 2500 K oven at a probable speed of nearly 8×10^5 cm/s, a $2P$ atom would not likely travel more than a few microns. There were also a large number of systematic effects, most traceable to the presence of the magnetic field, which distorted and/or shifted the resonance curves and required correction. Such effects included the variation of the transition matrix elements with magnetic field, the production (by the Lorentz transformation) of a 'motional' electric field in the atom rest frame with resulting displacement of the line center (Stark effect) and selected quenching of hyperfine states, the unsymmetrical distribution of the hyperfine levels about the mean fine structure energy caused by incomplete decoupling of the nuclear and electron angular momenta (Back–Goudsmit effect), and 'curvature' of the Zeeman levels due to partial decoupling of the electron orbital and spin angular momenta (Paschen–Back effect). In the words of Lamb [126]: "A lengthy programme of calculations and measurements is required to allow for all such sources of error."

To investigate non-metastable excited states, Lamb and his co-workers later introduced a new radio-frequency–optical technique in which the use of an atomic beam was abandoned for a resonance cell ('bottle' experiment) [127]. The method involved the excitation of hydrogen atoms to $3S$, $3P$, and $3D$ states by a low energy electron beam, simultaneous irradiation by a rf field of fixed frequency, and detection of atomic transitions by monitoring the variation in Balmer α emission as a function of magnetic field strength. As is characteristic of a bottle experiment, production, irradiation, and detection all took place in the same region for which the "understanding of the origin and magnitudes of electric fields ... was distinctly incomplete", in the words of Lamb and Sanders [127]. Electrical perturbations possibly due to space charge within the electron beam, charges on the walls of the glass vessel, motion of excited atoms across the magnetic field, and fluctuating fields of neighboring ions and electrons gave rise to line shifts and signal variations as conditions within the vessel changed with time. Since the sensitivity of hydrogen to electric perturbations increases rapidly with principal quantum number (the Stark shift, for example, varies roughly as n^6), the bottle method – in the period before quantum beat spectroscopy – promised serious difficulties as a general technique to probe excited atomic states.

If I have dwelt rather long upon Lamb's experimental methods, it is because – problems notwithstanding – they are the ingenious prototypes which, like Rabi's, inspired a variety of succeeding methods to expose the inner workings of the atom by approaching ever more closely the ideal of a two-level quantum system in a purely oscillating field. It is clear from the foregoing remarks that reaching such an ideal would be greatly facilitated if:

- atomic states were produced in quantity at one place and rapidly transported to a separate region for examination,
- state selection and spectroscopy were performed in the absence of an external static magnetic field.

One notably successful way this has been accomplished is by the electrostatic acceleration of atoms to a high (but nonrelativistic) speed and subsequent mapping out of resonance line shapes by the originally desired, and ideally simplest, procedure of sweeping the frequency ω of an applied rf or microwave field in an otherwise field-free environment. A representative experimental configuration is illustrated in Fig. 5.1. Typically, the signal

$$\text{signal}(\omega) = \frac{N(0) - N(\omega)}{N(0)} \tag{5.1}$$

is obtained by counting photons in the decay radiation for a set period of time with the oscillating field off $[N(0)]$ and on $[N(\omega)]$, respectively. Since changes in the resonant response of the atoms with frequency modify the intensity and polarization of spontaneous emission and can therefore be detected optically, I have designated this method fast-beam optical electric resonance or simply OER [128]. The basic principles of OER spectroscopy, delineated in the early 1970s [129], still provide one of the most straightforward ways of probing atomic structure [56]. A comparative summary of slow (thermal) beam, bottle, and accelerator-based OER methods is given in Table 5.1.

Although the direct acceleration of neutral atoms to an energy of tens of keV is not feasible, the conversion of accelerated protons by charge-capture collisions with gas or carbon-foil targets can provide a means of generating a beam of hydrogen atoms moving at speeds on the order of 10^8 cm/s, or nearly one hundredth the speed of light. As a consequence of the very short impact time, the fast-moving hydrogen atoms emerge from the target distributed over a large number of excited states with only a small loss in translational energy. The beam subsequently passes through one or more rf interaction regions in which transitions of interest are driven, while use of a precise power monitor ensures a constant rf power across the resonance line. Downstream from the rf interaction region the fluorescence of the atoms is detected as a function of the rf frequency. No external magnetic fields are applied, and the field of the

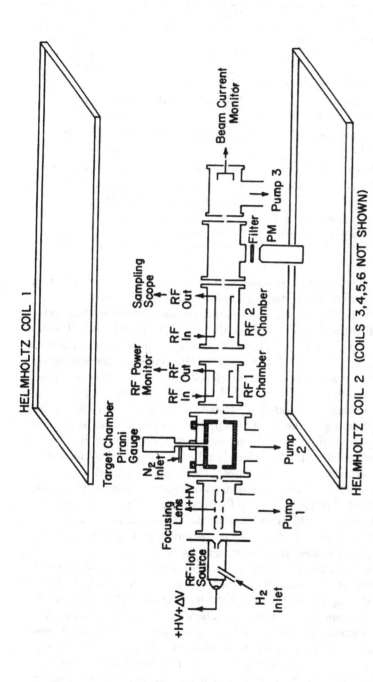

Fig. 5.1. Schematic diagram of a fast-beam optical electric resonance apparatus. Protons, extracted at high potential from the radiofrequency (rf) ion source, are focused by an electrostatic lens into a target chamber where some capture electrons and emerge as an accelerated beam of neutron hydrogen atoms. The atoms pass through a spectroscopy chamber where desired transitions are induced by a rf field, then a state-selection chamber where undesired states are removed by a second rf field, and finally a detection chamber where decay photons are monitored by a photodetector (PM). The residual proton current is received in a Faraday cup. Three mutually orthogonal pairs of Helmholtz coils shield the apparatus from the Earth's magnetic field. (Adapted from [130])

Table 5.1. Comparison of spectroscopic methods

	Slow beam	Bottle	Fast beam OER
Source	Thermal dissociation	Electron bombardment	Charge exchange conversion of a fast ion beam
	Electron bombardment		
Spectroscopy	RF or static electric field	RF electric field	Swept RF electric field
	Swept magnetic field	Swept magnetic field	Zero magnetic field
Detection	Metastable current	Decay radiation	Decay radiation
Limitations	Long-lived states	Complex electrical perturbations	Specifications of RF field
	Systematic errors from magnetic field	Variation in bottle environment	

Earth, itself, can be nulled by surrounding the apparatus with three mutually orthogonal sets of current-carrying coils known as Helmholtz coils.

An example of one of the first panoramic sweeps through the level structure of hydrogen is illustrated in Fig. 5.2 for the $n = 4$ manifold [130]. The accompanying level diagram shows in broad outline the types of single-quantum electric dipole transitions that can occur. In the event that transitions between two different pairs of fine structure levels have neighboring resonance frequencies – as in the case of $4^2S_{1/2}$–$4^2P_{3/2}$ and $4^2P_{1/2}$–$4^2D_{3/2}$ transitions – the use of two or more separated rf interaction chambers can often prove helpful, one field serving as the spectroscopy field and the others as quenching fields to drive atoms in unwanted long-lived states into shorter-lived states that decay before reaching the detection chamber. Figure 5.2 illustrates state selection for the preceding pair of transitions. Generated by a single oscillating field, the $4^2P_{1/2}$–$4^2D_{3/2}$ resonance is seen as a poorly delineated plateau in the high-frequency tail of the $4^2S_{1/2}$–$4^2P_{3/2}$ resonance curve. However, with a quenching field set at 1225 MHz close to the resonance frequency of the $4^2S_{1/2}$–$4^2P_{3/2}$ transitions, the $4^2S_{1/2}$ states are eliminated, and the profile of the $4^2P_{1/2}$–$4^2D_{3/2}$ resonance curve shows up clearly.

Besides overlapping fine-structure resonances, the hydrogen spectrum is further complicated by the magnetic interaction between electron and nuclear spins which divides each fine structure level into two hyperfine components with total angular momentum quantum numbers $F = J \pm 1/2$ and associated substates with magnetic quantum numbers $-F \leq m_F \leq +F$. Thus, transitions at three different resonant frequencies corresponding to quantum jumps with $\Delta F = \pm 1, 0$ can be induced between a pair of fine structure levels of opposite parity. In such a case sequential rf fields can be employed again to

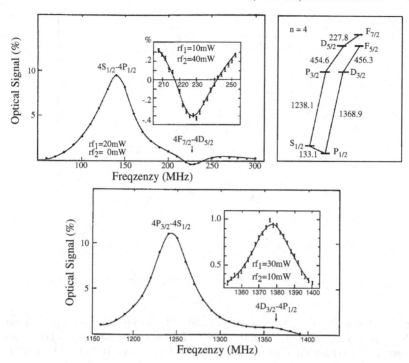

Fig. 5.2. Panorama of fine structure resonances in the $n = 4$ manifold of atomic hydrogen. *Insert* A shows the level structure and frequencies of allowed transitions. *Inserts* in part B show line shapes resolved by rf quenching of overlapping transitions. (Adapted from [130])

remove populations of atoms in unwanted hyperfine states. An example of hyperfine state selection is shown in Fig. 5.3 where the $4^2S_{1/2}$–$4^2P_{1/2}$ resonance curve is simplified by total quenching of atoms in the $4^2S_{1/2}(F = 0)$ state.

In addition to furnishing resonance frequencies – and therefore details of the internal atomic structure – electric resonance spectroscopy is also sensitive to the initial relative populations of the coupled states, and consequently can provide significant information regarding the charge-changing interactions that created the atom. This information is coded in the exact shape of the resonance profile. Depending on the initial populations, the transitions induced by an oscillating field can lead to either a greater or lesser fluorescent emission than with the field turned off. As illustrated in Fig. 5.2, most resonances are above the baseline; the oscillating field has driven the atoms from relatively long-lived states like $4^2S_{1/2}$ (lifetime 23 μs) to short-lived states like $4^2P_{1/2}$ (lifetime 12 ns) thereby diminishing the number of atoms that reach the detection chamber. The signal (5.1) is then positive. However, transitions induced between initially populated states of comparable lifetime can suppress the rate of spontaneous decay, as in the case of coupled $4^2D_{5/2}$ (lifetime

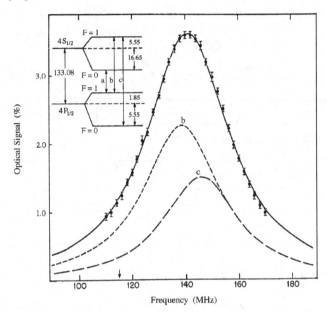

Fig. 5.3. Example of hyperfine state selection. Hydrogen $4^2S_{1/2}$–$4^2P_{1/2}$ resonance with complete removal of atoms in the $4^2S_{1/2}$ ($F = 0$) hyperfine state and consequent suppression of transition a as ascertained by a theoretical fit to the data of three Lorentzian line shapes. Numerical values for the level separations are in units of MHz. (Adapted from [128])

36 ns) and $4^2F_{7/2}$ (lifetime 73 ns) states, giving rise to a resonance curve below the baseline. Since the net rate and direction of transitions between coupled states depend on the initial occupation probabilities, one would expect to see a variation in line shape with the choice of target atom from which an incoming proton captures an electron to form a neutral hydrogen atom. An example of this is illustrated in the upper row of Fig. 5.4 for the overlapping transitions $4^2D_{5/2}$–$4^2P_{3/2}$ and $4^2F_{5/2}$–$4^2D_{3/2}$. The lower row shows computer simulations of these spectra for different initial state assignments according to a spin-independent Coulomb interaction model of the electron capture process [56, 128].

Traditional charge transfer experiments employing fast particle detection are based on measurements of either the total current of ions produced in various charge states, the attenuation of the primary beam, or the angular distribution of products scattered by violent close-encounter collisions. Such experiments yield information on the total electron capture cross-sections and cannot distinguish scattering events which lead to different excited states of products of the same charge. Subsequent methods depending on electric field ionization are limited to manifolds of large principal quantum number and provide a sum over the cross-sections of the component orbital sublevels. Similarly, methods that depend on the resolution of multi-state radiative decay

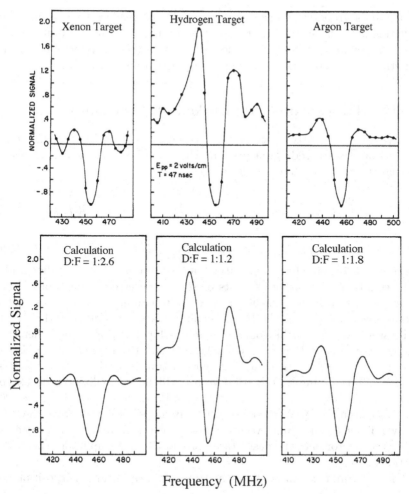

Fig. 5.4. Sensitivity of electric resonance line shapes to initial state populations: hydrogen $4^2D_{3/2}$, $4^2D_{5/2}$, and $4^2F_{5/2}$ states prepared by electron capture collisions with different targets. Experimental spectra are shown in the *top row*; theoretical spectra based on different initial D and F state amplitudes are shown in the *bottom row*. (Adapted from [129], Part III)

curves are incapable of resolving different states of the same lifetime such as fine or hyperfine structure states of the same orbital quantum number. By contrast, electric resonance line shapes differ in structure and occur in different portions of the frequency spectrum even for states of the same lifetime.

To extract both level structure and relative populations from an electric resonance line shape requires a detailed understanding of the interaction of a decaying multi-level atom with an applied oscillating field (and of the subsequent optical detection process as well). The analysis of so broad a problem goes well beyond the scope of this chapter and must be left to the cited literature. It is grounded, however, in the study of the two-level atom in an oscillating field whose solution we will take up next, for, in addition to spectroscopy, it is an important component of several experimental configurations exhibiting novel aspects of quantum superposition and interference.

5.2 The Two-Level Atom Looked at Two Ways

The dynamics of the two-level atom with energy eigenstates $|1\rangle$, $|2\rangle$ is deducible from the Schrödinger equation, which can be conveniently expressed in a matrix representation as follows

$$H\Psi = i\frac{\partial \Psi}{\partial t} \implies \begin{pmatrix} h_{11} & h_{12} \\ h_{21} & h_{22} \end{pmatrix} \begin{pmatrix} c_1 \\ c_2 \end{pmatrix} = i \begin{pmatrix} \partial c_1/\partial t \\ \partial c_2/\partial t \end{pmatrix}, \qquad (5.2)$$

where

$$h_{ij} = \langle i|H|j\rangle, \qquad c_i = \langle i|\Psi\rangle \qquad (i,j=1,2). \qquad (5.3)$$

The constant $\hbar$ is not shown explicitly, but has been divided out of both sides of the equation so that the elements of H are expressed in units of angular frequency. The diagonal elements of the Hamiltonian matrix (to be designated by the same symbol H as the Hamiltonian operator) characterize the energy of the states which can be complex numbers for decaying states where the real part is the actual energy (divided by $\hbar$) and the imaginary part is the decay rate. The off-diagonal elements govern transitions between the states. To keep notation simple, the state vector will be symbolized by Ψ except where it is part of a scalar product operation in which case the Dirac ket $|\Psi\rangle$ or corresponding bra will be used. In a matrix representation Ψ takes the form of a two-dimensional column vector whose elements c_1 and c_2, the projection of the state vector onto the basis states, are probability amplitudes (i.e., wave functions).

Any 2×2 matrix can be uniquely expressed as a linear superposition

$$H = h_0 1 + \sum_{i=1}^{3} h_i \sigma_i = h_0 1 + \boldsymbol{h} \cdot \boldsymbol{\sigma} \qquad (5.4)$$

of the unit 2×2 matrix 1 and the Pauli matrices σ_i $(i = 1, 2, 3)$ defined by the algebraic properties

$$\sigma_i \sigma_j = \sum_{i=1}^{3} \varepsilon_{ijk} \sigma_k + \delta_{ij} 1, \qquad (5.5)$$

where ε_{ijk} is the completely antisymmetric tensor or Levi-Civita symbol, and δ_{ij} is the Kronecker delta symbol.[1] In the second equality of (5.4), $\boldsymbol{\sigma}$ is a three-

---

[1] The symbol ε_{ijk} where i, j, k, can each take values 1,2,3, is ± 1 for respectively even and odd permutations of the sequence 1,2,3, and is 0 if any two indices are equal. Thus, for example, $\varepsilon_{123} = \varepsilon_{312} = +1$, $\varepsilon_{213} = \varepsilon_{132} = -1$, and $\varepsilon_{121} = 0$. The symbol δ_{ij} is $+1$ if $i = j$, and 0 if $i \neq j$.

dimensional vector whose elements are the Pauli matrices. There are infinitely many representations of the Pauli matrices (all related by unitary transformations) of which perhaps the most widely used is the following

$$\sigma_1 = \begin{pmatrix} 0 & 1 \\ 1 & 0 \end{pmatrix}, \qquad \sigma_2 = \begin{pmatrix} 0 & -i \\ i & 0 \end{pmatrix}, \qquad \sigma_3 = \begin{pmatrix} 1 & 0 \\ 0 & -1 \end{pmatrix}. \qquad (5.6)$$

Adopting the above representation leads to the following elements of the Hamiltonian four-vector $H = (h_0, \boldsymbol{h})$ whose components are the coefficients of the basis vectors $1, \boldsymbol{\sigma}$

$$h_0 = \frac{1}{2}(h_{11} + h_{22}), \quad h_1 = \frac{1}{2}(h_{12} + h_{21}),$$
$$h_2 = \frac{1}{2i}(h_{21} - h_{12}), \quad h_3 = \frac{1}{2}(h_{11} - h_{22}). \qquad (5.7)$$

If the Hamiltonian is Hermitian, $H^\dagger = H$, then $h_{12} = h_{21}^*$ and h_1, h_2 respectively are seen to be the real and imaginary parts of h_{12}.

Those familiar with physical optics will recognize the components h_0, $\boldsymbol{h}$ as an analogue of the Stokes parameters[2] that uniquely characterize the polarization of a light beam [7]. The analogy follows because an arbitrarily polarized light wave can be described in terms of two mutually orthogonal basis states-for example, vertical and horizontal linear polarizations or left and right circular polarizations; light is consequently an example of a two-state system. Since the quantum description of light is formulated in terms of photons, which are spin-1 bosons, one might wonder why there are only two components and not three, as is ordinarily the case for spin-1 particles. The answer, expounded in a delightful essay by Wigner [131], is intimately related to the fact that the photon has zero rest mass.

In the above formalism the Schrödinger equation (5.2) takes the succinct form

$$\frac{d\Psi}{dt} = -i(h_0 1 + \boldsymbol{h} \cdot \boldsymbol{\sigma})\Psi, \qquad (5.8)$$

which can be integrated immediately

$$\Psi(t) = e^{-iH_0 t}\Psi(0) = e^{-ih_0 t}e^{-i\boldsymbol{h} \cdot \boldsymbol{\sigma} t}\Psi(0), \qquad (5.9)$$

if the elements of H are independent of time. The first exponential factor in the second equality is an unimportant global phase factor that depends on the (arbitrarily adjustable) mean energy of the states and does not appear in expressions related to measurable quantities. (Note, however, that what may initially appear to be a global phase factor can have experimental consequences when the two coupled states must actually be regarded as a part of

---

[2] The Stokes parameters (I, U, V, Q) are related to the total light intensity (I), the angular orientation and eccentricity of the elliptical motion traced out by the electric vector of the light wave (U and Q), and the handedness or sense of circulation of the electric vector (V).

a larger quantum system. We will return to this point later in the chapter.) It is the second exponential factor of (5.9) upon which the time-evolution of the system depends significantly. Operationally, expressions of the form of (5.9) that involve exponential powers of matrices are defined by a Taylor series expansion of the exponential function

$$\mathrm{e}^{-\mathrm{i}\boldsymbol{h}\cdot\boldsymbol{\sigma}t} \equiv \sum_{n=0}^{\infty} \frac{(-\mathrm{i}\boldsymbol{h}\cdot\boldsymbol{\sigma}t)^n}{n!} = 1\cos(ht) - \mathrm{i}\frac{\boldsymbol{h}\cdot\boldsymbol{\sigma}}{h}\sin(ht) ,\qquad(5.10)$$

in which the scalar h (not to be confused with Planck's constant which shall rarely be needed in this chapter) is the magnitude of the vector $\boldsymbol{h}$

$$h = \sqrt{\boldsymbol{h}\cdot\boldsymbol{h}} = \sqrt{h_1^2 + h_2^2 + h_3^2} = \frac{1}{2}\sqrt{(h_{11}-h_{22})^2 + 4h_{12}h_{21}} .\qquad(5.11)$$

The reduction to the closed-form expression of (5.10) follows straightforwardly from the algebraic properties of the Pauli matrices summarized in (5.5) and from the Taylor series for sine and cosine.

One can also evaluate expressions of the form of (5.9) by a more general procedure, the solution of an eigenvalue problem, that is independent of the dimension of the representation of H and therefore does not rely on the algebra of the Pauli matrices. It is instructive to examine this alternative method for it provides another mathematical interpretation of h_0 and $\boldsymbol{h}$. The idea is to diagonalize H – i.e., to transform it to a matrix η containing elements $\eta_{ij} = \eta_i\delta_{ij}$ only along the principal diagonal – since the exponential of a diagonal matrix e^{η} is readily shown to take the form of a diagonal matrix whose elements are e^{η_i}. (To show this one employs again the Taylor series representation of the exponential function.) Then the solution given by the first equality of (5.9) can be explicitly evaluated. The diagonalization of H is accomplished by first solving the eigenvalue problem

$$HX_i = \eta_i X_i ,\qquad(5.12)$$

which yields n eigenvalues η_i for an n-dimensional matrix, and relating H and η by the transformation

$$H = D\eta D^{-1} ,\qquad(5.13)$$

where η displays the eigenvalues of H along the principal diagonal in some designated order $\eta_1, \eta_2, \ldots, \eta_n$; the diagonalizing matrix D is then constructed by juxtaposing the corresponding eigenvectors X_i of H in the same order (from left to right)

$$D = X_1 X_2 \cdots X_n .\qquad(5.14)$$

By employing the Taylor series definition of the exponential yet again, one can demonstrate that

$$\mathrm{e}^{-\mathrm{i}Ht} = \mathrm{e}^{-\mathrm{i}D\eta D^{-1}t} = D\mathrm{e}^{-\mathrm{i}\eta t}D^{-1} ,\qquad(5.15)$$

from which the solution to the Schrödinger equation in the case of a two-level system can be written explicitly

$$\Psi(t) = e^{-iH_0 t}\Psi(0) = D \begin{pmatrix} e^{-i\eta_1 t} & 0 \\ 0 & e^{-i\eta_2 t} \end{pmatrix} D^{-1}\Psi(0) . \tag{5.16}$$

The equivalence of solutions (5.9) and (5.16) is apparent once the eigenvalue problem is actually solved. In the case of the two-level atom, the two eigenvalues of H deducible from solution of a quadratic secular equation are

$$\eta_{1,2} = \frac{1}{2}\left[(h_{11} + h_{22}) \pm \sqrt{(h_{11} - h_{22})^2 + 4h_{12}h_{21}}\right] . \tag{5.17}$$

The components of the four-vector $(h_0, \boldsymbol{h})$ are therefore related to the eigenvalues of H by

$$\eta_{1,2} = h_0 \pm h . \tag{5.18}$$

If the elements of H depend on time, then the resulting solution does not, in general, take the simple form of relations (5.9). Indeed, as pointed out at the beginning of the chapter, there may be no known exact analytical solution at all even in the simplest case of a two-level atom interacting with a harmonically oscillating field. Where a closed-form analytical solution is not achievable, one must then either seek an approximate solution to the original equations or simplify the model further to obtain an analytical solution to reduced equations. In the latter case the strategy is to remove from H by means of appropriate transformations as much of the time-dependence as possible, so that the solution to the transformed Schrödinger equation can be cast in the form of (5.9). One customary way in which this has been done is to assume (although it is ordinarily not the case experimentally) that the atom is interacting with a *rotating* field, rather than with an *oscillating* field. In the next section we will discuss this procedure and a more general one, designated the 'oscillating field theory' [132], which produces a closed-form analytical solution almost identical to the exact solution obtainable by numerical integration of the Schrödinger equation.

Before examining the specific problem of an external radiofrequency or microwave field, it is conceptually useful to re-examine the dynamics of a two-level system from the perspective of a second general mathematical formalism, that of the density operator ρ whose elements in a matrix representation enter directly into theoretical expressions for observable quantities. Consider first a system with two coupled *stable* states for which the Hamiltonian matrix is necessarily Hermitian. By multiplying the two sides of the Schrödinger equation (5.2) by Ψ on the right, and subtracting from the result the Hermitian conjugate of the Schrödinger equation multiplied by $\Psi^\dagger$ on the left, one obtains the equation of motion in the Heisenberg form

$$\frac{d\rho}{dt} = -i[H, \rho] \tag{5.19}$$

(which we used, but did not derive, in the discussion of laser-induced quantum beats). Employing again the decomposition in (5.4), we can express he two-dimensional matrix representation of the density operator as

$$\rho \equiv \Psi\Psi^{\dagger} = \begin{pmatrix} \rho_{11} & \rho_{12} \\ \rho_{21} & \rho_{22} \end{pmatrix} = \rho_0 1 + \boldsymbol{\rho} \cdot \boldsymbol{\sigma} , \tag{5.20}$$

where the expressions of ρ_0 and $\boldsymbol{\rho}$ in terms of the elements ρ_{ij} are formally the same as those given by (5.5) and (5.6) for the elements of H. The diagonal elements represent the relative populations of each level, and hence the trace, $\text{Tr}(\rho) = \rho_{11} + \rho_{22}$, to which ρ_0 is proportional, is a measure of the total probability of finding the atom in one or the other of the two available states. The off-diagonal elements characterize the coherence properties of the system, i.e., the capacity to produce quantum interference effects. Note that in the definition of the density operator the order of the state vector and its Hermitian conjugate is significant; the reverse order $\Psi^{\dagger}\Psi$ produces a number $[\text{Tr}(\rho)]$ and not a matrix.

Substitution of the expansion of ρ in (5.20) into the commutator of (5.19), use of the commutation properties of the Pauli matrices

$$[\sigma_i, \sigma_j] = 2\mathrm{i} \sum_{k=1}^{3} \varepsilon_{ijk}\sigma_k \tag{5.21}$$

that follow from (5.5), and extraction of the coefficients of the bases 1 and $\boldsymbol{\sigma}$ lead to the scalar and vector equations of motion

$$\frac{\mathrm{d}\rho_0}{\mathrm{d}t} = 0 , \tag{5.22}$$

$$\frac{\mathrm{d}\boldsymbol{\rho}}{\mathrm{d}t} = 2(\boldsymbol{h} \times \boldsymbol{\rho}) \equiv \boldsymbol{\Omega}_{\mathrm{r}} \times \boldsymbol{\rho} . \tag{5.23}$$

In view of the interpretation of ρ_0, the first relation (5.22) is a statement of the conservation of probability for a closed system. If the two coupled states of interest were not stable, and the atom could decay radiatively to lower states whose dynamics were not taken into account, then ρ_0 would not be constant in time. We will see shortly how this case can be treated. The dynamics of the stable two-state system is effectively contained in relation (5.23) which formally resembles the equation of motion of an angular momentum vector precessing at an angular velocity $\boldsymbol{\Omega}_{\mathrm{r}} = 2\boldsymbol{h}$. The relationship is only a formal one – nothing in the two-level atom need actually be precessing – but it provides a useful classical picture of the time evolution of the system.

The connection to rotation may also be seen directly in the solution for the state vector, (5.9) since the unitary operator $U(\theta)$ for rotating a spin-1/2 system with angular momentum $\boldsymbol{S} = \boldsymbol{\sigma}/2$ (in units of $\hbar$) by an angle θ about the unit normal vector $\boldsymbol{n}$ is

$$U(\theta) = \mathrm{e}^{-\mathrm{i}\boldsymbol{\sigma} \cdot \boldsymbol{n}\theta/2} . \tag{5.24}$$

Comparison of (5.9) and (5.24) shows that the rotation axis is

$$n = \frac{h}{h} \, , \tag{5.25}$$

with angle of rotation

$$\theta = 2ht \equiv \Omega_r t \, . \tag{5.26}$$

Thus, one may imagine an ensemble of precessing dipoles as an analogue of the transitions between two quantum states. When the atoms are subjected to the interaction h over a time t, the corresponding dipoles precess through an angle $\theta = 2ht$.

If the elements of H are independent of time, it follows from relations (5.9) and (5.20) that the time-evolution of the density matrix takes the form

$$\rho(t) = e^{-ih\cdot\sigma t}\rho(0)e^{ih\cdot\sigma t} \, . \tag{5.27}$$

Upon use of (5.10) to evaluate (5.27) explicitly, one finds that the atomic system returns to its initial state described by $\rho(0)$ when the interaction is of such strength and duration that the precession angle θ equals π radians. One might have expected that the periodicity of θ should be 2π radians, but this is not the case for a two-level system. Although $\rho(\theta = \pi) = \rho(0)$, the corresponding state vectors differ by a sign: $\Psi(\theta = \pi) = -\Psi(0)$. As I mentioned previously, global phase factors such as $e^{i\pi} = -1$ ordinarily have no observable consequences. However, if the two coupled states should subsequently be regarded as part of a larger system of states, then the once global phase may actually become a relative phase that can be manifested through a quantum interference experiment [133]. We shall examine this possibility in more detail in a subsequent section.

In analyzing the two-level atom – or, indeed, a quantum system with arbitrary number of levels – it is often convenient to remove at the outset the highest frequency terms in the diagonal elements of H corresponding to the energy eigenvalues. The precession of the system vector ρ then takes place more slowly under the influence of the weaker interactions, both internal and external, which are of principal interest. (It is assumed that the unperturbed energies have already been determined by solution of the appropriate eigenvalue problem.) This simplification is effected by decomposing the Hamiltonian into a diagonal part H_0 yielding the unperturbed energies

$$H_0 = \begin{pmatrix} \omega_1 & 0 \\ 0 & \omega_2 \end{pmatrix} \tag{5.28}$$

and an interaction V given by

$$V = \begin{pmatrix} V_{11} & V_{12} \\ V_{21} & V_{22} \end{pmatrix} \tag{5.29}$$

(all matrix elements expressed in units of angular frequency) and then transforming to the so-called interaction representation by substituting

$$\Psi = e^{-iH_0 t}\Psi_I \tag{5.30}$$

into the Schrödinger equation to obtain the transformed equation

$$H_I\Psi_I = i\frac{\partial\Psi_I}{\partial t} , \tag{5.31}$$

with interaction Hamiltonian

$$H_I = U_I^\dagger H U_I - iU_I^\dagger\frac{\partial U_I}{\partial t} = U_I^\dagger V U_I . \tag{5.32}$$

If V does not contribute to the level energies – i.e., if V serves only to couple different states – then the resulting interaction Hamiltonian

$$H_I = \begin{pmatrix} 0 & v_{12}e^{i\omega_0 t} \\ v_{21}e^{-i\omega_0 t} & 0 \end{pmatrix} , \tag{5.33}$$

with energy interval

$$\omega_0 \equiv \omega_1 - \omega_2 \tag{5.34}$$

has no elements along the principal diagonal.

Under the preceding conditions where $v_{11} = v_{22} = 0$, the interaction vector V – in the four-vector representation $V = (V_0, \boldsymbol{V})$ – has a null component V_3. If, in addition, $\boldsymbol{V}$ is purely real-valued, then $v_{12} = v_{21}$, and component $V_2 = \mathrm{Im}(v_{21})$ also vanishes. The angular velocity $\boldsymbol{\Omega}_r = 2\boldsymbol{h}$ then has components $(\boldsymbol{\Omega}_r)_1 = 2v_{12}$ and $(\boldsymbol{\Omega}_r)_3 = \omega_0$. For transitions induced by an optical interaction between the atomic ground state and an electronic excited state, the frequency ω_0 – if atomic energy levels are to have any meaning – must be much greater than the corresponding frequency $2v_{12}$ of the interaction. In that case, one could regard the system vector as precessing around the '3-axis' at the rate ω_0 with a much slower nutational motion about the '1-axis' at the frequency $2v_{12}$. However, for the example of rf coupling of atomic fine or hyperfine states, the frequencies ω_0 and $|\boldsymbol{V}|$ can be of comparable magnitude. Let us examine this point quantitatively.

The energy of an electron with principal quantum number n in a hydrogenic atom with atomic number Z

$$\varepsilon_n \sim \frac{Z^2}{n^2}\mathrm{Ry} \tag{5.35}$$

is proportional to the Rydberg unit $\mathrm{Ry} \equiv e^2/2a_0$, where e is the electron charge and a_0 is the Bohr radius. The splitting of this level arising from the interaction of the electron spin and orbital motion (fine structure) is smaller by the square of the fine structure constant $\alpha_f = e^2/\hbar c \sim 1/137$, or more precisely

$$\omega_0 \sim \frac{Z^2\alpha_f^2}{n}\varepsilon_n \sim \frac{Z^4\alpha_f^2}{n^3}\text{Ry} . \tag{5.36}$$

Since the orbital radius $r(n, Z)$ of the electron varies in the Bohr model as n^2/Z, the electric dipole interaction between the electron and an external oscillating electric field of amplitude E_0 can be estimated by

$$v_{12} = er(n, Z)E_0 \sim \frac{en^2a_0}{Z}E_0 . \tag{5.37}$$

Comparing expressions (5.36) and (5.37), one has

$$\frac{v_{12}}{\omega_0} \sim \frac{n^5E_0}{\alpha_f^2 Z^5 E_a} , \tag{5.38}$$

where the atomic unit of electric field strength is

$$E_a = \frac{e}{a_0^2} = 5.14 \times 10^9 \text{ V cm}^{-1} . \tag{5.39}$$

Thus, in the case of the hydrogen atom ($Z = 1$) subjected to an external field of $E_0 \sim 10 \text{ V cm}^{-1}$ for example, the ratio of precession frequencies v_{12}/ω_0 would be about 10^{-3} in the level $n = 2$ and slightly greater than 1 in $n = 8$. The fact that the interaction with an external rf field can be comparable to, or even exceed, the internal interactions that split the coupled levels makes the analysis of the rf resonance problem in some ways more difficult than that of optical resonance for which the ratio in (5.38) is ordinarily very small.

We conclude this section by taking account of the radiative decay of the coupled states, for it is by atomic fluorescence that the rf-induced transitions are observed in optical electric resonance experiments. As I indicated previously, unstable levels can be described by a complex frequency of the form $\omega - i\gamma$ where $\hbar\omega$ is the energy eigenvalue of the state in the absence of spontaneous emission and γ is the decay rate, or reciprocal of the mean lifetime τ. One then adds to the Hamiltonians (5.28) and (5.29) for stable levels a pure imaginary diagonal decay operator $-i\hbar\Gamma/2$ for which the representation of Γ as a 2×2 matrix takes the form

$$\Gamma = \begin{pmatrix} \gamma_1 & 0 \\ 0 & \gamma_2 \end{pmatrix} = \Gamma_0 1 + \boldsymbol{\Gamma} \cdot \boldsymbol{\sigma} . \tag{5.40}$$

The total Hamiltonian is no longer Hermitian, and the procedure by which the Heisenberg form of the equation of motion (5.19) was derived now leads to an equation of motion

$$\frac{d\boldsymbol{\rho}}{dt} = -i[H, \boldsymbol{\rho}] - \{\Gamma, \boldsymbol{\rho}\} , \tag{5.41}$$

in which curly brackets signify the anticommutator $\{A, B\} \equiv AB + BA$ as before. The components of the density matrix in the 1, $\boldsymbol{\sigma}$ basis then satisfy the coupled relations

$$\frac{\mathrm{d}\rho_0}{\mathrm{d}t} = -\Gamma_0\rho_0 - \boldsymbol{\Gamma} \cdot \boldsymbol{\rho} , \tag{5.42}$$

$$\frac{\mathrm{d}\boldsymbol{\rho}}{\mathrm{d}t} = \boldsymbol{h} \times \boldsymbol{\rho} - \Gamma_0\boldsymbol{\rho} - \rho_0\boldsymbol{\Gamma} , \tag{5.43}$$

where the vector $\boldsymbol{h}$ comes from the Hermitian part of the Hamiltonian $H_0 + V$. From (5.42) it is seen that probability of finding the atom in one or the other of the two specified levels is no longer conserved, but decays exponentially. The level instability also influences the precession of the system vector $\boldsymbol{\rho}$ in accordance with (5.43).

5.3 Oscillating Field Theory

The foregoing equations of motion of a two-level decaying atom, whether expressed in terms of amplitudes or density matrix elements, have no known exact analytical solution for the case of a pure oscillatory interaction. To the extent that high accuracy is not required in the study of experimental resonance line shapes, an exact solution can be obtained for a simplified interaction, that of the rotating-field or rotating-wave approximation, first employed by Rabi in his derivation of the 'flopping formula'. The underlying physical idea, expressed in classical imagery, is that a linear oscillation can be decomposed into a sum of two counter-rotating motions, one of which will be nearly resonant with the precessing dipole moment (representative of the quantum transitions between two states), and the other anti-resonant, i.e., rotating in a sense opposite that of the dipole precession. From a classical perspective the component of the applied field rotating with the dipole exerts a steady torque whose effect is cumulative over many precession periods. The torque exerted by the counter-rotating component, however, reverses itself at a rate 2ω and therefore might be expected to have no significant long-time effect. The rotating-wave approximation (RWA) consists of neglecting all terms at the counter-rotating frequency.

Unless, of course, one actually has a rotating field, neglect of the anti-resonant interaction has theoretical consequences. The first consideration of these effects was given by Bloch and Siegert [134] and by Stevenson [135] who showed that the resonance maximum undergoes a small displacement. These results, derived for the case of magnetic resonance in a stable spin-1/2 system, are not a priori applicable in the case of decaying states. In this section the problem of a two-level atom subjected to an oscillating field is examined more generally by a procedure I have designated the oscillating field (OF) approximation [132]. By means of several matrix transformations a representation is found in which the major effects of the applied field are independent of time – so that the procedure of the foregoing section is applicable – and the residual time-dependent portion contributes negligibly to observable atom–field interactions. It will be seen that, besides a frequency shift, the linearly oscillating field alters as well the difference in level decay rates, and indeed can affect the details of the overall line shape. The oscillating field approximation yields results that are virtually identical to those obtained by exact numerical integration of the Schrödinger equation.

Our quantum system consists of two quasi-stationary states $|1\rangle$, $|2\rangle$ interacting through the electric dipole moment $\boldsymbol{\mu}_{\mathrm{E}}$ with a classical oscillating field $\boldsymbol{E}_0 \cos(\omega t)$ of well defined phase (here chosen to be zero). These atomic states are quasi-stationary in the sense that, although they are energy eigenfunctions of a Hermitian Hamiltonian, their interaction with the vacuum electromagnetic field results in a finite lifetime and hence level width. For well-defined states to exist, however, the level width must be much smaller than the energy eigenvalue, or $\omega_i \gg \gamma_i$ (for $i = 1, 2$). The dynamics of the atom-rf field system is described by the Hamiltonian operator

$$H = H_0 - \frac{1}{2}i\hbar\Gamma - \boldsymbol{\mu}_{\mathrm{E}}{\cdot}\boldsymbol{E}_0 \cos(\omega t) , \qquad (5.44)$$

which leads to a matrix differential equation of the form of (5.2)

$$\begin{pmatrix} -\left(i\omega_1 + \frac{1}{2}\gamma_1\right) & -iV_{12}\left(e^{i\omega t} + e^{-i\omega t}\right) \\ iV_{12}\left(e^{i\omega t} + e^{-i\omega t}\right) & -\left(i\omega_2 + \frac{1}{2}\gamma_2\right) \end{pmatrix} \begin{pmatrix} c_1(t) \\ c_2(t) \end{pmatrix} = \frac{\mathrm{d}}{\mathrm{d}t}\begin{pmatrix} c_1(t) \\ c_2(t) \end{pmatrix} , \qquad (5.45)$$

in which

$$V_{12} = \frac{\langle 1| - \boldsymbol{\mu}_{\mathrm{E}}{\cdot}\boldsymbol{E}_0|2\rangle}{2} \qquad (5.46)$$

is the interaction matrix element in units of angular frequency.

Upon transforming to the interaction representation, as described in the previous section (5.30)–(5.32), one eliminates from (5.45) the high-frequency eigenvalues of H_0 to obtain the equation

$$\begin{pmatrix} -\gamma_1/2 & -iV_{12}\left(e^{i\tilde{\Omega}t} + e^{-i\Omega t}\right) \\ -iV_{12}\left(e^{-i\tilde{\Omega}t} + e^{i\Omega t}\right) & -\gamma_2/2 \end{pmatrix} \Psi_{\mathrm{I}} = \frac{\mathrm{d}}{\mathrm{d}t}\Psi_{\mathrm{I}} , \qquad (5.47)$$

where

$$\Omega = \omega - \omega_0 \qquad (5.48)$$

is the deviation of the applied frequency from the unperturbed resonance frequency ω_0 and

$$\tilde{\Omega} = \omega + \omega_0 \qquad (5.49)$$

is the frequency of the anti-resonant component.

At this point the rotating-wave approximation is usually invoked by neglecting all terms containing $\tilde{\Omega}$. It is instructive for purposes of comparison to complete the RWA solution for decaying states before taking up the more accurate oscillating-field solution. The problem is in fact solved by (5.9) once a transformation is found to remove the residual resonant time dependence from the matrix of (5.47). This task is accomplished by the unitary transformation

$$\Psi_{\mathrm{R}} = \begin{pmatrix} e^{i\Omega t/2} & 0 \\ 0 & e^{-i\Omega t/2} \end{pmatrix} \Psi_{\mathrm{I}} , \tag{5.50}$$

which leads to the time-independent Schrödinger equation in the rotating frame

$$\begin{pmatrix} -\dfrac{1}{2}(\gamma_1 - i\Omega) & -iV_{12} \\ -iV_{12} & -\dfrac{1}{2}(\gamma_2 + i\Omega) \end{pmatrix} \Psi_{\mathrm{R}} = \dfrac{\mathrm{d}\Psi_{\mathrm{R}}}{\mathrm{d}t} . \tag{5.51}$$

Upon integrating (5.51) and making the subsequent inverse transformations back to the original representation of (5.45), one obtains the complete solution

$$\Psi(t) = \begin{pmatrix} e^{-(\frac{1}{2}\gamma_1 + i\omega_1)t} & 0 \\ 0 & e^{-(\frac{1}{2}\gamma_2 + i\omega_2)t} \end{pmatrix} \begin{pmatrix} I_{11} & I_{12} \\ I_{21} & I_{22} \end{pmatrix} \Psi(0) , \tag{5.52}$$

where the elements I_{ij} of the interaction matrix are

$$I_{11} = \exp\left[\dfrac{(G - i\Omega)t}{2}\right] \left[\cos(\nu t) - \dfrac{G - i\Omega}{2\nu} \sin(\nu t)\right] , \tag{5.53}$$

$$I_{12} = -2i \exp\left[\dfrac{(G - i\Omega)t}{2}\right] \dfrac{V_{12}}{2\nu} \sin(\nu t) , \tag{5.54}$$

$$I_{21} = -2i \exp\left[-\dfrac{(G - i\Omega)t}{2}\right] \dfrac{V_{12}}{2\nu} \sin(\nu t) , \tag{5.55}$$

$$I_{22} = \exp\left[-\dfrac{(G - i\Omega)t}{2}\right] \left[\cos(\nu t) + \dfrac{G - i\Omega}{2\nu} \sin(\nu t)\right] , \tag{5.56}$$

with 'precession frequency' ν given by

$$\nu = \sqrt{\dfrac{(\Omega + iG)^2}{4} + V_{12}^2} \tag{5.57}$$

and system decay constant

$$G = \dfrac{1}{2}(\gamma_1 - \gamma_2) . \tag{5.58}$$

From relations (5.53)–(5.56) and (5.57) one can infer that $I_{ij}(-\Omega) = I_{ij}(\Omega)^*$, and therefore the magnitude of each element I_{ij} is symmetric about the resonance center $\Omega = 0$; the predicted resonance frequency is exactly ω_0. For the case where one state is initially populated and the other unpopulated, the occupation or transition probabilities $|c_i(t)|^2$ $(i = 1, 2)$ as a function of frequency produce symmetric oscillatory line shapes. Note that only the difference in decay rates (as expressed by G), and not the individual decay constants, enter each element I_{ij}. Consequently, the RWA resonance line shape for transitions induced between two decaying states of the same lifetime is identical to that for stable states except for the diminution in overall intensity due to the decay matrix. This is, of course, not the case if the coupled states have different lifetimes.

Let us return to the initial problem of solving the Schrödinger equation for a two-level atom in an oscillating field and determine the effects of the anti-resonant component. We start again with the exact Schrödinger equation (5.47), in the interaction representation and transform it – *not* into the 'rotating' reference frame – but into the frame of the counter-rotating component of the oscillating field

$$\Psi_A = \begin{pmatrix} e^{-i\tilde{\Omega}t/2} & 0 \\ 0 & e^{i\tilde{\Omega}t/2} \end{pmatrix} \Psi_I , \tag{5.59}$$

by a transformation analogous to (5.50). This leads to a Schrödinger equation in which the Hamiltonian can be decomposed into a time-independent part and a part that oscillates at 2ω. Following the procedure outlined in the previous section, we next diagonalize the time-independent part and re-express the equation in a basis of its eigenstates. At this point, the substantive part of the anti-resonant interaction has been captured, so to speak, in the time-independent eigenvalues and eigenvectors, and one transforms the Schrödinger equation to the reference frame of the rotating component by setting

$$\Psi_R = \begin{pmatrix} e^{i\omega t} & 0 \\ 0 & e^{-i\omega t} \end{pmatrix} \left(C_A^{-1} \Psi_A \right) , \tag{5.60}$$

where the matrix

$$C_A = \begin{pmatrix} 1 & -\kappa_A \\ \kappa_A & 1 \end{pmatrix} \tag{5.61}$$

executes the diagonalization in the anti-rotating frame. The element κ_A, defined by

$$\kappa_A = \frac{\vartheta}{\sqrt{1 + \vartheta^2} + 1} , \tag{5.62}$$

where

$$\vartheta = \frac{2iV_{12}}{G + i\tilde{\Omega}} , \tag{5.63}$$

will ordinarily be a small quantity (since $V_{12} \ll \omega + \omega_0 \sim 2\omega_0$) and serves as a weak coupling parameter characterizing the interaction of the atom with the anti-resonant part of the oscillating field. The resulting Schrödinger equation in the rotating frame now takes the form

$$\frac{\mathrm{d}}{\mathrm{d}t}\Psi_{\mathrm{R}} \tag{5.64}$$

$$= \left[\begin{pmatrix} \varepsilon_{\mathrm{A}}^- + \mathrm{i}\omega & -\dfrac{\mathrm{i}V_{12}}{1+\kappa_{\mathrm{A}}^2} \\[2ex] -\dfrac{\mathrm{i}V_{12}}{1+\kappa_{\mathrm{A}}^2} & \varepsilon_{\mathrm{A}}^+ - \mathrm{i}\omega \end{pmatrix} + \begin{pmatrix} -\dfrac{2\mathrm{i}V_{12}\kappa_{\mathrm{A}}}{1+\kappa_{\mathrm{A}}^2}\cos(2\omega t) & \dfrac{\mathrm{i}V_{12}\kappa_{\mathrm{A}}^2}{1+\kappa_{\mathrm{A}}^2}\mathrm{e}^{4\mathrm{i}\omega t} \\[2ex] \dfrac{\mathrm{i}V_{12}\kappa_{\mathrm{A}}^2}{1+\kappa_{\mathrm{A}}^2}\mathrm{e}^{-4\mathrm{i}\omega t} & \dfrac{2\mathrm{i}V_{12}\kappa_{\mathrm{A}}}{1+\kappa_{\mathrm{A}}^2}\cos(2\omega t) \end{pmatrix} \right]\Psi_{\mathrm{R}},$$

in which

$$\varepsilon_{\mathrm{A}}^{\pm} = -\frac{1}{4}(\gamma_1 + \gamma_2) \mp \frac{1}{2}\sqrt{(G + \mathrm{i}\tilde{\Omega})^2 - 4V_{12}^2} \tag{5.65}$$

are the eigenvalues obtained in the anti-rotating frame.

It should be emphasized that the OF analysis to this point is exact; no approximations have yet been made, and (5.64) contains all the information of the initial Schrödinger equation (5.45). The virtue of (5.64) lies in the following considerations. Not only is the parameter κ_{A} usually small-diminishing steadily in magnitude across the resonance profile roughly as the inverse of the applied frequency – but it multiplies highly oscillatory terms in 2ω and 4ω which have a negligible effect on the solution close to resonance. Thus, to a good approximation, one can ignore the time-dependent part of (5.64) and apply the theory of the preceding section to solve the residual time-independent equation. The details of obtaining the resulting closed form solution, which is somewhat complicated, are left to the original literature, and the solution, itself, designated by the interaction elements J_{ij} $(i, j = 1, 2)$ corresponding to the RWA expressions (5.53)–(5.56), is given in Appendix 5A.

Let us examine, however, the precession frequency μ,

$$\mu = \sqrt{\left[\frac{\tilde{\Omega} - \mathrm{i}G}{2}(1 + \vartheta^2)^{1/2} - \omega\right]^2 + \frac{V_{12}^2}{(1 + \kappa_{\mathrm{A}}^2)}}, \tag{5.66}$$

obtained by diagonalizing the time-independent part of (5.64). This is the frequency at which transitions between the two states are induced and from which the observed central frequency of an electric resonance line shape can be determined. By expanding the inside radical in a Taylor series to order ϑ^2 and neglecting the small term κ_{A}^2 in the denominator of the second term, one can reduce (5.66) to the form

$$\mu \sim \sqrt{\frac{\left[(\Omega - \delta\Omega) + \mathrm{i}(G - \delta G)\right]^2}{4} + V_{12}^2}, \tag{5.67}$$

directly comparable to the corresponding expression for ν, (5.57) of the rotating-field theory. One would then find that the oscillating field gives rise to an apparent shift in resonance frequency of

$$\delta\Omega = \frac{2V_{12}^2\tilde{\Omega}}{\tilde{\Omega}^2 + G^2} , \tag{5.68}$$

and change in the decay rate difference of

$$\delta G = \frac{2V_{12}^2 G}{\tilde{\Omega}^2 + G^2} . \tag{5.69}$$

At resonance ($\tilde{\Omega} = 2\omega_0$) the frequency shift (5.68) reduces in the case of stable states to (V_{12}^2/ω_0) as derived by Bloch and Siegert. [Equation (5.68) differs from the result derived by Willis Lamb [136], but has been confirmed by exact numerical integration of the Schrödinger equation [137].]

An oscillatory field can do more than just shift the resonance frequency and decay rate difference, however. Particularly in the case of transitions induced between unstable states, the entire line shape – i.e., either the occupation or transition probability as a function of frequency – can exhibit additional structure, a high-frequency modulation of the basic line shape derived by the rotating-field approximation. An example of this is illustrated in Fig. 5.5 for transitions induced between an initially populated hydrogen $4^2S_{1/2}$ state [$c_1(0) = 1$] and initially unpopulated $4^2P_{1/2}$ state [$c_2(0) = 0$], where the lifetimes are respectively $\tau_1 = 232$ ns and $\tau_2 = 12.4$ ns. In the case of the longer-lived S state, the resonance profile (not shown), calculated from the expressions

$$P_1(t) = \left|\langle 1|\Psi_{\text{OF}}(t)\rangle\right|^2 = |J_{11}(t)|^2 e^{-\gamma_1 t} , \tag{5.70}$$

$$P_1^0(t) = \left|\langle 1|\Psi_{\text{RWA}}(t)\rangle\right|^2 = |I_{11}(t)|^2 e^{-\gamma_1 t} , \tag{5.71}$$

is virtually the same for the RWA and OF solutions. This is not the case in a corresponding comparison of the OF and RWA profiles of the short-lived P state

$$P_2(t) = \left|\langle 2|\Psi_{\text{OF}}(t)\rangle\right|^2 = |J_{21}(t)|^2 e^{-\gamma_2 t} , \tag{5.72}$$

$$P_2^0(t) = \left|\langle 2|\Psi_{\text{RWA}}(t)\rangle\right|^2 = |I_{21}(t)|^2 e^{-\gamma_2 t} , \tag{5.73}$$

which differ markedly outside the immediate vicinity of resonance. In all cases the resonance line shapes determined from the oscillating-field solution and numerical integration of the Schrödinger equation are nearly indistinguishable.

The modification of the RWA line shape by the counter-rotating component of the oscillating field can be approximated by retaining in the expression for J_{21} [see (5A.4) in Appendix 5A] only factors linear in κ_{A}. This leads to the approximate relation

$$P_2(t) = P_2^0(t) + \frac{1}{\tilde{\Omega}}\text{Im}\left\{(1 - e^{-2i\omega t})\left[\frac{2\nu\cos(\nu t)}{\sin(\nu t)} - G + i\Omega\right]\right\} P_2^0(t) . \tag{5.74}$$

Thus, the total line shape is the sum of the RWA line shape and a smaller term inversely proportional to $\omega + \omega_0$ and modulated at the frequency 2ω [which accounts for the separation $\Delta\omega$ between modulation maxima in Fig. 5.5 given by $(\Delta\omega)t = 1/2$].

Those familiar with experimental electric or magnetic resonance line shapes may wonder why the high-frequency modulation is not usually evident. First, this additional structure is most pronounced in the case of short-lived states, whereas the observed signal ordinarily results from transitions between long-lived states, as, for example, in the pioneering experiments of Rabi and Lamb. Second, if there is a sufficiently wide dispersion in the atomic velocities, as in the case of a thermal beam, the transition probability must be averaged over the interaction time and this tends to smooth out the resonance line shape. It is worth noting, however, that the modulation is largely unaffected

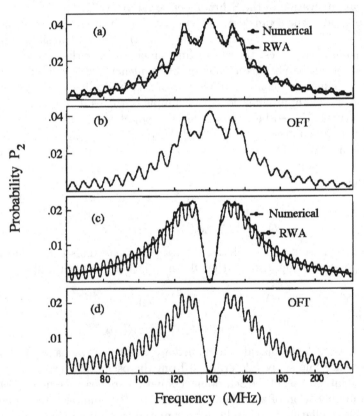

Fig. 5.5. Comparison of oscillating-field theory (OFT), rotating-wave approximation (RWA) and exact numerical integration. Occupation probability of a short-lived state $|2\rangle = |4P\rangle$ coupled to a long-lived state $|1\rangle = |4S\rangle$ as a function of driving frequency for theoretical parameters: $V_{12}/\omega_0 = 5/140$; $c_1(0) = 1$, $c_2(0) = 0$; $\gamma_1 = 4.35 \times 10^6 \text{ s}^{-1}$, $\gamma_2 = 80.6 \times 10^6 \text{ s}^{-1}$; interaction time $t = 80$ ns [line shapes (**a**) and (**b**)], 125 ns [line shapes (**c**) and (**d**)]. (Adapted from [132])

by the phase of the field. In the preceding analysis the initial phase of the oscillating field was arbitrarily chosen to be zero. For an experimental configuration such as an atomic beam, however, where atoms are continually passing through the interaction region, the initial phase of the field sampled by different atoms spans the full range of possibilities from 0 to 2π. To correspond to the experimental signal, therefore, theoretical expressions for occupation or transition probabilities must be averaged over the phase of the field. The phase, as will be shown in Sect. 5.5, appears in a phase factor multiplying the off-diagonal elements of the matrix (I_{ij}) or (J_{ij}). The probability of transition out of an initially pure state is totally insensitive to the phase of the field. This is not to say, however, that phase plays no role. In the following sections important experimental consequences of a sharp relative phase will be explored further in experimental configurations giving rise to quantum interference.

A detailed examination of the variation of the oscillating field solution with diverse experimental parameters (frequency and power of the applied field and duration of interaction) must again be left to the cited literature. However, it is worth mentioning one further unusual, or perhaps unanticipated, fact concerning the response of the atom to an increasing applied field strength (as represented by the matrix element V_{12}) in the case of decaying states.

First note that in the absence of spontaneous decay the occupation probabilities at resonance predicted by the rotating-field approximation reduce to the familiar results

$$P_1(t) = \cos^2(V_{12}t) \,, \tag{5.75}$$

$$P_2(t) = \sin^2(V_{12}t) \,, \tag{5.76}$$

which oscillate between 0 and 1 with a recurrence relation $(\Delta V_{12})t = \pi$. From these expressions it is obvious that, for a fixed interaction time, the maximum probability for transition *out* of one state occurs at the field strength at which the transition *into* the other state is also maximum. This must be so, for there are but two states, and it is obvious from relations (5.75) and (5.76) that total probability is conserved.

Figure 5.6 shows the variation in RWA probabilities $P_1(t)$ and $P_2(t)$ at resonance ($\omega = \omega_0$) as a function of coupling strength for the coupling of hydrogen $4^2S_{1/2}$ and $4^2P_{1/2}$ states whose lifetimes were given earlier. The duration of the interaction of the atoms with the field is 50 ns. Since the numerically exact and OF calculations yield virtually identical results, they are not shown. When $V_{12} = 0$, and therefore no transitions are induced, the probabilities are seen to be $P_1 = 0.8$ and $P_2 = 0$ as determined by the initial conditions. With increasing strength of the applied field, the probability of remaining in the longer lived state 1 rapidly decreases and then oscillates between zero and a constant maximum value less than unity with the same recurrence relation $(\Delta V_{12})t = \pi$. The occupation probability of the initially unpopulated state 2 reaches a maximum at a certain value of V_{12}, then exhibits analogous oscillatory behavior although with a periodicity that depends on the decay parameter G.

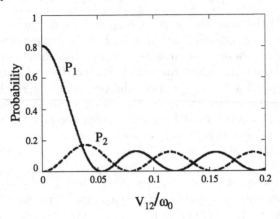

Fig. 5.6. Variation of resonant occupation probabilities P_1 and P_2 with the ratio of the strength of atom–field coupling to the level separation for a long-lived $4S$ state and short-lived $4P$ state, respectively. The interaction time is 50 ns; initial amplitudes and decay rates are the same as for Fig. 5.5

As clearly illustrated in the figure, however, for transitions between unstable states a maximum quenching of one state need no longer occur at the same field strength as a maximum pumping of the other. This can be readily inferred from the theoretical resonance expressions analogous to (5.75) and (5.76)

$$P_1(t) = e^{-(\gamma_1+\gamma_2)t/2} \left(\cos Xt - \frac{G}{2X} \sin Xt \right)^2 , \tag{5.77}$$

$$P_2(t) = e^{-(\gamma_1+\gamma_2)t/2} \left(\frac{V_{12}}{X} \right)^2 \sin^2 Xt , \tag{5.78}$$

where

$$X = \sqrt{V_{12}^2 - (G/2)^2} . \tag{5.79}$$

The reason for this is that an unstable two-level atom is not really a two-level system, being coupled by the vacuum electromagnetic field to all states of lower energy and appropriate symmetry. Indeed, as readily verified from Fig. 5.6 and (5.77) and (5.78), or (5.42), the sum of the probabilities at resonance – or indeed at any frequency – is not unity.

5.4 Resonance and Interference: Tell-Tale Mark of a Quantum Jump

All physicists at some time during their study of the quantum theory of angular momentum undoubtedly encounter the seemingly peculiar property of spinors that a rotation of 4π, rather than 2π, radians is required to return them to their original state. A rotation of 2π radians, which intuitively ought to correspond to no rotation at all, multiplies the spinor wave function by -1, or equivalently by the phase factor $e^{i\pi}$. Theoretically, this follows from the form of the unitary operator $U(\theta) = e^{-i\boldsymbol{J}\cdot\boldsymbol{n}\theta}$, which rotates a state vector of angular momentum $\boldsymbol{J}$ (in units of $\hbar$) by an angle θ about the direction specified by unit vector $\boldsymbol{n}$. If the magnitude of $\boldsymbol{J}$ is an odd half integer (e.g., $J = 1/2, 3/2$, etc.), then $U(2\pi) = -1$ independent of the rotation axis.

In so far as one is discussing the properties of an abstract mathematical object, the above rotational property is not at all disturbing. Mathematics need satisfy no criteria imposed by the real world for its justification. But physics clearly must, and if spinors are to be suitable representations of fermionic systems, then it becomes a legitimate question to ask whether the consequences of rotating a spinorial system by 2π are observable.

The pedagogical literature has not been encouraging in this regard. Dirac, himself, asserted as a 'general result' in his classic treatise on the principles of quantum mechanics that [138]: "[...] the application of one revolution about any axis leaves a ket unchanged or changes its sign. A state, of course, is always unaffected by the revolution, since a state is unaffected by a change of sign of the ket corresponding to it." This sentiment has been repeated often in physics textbooks. One finds, for example, in one well-known book [139]: "That the rotation of a ket through 2π does not give the same ket raises no difficulty of principle so long as no observable effect is produced." Underlying all such remarks is the basic idea that the result of any measurement is representable by expectation values bilinear in the wave function. Therefore two wave functions differing only in overall sign cannot lead to different physical predictions.

There is, in fact, nothing incorrect in the above assertions. Indeed, the rotation of an isolated system – measuring apparatus included – does not lead to experimental consequences. This is not, however, what one ordinarily means by a rotation. Only a part of a system can be rotated; part must remain fixed to provide a reference against which the rotation is to be measured. The global phase change resulting from the former process has no experimental counterpart and therefore no physical implications. The latter process, the rotation of a spinor-characterized portion by 2π radians with respect to a fixed portion of an encompassing larger system, *does* have physical implications, as demonstrated in particular by a number of clever experiments.[3]

In the first, and perhaps best known, of these experiments the rotation of neutrons, which are spin-1/2 particles, was observed by means of neutron interferometry [140]. In basic outline, a beam of unpolarized neutrons was coherently divided at a single-crystal beam splitter, one component passing between the poles of a magnet, the other component passing directly through field-free space to the analyzer where both components were subsequently

---

[3] For a discussion of experimental confirmations of the properties of spinor rotation see [123], and my book [114], Chap. 2, or [82], Chap. 4.

recombined. As a result of the two-beam interference, the neutron intensity transmitted by the analyzer exhibited oscillations as a function of the magnetic field strength.

Considered classically, each neutron is a small magnetic dipole and therefore undergoes Larmor precession in a magnetic field. For appropriate values of magnetic field and interaction time, any desired angle of precession can be realized. Quantum mechanically, the precession angle corresponds to the relative phase shift between the two components of the total wave function. Thus, from the periodicity of the observed oscillations in neutron intensity, such as shown in Fig. 5.7, one could conclude that a neutron returned to its original state after precessing through 4π (rather than 2π) radians.

Although seemingly straightforward, the neutron rotation experiments gave rise to some conceptual difficulties of interpretation [141]. At the root of the difficulty was the Heisenberg uncertainty principle; for fermions the simultaneous observation of the relative rotation of the spins in the two beams

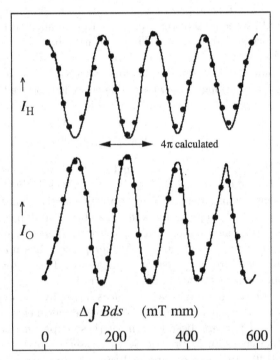

Fig. 5.7. Demonstration of spinor rotation by neutron interferometry. A single-crystal interferometer splits an incident neutron beam into two components, one of which passes between the poles of a magnet. The two beams recombine at the rear face of the interferometer and emerge in a forward beam (O) and a deviated diffracted beam (H). The intensity of each, determined by neutron counting, manifests oscillations with a 4π periodicity as a function of magnetic field. (Adapted from Rauch et al. [140])

and of the interference pattern is incompatible. In the context of an interference experiment one can never know whether a particular neutron followed a classical path through the magnetic field or the field-free region. Although one can measure the rotation of neutrons along one path relative to the rotation of neutrons along the other, an intrusive measurement of this kind would destroy the interference pattern. Thus, according to Byrne, who first called attention to this complication, "[...] the notion of relative rotation ceases to have a meaning as it corresponds to nothing which is measurable".

The objection derives from examination of the density matrix of the beam of neutrons. For unpolarized neutrons the experimental possibilities are exhausted by observation of the interference pattern and determination of the fringe visibility. If the neutrons are polarized, one can determine the degree of polarization parallel to and transverse to the magnetic field. The former, the longitudinal polarization, is unchanged by division and subsequent recombination of the wave function. From Byrne's analysis, it is the rotation of the transverse polarization that is measured by neutron interferometry, and this observable corresponds to either a rotation or nutation of the particles in the *recombined* beam relative to those in the *incident* beam. In no case, however, does one observe the relative rotation of particles in the *split* beams.

The reinterpretation of the neutron interferometry experiments may at first glance seem like quibbling over semantics. It is not, however, since a similar situation does not arise in the case of massive bosons which always have a transverse spin component ($m_s = 0$) whose complex amplitude is unaffected by passage through the magnetic field. The transverse polarization in this case is expressible as the incoherent sum of a fixed polarization and a rotated polarization. Experimentally, the observed angle of rotation corresponds to the angle between *two* orientations of the analyzing axis of a filter giving maximum transmission of the particles – in contrast to fermions for which there is but *one* transmission axis. Thus, the concept of relative spin rotation retains a meaning for massive bosons in the context of an interference experiment. The rotation of a boson, however, is not of particular conceptual interest, since it returns to its original state with the usual angular periodicity of 2π.

In the preceding chapter we have examined various facets of the physics of laser-induced quantum beats. It will be recalled that, from the standpoint of spectroscopy, one of the principal advantages of this technique was to eliminate the need for external radiofrequency or microwave fields. Spectroscopy aside, however, the application of an oscillating field to states prepared in a coherent linear superposition permits one to determine the relative phase of different components of the wave function – a result of interest in its own right. One noteworthy application pertinent to the above discussion of spinors is that under appropriate circumstances a quantum beat resonance experiment can reveal the 4π periodicity of spinor rotation. Actually, the experiment is of more general scope, for, it does not require two beams of spin-1/2 particles. Furthermore, since particles are neither physically rotated nor spatially diffracted, no conceptual difficulties arise concerning the classical interpretation of rotation in a quantum mechanical context. Indeed, examined from a purely quantum perspective, what the experiment directly shows is the perhaps more surprising fact that an atom which has undergone a transition from one state to a second and then back again to the first may still be experimentally distinguishable from an identical atom which has undergone no transition at all [142].

Let us consider a beam of atoms with ground state $|0\rangle$ and three close-lying excited states $|1\rangle$, $|2\rangle$, $|3\rangle$ labeled in order of increasing energy (expressed in units of $\hbar$) $\omega_1 < \omega_2 < \omega_3$ such that the energy (i.e., angular frequency) intervals ω_{32} and ω_{21} (where $\omega_{ij} = \omega_i - \omega_j$) are unequal. The configuration of states is shown in Fig. 5.8. At a fixed time, $t = 0$, a laser pulse of mean frequency ν_0 where $\omega_1 < \nu_0 < \omega_2$ and bandwidth $\Delta\nu$ satisfying $\omega_{21} < \Delta\nu < \omega_{32}$ prepares the beam in a linear superposition of states $|1\rangle$ and $|2\rangle$ with respective amplitudes a_1 and a_2. Were the atoms to evolve freely in time under the Hamiltonian H_0 whose eigenvalues are the frequencies ω_i, the state vector at some subsequent time t would take the form

$$\Psi(t) = \mathrm{e}^{-\mathrm{i}H_0 t}\Psi(0) = a_1\mathrm{e}^{-\mathrm{i}\omega_1 t}|1\rangle + a_2\mathrm{e}^{-\mathrm{i}\omega_2 t}|2\rangle \ . \tag{5.80}$$

In what follows the ground state plays no role and has been eliminated from the above superposition. Also, each amplitude should in principle be multiplied by an exponential damping factor of the form $\mathrm{e}^{-t/2\tau_i}$, where τ_i is the mean lifetime of the ith state. However, in the case of a sufficiently long, although finite, lifetime for which $\tau^{-1} \ll \omega_{ij}$, so many oscillations between superposed states occur within one e-folding time, that it is permissible to simplify the ensuing mathematics without sacrificing essential physical ideas by omitting the decay factors.

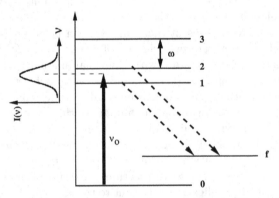

Fig. 5.8. Diagram of a quantum beat resonance experiment illustrating the observable effect of a cyclic quantum transition. A laser pulse of intensity $I(\nu)$ centered on frequency ν_0 excites an atom into a linear superposition of states 1 and 2. A rf field of frequency ω induces the cyclic transition $2 \to 3 \to 2$. The associated phase change is observable in the quantum beats of the radiative decay to lower level f

Following the analysis of Chap. 4, one can express the radiation intensity of polarization $\boldsymbol{\varepsilon}$ emitted at time t as a result of a spontaneous transition to a lower state $|f\rangle$ by

$$I(t) = \left[\langle f|\boldsymbol{\mu}_{\mathrm{E}}\cdot\boldsymbol{\varepsilon}|\Psi(t)\rangle\right]^2 , \tag{5.81}$$

where $\boldsymbol{\mu}_{\mathrm{E}}$ is the electric dipole operator. Thus, substitution of (5.80) into (5.81) leads to an optical signal with harmonic component, the quantum beat, at frequency ω_{21}. For the production of the beat it is essential that the transition matrix elements $D_{ij} \equiv \langle i|\boldsymbol{\mu}_{\mathrm{E}}\cdot\boldsymbol{\varepsilon}|j\rangle$ of both initial states ($i = 1, 2$) to the same final state ($j = f$) be non-zero.

The atoms in the beam, however, are *not* permitted to evolve freely, but are subjected instead to an oscillating electric field of adjustable frequency ω and arbitrary phase ϕ giving rise to the perturbation $V(t)$ (in units of angular frequency). The oscillation frequency is chosen so as to be in the vicinity of ω_{32} and to differ appreciably from ω_{21}, i.e., to couple states $|3\rangle$ and $|2\rangle$ without appreciably coupling states $|2\rangle$ and $|1\rangle$. Alternatively, the interaction $V(t)$ may be of such symmetry that only the element $V_{32} = \langle 3|V_0|2\rangle$ is non-null [where V_0 is the amplitude of the oscillatory potential $V(t)$]. Thus, in regard to this interaction, the atom behaves like a two-state system, since transitions can be induced only between states $|3\rangle$ and $|2\rangle$. The state of the atom at time t, whose initial state is given by (5.80) with $t = 0$, is obtained from solution of the time-dependent Schrödinger equation with potential $V(t)$. The equation immediately reduces to two uncoupled equations: one for a_1, whose time dependence is already given in (5.80), and a second for the coupled amplitudes a_2 and a_3.

For a weak perturbation, i.e., where the matrix elements of V are small compared to state eigenfrequencies, one can apply the rotating-field approximation developed in the preceding section. Then, with neglect of the small shift in resonance frequency and radiative decay rates during passage through the field, the solution to the two-state equation can be written in the form of a spinor rotation followed by free evolution

$$\begin{pmatrix} b_1 \\ b_2 \end{pmatrix} = \mathrm{e}^{-\mathrm{i}H_0 t}\mathrm{e}^{\mathrm{i}\sigma_3 \Omega t/2}\mathrm{e}^{-\mathrm{i}\boldsymbol{n}\cdot\boldsymbol{\sigma}\theta/2} \begin{pmatrix} a_2 \\ 0 \end{pmatrix} . \tag{5.82}$$

Here, $\boldsymbol{\sigma} = (\sigma_1, \sigma_2, \sigma_3)$ is again the vector of Pauli spin matrices,

$$\Omega = \omega - \omega_{32} \tag{5.83}$$

is the deviation from resonance, and

$$\theta = 2\left(|V_{32}|^2 + \frac{1}{4}\Omega^2\right)^{1/2} t \equiv 2\Omega_{\mathrm{r}}t \tag{5.84}$$

is the 'rotation' or 'precession' angle with corresponding angular frequency Ω_r. The unit vector n specifying the rotation axis has components (n_1, n_2, n_3) given by

$$n = \left(\frac{V_{32} \cos \phi}{\Omega_r}, \frac{V_{32} \sin \phi}{\Omega_r}, \frac{\Omega}{2\Omega_r} \right) . \qquad (5.85)$$

For simplicity V_{32} has been taken to be real-valued. Note, too, that the arbitrary phase ϕ of the rf field is contained only in the components n_1 and n_2, and this feature is unchanged should V_{32} be complex-valued.

At resonance, $\Omega = 0$, and the rotational parameters reduce to the simple expressions

$$n = (\cos \phi, \sin \phi, 0) \mathrm{sgn}(V_{32}) , \qquad (5.86)$$

$$\theta = 2|V_{32}|t . \qquad (5.87)$$

The signum function $\mathrm{sgn}(x)$ returns the sign of its argument. By adjustment of the oscillating field strength (to which V_{32} is proportional) and duration t of the atom–field interaction, the angle θ can be appropriately selected. Equations (5.86) and (5.87) show that under resonant conditions a variation in the field strength does not alter the rotation axis, since n is independent of $|V_{32}|$. Moreover, for rotations of an integer multiple of 2π radians, the state of the system is independent of n and therefore of the arbitrary phase ϕ.

From (5.82) through (5.87) – with application of the identity in relation (5.10) – it follows that at resonance a rotation of $\theta = \pi$ results in all atoms initially in state $|2\rangle$ being driven into state $|3\rangle$

$$\begin{pmatrix} b_2 \\ b_3 \end{pmatrix}_{\Omega=0, \theta=\pi} = e^{-iH_0 t} \begin{pmatrix} 0 \\ -ia_2 e^{i\phi} \end{pmatrix} , \qquad (5.88)$$

whereas a rotation of $\theta = 2\pi$ corresponds to a transition from $|2\rangle$ to $|3\rangle$ and back again to $|2\rangle$

$$\begin{pmatrix} b_2 \\ b_3 \end{pmatrix}_{\Omega=0, \theta=2\pi} = -e^{-iH_0 t} \begin{pmatrix} a_2 \\ 0 \end{pmatrix} . \qquad (5.89)$$

The occurrence of the minus sign is normally not observed since it is not revealed by any bilinear products of the wave function. In the present experiment, however, this sign change is important and will be seen to have experimental consequences. A minimal resonant rotation of $\theta = 4\pi$, which corresponds to two cyclic transitions between $|2\rangle$ to $|3\rangle$, is required to return the system to the same state as if no transition had occurred.

We now return to a description of the full three-state system. At the time of emergence from the oscillating field, the state vector of the atom is generally given by

$$\Psi(t) = e^{-iH_0 t} \left[a_1 |1\rangle + e^{i\sigma_3 \Omega t/2} e^{-in \cdot \sigma \theta/2} \begin{pmatrix} a_2 \\ 0 \end{pmatrix} \right] , \qquad (5.90)$$

which, when expanded, takes the explicit form

$$\Psi(t) = e^{-iH_0 t}\left[a_1|1\rangle + a_2 e^{i\Omega t/2}\left(\cos\frac{\theta}{2} - in_3 \sin\frac{\theta}{2} \right)|2\rangle \right. \tag{5.91}$$

$$\left. - ia_2 e^{-i\Omega t/2}(n_1 + in_2)\sin\frac{\theta}{2}|3\rangle \right] .$$

The operator $e^{-iH_0 t}$ acts on each basis ket $|i\rangle$ to generate the phase factor $e^{-i\omega_i t}$ which occurs under free evolution. Thus, after passage through the oscillating field, free evolution for a time T is accounted for by replacing t by $t + T$ in the first factor of (5.91). One can also easily verify that the total state vector is still properly normalized to $|\Psi|^2 = 1$ as expected (spontaneous decay having been neglected).

At an arbitrary time T after emergence from a *resonant* field the atom is characterized by the state vector

$$\Psi_{\Omega=0}(t + T) = e^{-iH_0(t+T)}\left(a_1|1\rangle + a_2\cos\frac{\theta}{2}|2\rangle - ia_2 e^{i\phi}\sin\frac{\theta}{2}|3\rangle \right) . \tag{5.92}$$

At this point the spontaneous emission from the atom plays a role, for it furnishes the optical signal to be observed. The general expression for the resonant atomic fluorescence of polarization ε emitted at time $t + T$, obtained by substituting (5.92) into (5.81), is somewhat complicated (with quantum beats at frequencies ω_{21}, ω_{31}, and ω_{32}) and need not be given explicitly. Of particular significance, however, is the result that for 'rotations' of 0 and 2π, the quantum beat signals at frequency ω_{21} are independent of the arbitrary initial phase of the oscillating field and differ in relative phase by 180° as shown in the expression

$$I_{\Omega=0}^{0,2\pi}(t + T) = |a_1|^2|D_{f1}|^2 + |a_2|^2|D_{f2}|^2 \tag{5.93}$$

$$\pm 2\mathrm{Re}\left\{ a_1 a_2^* D_{f1} D_{f2}^* \exp\left[i\omega_{21}(t + T)\right] \right\} ,$$

where the upper and lower signs correspond to 0 and 2π, respectively.

Experimentally, one can measure the above fluorescent intensities in the traditional way as a function of T for fixed amplitude of the oscillating field. The effect of the sign change can be enhanced by taking the difference signal

$$S(t+T) = I_{\Omega=0}^0(t+T) - I_{\Omega=0}^{2\pi}(t+T) = 4|a_1 a_2^*||D_{f1}D_{f2}^*|\cos(\omega_{12}t+\beta) \tag{5.94}$$

to obtain ideally a harmonic transient with 100% contrast. (The signal, of course, is exponentially damped when state decay is explicitly included in the analysis.) The phase β depends on the interaction time t and on constant phase factors arising from possibly complex-valued matrix elements and initial amplitudes. Thus, the distinction between a cyclic transition between two states ($\theta = 2\pi$) and no transition at all is revealed by a tell-tale minus sign in the wave function.

To show explicitly the 4π periodicity of the transition one can modify the foregoing procedure to measure, at fixed time $t + T$, the fluorescence as a function of the oscillating field strength or equivalently θ. Since the atoms passing through the field experience all values of the initial phase on which the signal in general depends, one must in this case average $I(t + T)$ over the range $0 \leq \phi \leq 2\pi$. Assuming that the atomic states have a well-defined parity – which is valid upon neglect of small effects on the electron wave function due to weak nuclear interactions – the states $|2\rangle$ and $|3\rangle$ must have opposite parities to be coupled by an oscillating electric field. Thus, for an electric dipole decay transition to the lower state $|f\rangle$, one of the matrix elements D_{f2} or D_{f3} must be null. With $|f\rangle$ chosen so that $D_{f3} = 0$, for example, the calculated phase-averaged intensity at resonance can be shown to be

$$\langle I_{\Omega=0}(t+T)\rangle_\phi = |a_1|^2|D_{f1}|^2 + |a_2|^2|D_{f2}|^2 \cos^2 \frac{\theta}{2} \qquad (5.95)$$

$$+2\mathrm{Re}\left[a_1 a_2^* D_{f1} D_{f2}^* e^{i\omega_{21}(t+T)}\right] \cos \frac{\theta}{2} \; .$$

The signal represented by (5.95) is then processed by taking pointwise over a range of θ the difference

$$S(\theta) = \langle I_{\Omega=0}(\theta)\rangle_\phi - \langle I_{\Omega=0}(\theta + \pi)\rangle_\phi = 4\mathrm{Re}\left[a_1 a_2^* D_{f1} D_{f2}^* e^{i\omega_{21}(t+T)}\right] \cos \frac{\theta}{2} \; , \tag{5.96}$$

which results in a signal explicitly showing the 4π periodicity of spinor rotation.

It is worth noting expressly that the phase change under a cyclic transition is observable in this experiment precisely because it is *not* a global phase, but rather a relative phase between the components of the wave function characterizing states $|2\rangle$ and $|1\rangle$. The latter is unaffected by the external oscillating field and thereby assumes the role of a reference state in much the same way in which phase information of light scattered from an object is recorded holographically.

As a final consideration on spinors, I would like to point out an amusing, if not intellectually mystifying, demonstration that I have often used before lecture audiences to illustrate the spinor rotational property concretely by means of a classical object – rather than a purely quantum system like an electron or neutron. Invented originally by Dirac, as far as I am aware, and referred to as a 'spinor spanner', the apparatus consists of a spanner (or wrench) fastened by three parallel cords – one from each of its ends – to two vertical walls as shown in Fig. 5.9. If the spanner is rotated 360° about its axis, the cords become twisted, and no maneuvering of the object or cords – short of a counter-rotation or cutting of the cords – can undo the tangle. If the spanner is now turned another 360° in the *same* sense, the entanglement appears at first even worse than before. But the astonishing thing is that by looping the cords around the spanner in an appropriate manner – the

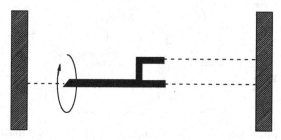

Fig. 5.9. The spinor spanner. Rotating the spanner around its axis by 360° entangles the cords which can be made to untangle by a further 360° rotation in the same sense. The initial impression following a 720° rotation, however, is that the cords are even more entangled than before

orientation of the spanner all the while kept fixed – one can untangle the snarl and recover the original configuration of the system as shown in the figure. Why a 4π rotation is equivalent to the identify operation in this case is an instructive exercise in topology [143] which cannot be pursued further in this book. The demonstration itself, however, is easy enough to construct and will undoubtedly fascinate the reader as it has for many years the author.

5.5 Quantum Interference in Separated Oscillating Fields

In the discussion of atomic state selection in Sect. 5.1, the various oscillators that furnished the distinct spectroscopy and quenching fields were uncorrelated, i.e., they had in general different frequencies and no well-defined relative phase. A configuration of two (or more) sequential oscillating fields, such as illustrated in Fig. 5.10, which *together* comprise a *single* spectroscopy region was proposed in 1949 by Norman Ramsey [144] and contributed to his award of the Nobel Prize for physics for 1989. In this configuration two spatially separated interaction regions created from the *same* oscillator have identical frequencies and a well-defined relative phase which is adjustable by a phase-shifter. A sensitive spectroscopic tool for resolving resonance line shapes [56, Chap. 7], this field configuration is of conceptual interest here because it leads to a different kind of particle self-interference in the time domain than discussed so far.

Under the usual circumstances where an atom in a beam is not observed until after having passed through both fields, an observer cannot know in which of the two interaction regions a transition between atomic states may have occurred. Thus, the entire configuration is again analogous to the electron two-slit experiment, except that in the present case the slits are not laterally spaced apertures, but longitudinally separated windows in time. All the same, the transition probability contains an interference term with the experimental

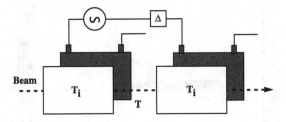

Fig. 5.10. Separated oscillating field configuration. Two rf fields driven by the same oscillator have the same frequency and well-defined relative phase Δ. The atoms in the beam interact with each field for a time T_{i} with a time of transition T between fields

consequence that the resonance curve exhibits an oscillatory profile narrower by nearly a factor of two than the width of the resonance produced by a single oscillatory field of the same overall length. One may have anticipated this by analogy with diffraction and interference in physical optics. The central diffraction peak of a single slit of given width is broader than the fringes of an interference pattern produced by two slits of equivalent total width. It is this interference narrowing that in part gives the separated oscillatory field configuration its spectroscopic utility, for it is difficult to determine a precise resonance frequency from a broad line shape.

There is a second interesting feature associated with the coherently oscillating field configuration when the coupled atomic states are unstable (a situation that Ramsey did not treat). As illustrated later in the section (in Fig. 5.12), the greater the separation between the two interaction regions, the more oscillatory is the resulting line shape – and hence the more sharply defined is the resonant transition frequency. This is a practical consequence of the uncertainty principle. Recall that the lifetime of a quantum state is a statistical measure of the duration of that state. In an ensemble of similarly prepared unstable systems, some decay sooner and others last longer than the mean lifetime. Of course, the probability of surviving much longer than the mean lifetime τ diminishes exponentially as $e^{-t/\tau}$. Thus, the probability that an atom remains in an unstable state for a period of five lifetimes before decaying is $e^{-5} \sim 6.7 \times 10^{-3}$, or approximately seven atoms out of every thousand. By contrast, the number surviving for 10 lifetimes is $e^{-10} \sim 4.5 \times 10^{-5}$ or approximately five out of one hundred thousand. Nevertheless, with a sufficiently intense initial beam, one can have a measurable population of atoms in an unstable excited state pass from the first to the second transition region if the separation between regions is not too great. The mean lifetime of this population, however, is now longer than that of the original unselected population of atoms in the same state. According to the version of the uncertainty principle concerning time intervals and energy, $(\Delta\varepsilon)(\Delta t) \geq \hbar$, the longer the time interval $\Delta t = \tau$ over which a state can last, the smaller is the uncertainty $\Delta\varepsilon$, and hence the narrower will be the corresponding resonance line shape.

Let us examine quantitatively the quantum mechanics of a two-level atom with energy interval $\omega_0 = \omega_1 - \omega_2$ traversing the configuration of two separated oscillating fields. As it passes the two interaction regions, the atom experiences the electric field $E_0 \cos(\omega t + \delta_1)$ for a time T_i, field-free space for a time T, and a second electric field $E_0 \cos(\omega t + \delta_2)$ again for a time T_i. (The more general case of different field amplitudes, frequencies, and interaction times is treated in [56, 128].) We take the two fields to be linearly polarized parallel to one another. At the time $t = 0$ of entry into the first field, the state of the atom is specified by the wave function $\Psi(0)$ which is representable as a vector, as in (5.2) and (5.3), where the upper component is the amplitude of state 1 and the lower component the amplitude of state 2. Upon leaving the second field a time interval $t = 2T_i + T$ later, the state of the atom is represented by a wave function $\Psi(t)$ which can be cast in the form

$$\Psi(t) = e^{-iXt} M \Psi(0) , \tag{5.97}$$

where X is the operator $H_0 - i\hbar\Gamma/2$ in (5.44) responsible for free evolution and spontaneous decay, and M is the transition matrix whose elements are sought. To determine M we utilize the results of the preceding section and calculate the state of the atom at the entry and exit of each of the interaction regions.

At the time of emergence T_i from the first field, the wave function has evolved according to

$$\Psi(T_i) = e^{-iXT_i} K(T_i, \delta_1) \Psi(0) , \tag{5.98}$$

where the matrix representation of X is diagonal with elements

$$X_j = \omega_j - i\frac{1}{2}\gamma_j \qquad (j = 1, 2) , \tag{5.99}$$

and the elements of $K(T_i, \delta_1)$, obtained in the rotating-field approximation from (5.53)–(5.56), take the form

$$K_{11} = I_{11} , \quad K_{22} = I_{22} , \quad K_{12} = I_{12} e^{-i\delta_1} , \quad K_{21} = I_{21} e^{i\delta_1} . \tag{5.100}$$

The phase δ_1 is a distributed variable, differing from one atom to the next that enters the first field.

The time-evolution of the wave function in field-free space is generated simply by the matrix operator e^{-iXT}. To account for passage through the second interaction region one merely repeats the application of (5.98), replacing the initial wave function $\Psi(0)$ by the expression

$$\Psi(T + T_i) = e^{-iXT} \Psi(T_i) , \tag{5.101}$$

to obtain

$$\Psi(t) = e^{-iXT_i} K(T_i, \phi_2) e^{-iX(T+T_i)} K(T_i, \delta_1) \Psi(0) . \tag{5.102}$$

The elements of the matrix $K(T_i, \phi_2)$ have the same form as those of (5.100) but with the phase δ_1 replaced by

$$\phi_2 = \omega(T + T_i) + \delta_2 , \tag{5.103}$$

which is also a distributed variable. [In general, the phase to be inserted in relations (5.100) for an atom entering the field $E_0 \cos(\omega t + \delta)$ at time t_0 is $\phi = \omega t_0 + \delta$.]

Upon inverting the order of the inside exponential and transition matrix $K(T_i, \phi_2)$ in expression (5.102), and taking account of their noncommutativity, one obtains (5.97) where the elements of M are

$$M_{11} = I_{11}^2 + I_{12} I_{21} \exp \left[G(T + T_i) - i\Theta \right] , \tag{5.104}$$

$$M_{12} = e^{-i\delta_1} \left\{ I_{11} I_{12} + I_{12} I_{22} \exp \left[G(T + T_i) - i\Theta \right] \right\} , \tag{5.105}$$

$$M_{21} = e^{i\delta_1} \left\{ I_{22} I_{21} + I_{21} I_{11} \exp \left[- G(T + T_i) + i\Theta \right] \right\} , \tag{5.106}$$

$$M_{22} = I_{22}^2 + I_{21} I_{12} \exp \left[- G(T + T_i) + i\Theta \right] , \tag{5.107}$$

and the phase Θ is

$$\Theta = \Omega(T + T_i) + (\delta_2 - \delta_1) . \tag{5.108}$$

G is the decay rate difference given in (5.58), and Ω is the displacement from resonance of (5.48). Although δ_1 and δ_2 are distributed quantities, the phase difference

$$\Delta \equiv \delta_2 - \delta_1 \tag{5.109}$$

is an experimentally adjustable, well-defined parameter in the case of two interaction regions driven by the same oscillator.

The density matrix of the two-level atom is now determined by averaging the product $\Psi(t)\Psi(t)^\dagger$ over the variable δ_1 while keeping Δ constant

$$\rho(t) = \langle \Psi(t)\Psi(t)^\dagger \rangle_{\delta_1} = \langle e^{-iXt} M \rho(0) M^\dagger e^{iX^\dagger t} \rangle_{\delta_1} . \tag{5.110}$$

Assuming for the sake of illustration – and because it is often the case experimentally – that the initial density matrix $\rho(0)$ represents an atom initially prepared in one of the two coupled states, for example the longer-lived state $|1\rangle$,

$$\rho(0) = \begin{pmatrix} 1 & 0 \\ 0 & 0 \end{pmatrix} \tag{5.111}$$

leads to the following occupation probabilities after passage of the atom through both interaction regions

$$\rho_{11}(t) = e^{-\gamma_1 t} \left| I_{11}^2 + I_{12} I_{21} e^{(G - i\Omega)(T+T_i)} e^{-i\Delta} \right|^2 , \tag{5.112}$$

$$\rho_{22}(t) = e^{-\gamma_2 t} \left| I_{22} I_{21} + I_{21} I_{11} e^{-(G - i\Omega)(T+T_i)} e^{i\Delta} \right|^2 . \tag{5.113}$$

The order of the interaction matrix elements I_{ij} in each term of (5.112) and (5.113) corresponds from right to left to the first and second interaction regions, respectively Thus, in $\rho_{11}(t)$, for example, one sees that the first term characterizes an atom in state $|1\rangle$ that remains in state $|1\rangle$ after passage through both fields, whereas the second term characterizes an atom that undergoes a transition from $|1\rangle$ to $|2\rangle$ in the first field and back again from $|2\rangle$ to $|1\rangle$ in the second field. Under the conditions of the experiment there is no way to distinguish which process produced an atom emerging in state $|1\rangle$ from the second field; the total amplitude is therefore the sum of the amplitudes for these two processes, and the occupation probability $\rho_{11}(t)$ [and likewise for $\rho_{22}(t)$] contains an interference term. In keeping with the earlier heuristic argument based on the uncertainty principle, note that the relative phase between the two terms depends on the field-free interaction time T; the larger T, the more oscillatory is the exponential phase factor. Under the conditions of resonance $(\Omega = 0)$ between two states of the same lifetime $(G = 0)$, however, the relative phase between the two terms is independent of T; the separation between the interaction regions then has no effect except to diminish the overall intensity if the states are unstable.

Let us examine the case of exact resonance between two states of the same lifetime more closely to see better the physical effect of the relative phase Δ. When the matrix element V_{12} is greater than $G/2$, it follows from (5.53)–(5.56) and (5.57) that the products of interaction matrix elements I_{ij} in (5.112) and (5.113) are real, and therefore the interference terms in (5.112) and (5.113) are proportional to $\cos \Delta$. When $G = 0$, the precession frequency ν is simply equal to the dipole matrix element V_{12}, and the occupation probability of state $|1\rangle$, for example, becomes

$$\rho_{11}(t) = e^{-\gamma_1 t} \left[\cos^4(V_{12}T_i) + \sin^4(V_{12}T_i) - 2\cos^2(V_{12}T_i)\sin^2(V_{12}T_i)\cos\Delta \right] .$$
$$(5.114)$$

For a choice of field strength and interaction time such that $V_{12}T_i = \pi/4$ radians, the above expression reduces to

$$\rho_{11}(t) = \frac{1}{2}e^{-\gamma_1 t}(1 - \cos\Delta) , \qquad (5.115)$$

and the interference term in ρ_{11} shows oscillations with 100% contrast. If the two coherently oscillating fields are in phase $(\Delta = 0)$, then all nondecaying atoms initially in state $|1\rangle$ emerge in state $|2\rangle$. On the other hand, if the two fields are 180° out of phase $(\Delta = \pi)$, all nondecaying atoms emerging from the second interaction region will be found in the original state $|1\rangle$.

Under conditions where the frequency is not exactly at resonance or the two states decay at different rates, the interference term still diminishes the $|1\rangle$ component of the wave function for in-phase oscillating fields and augments this component for fields oscillating out of phase. However, the contrast in these cases may be too low for different choices of Δ to be directly noticeable in the variation of ρ_{11} or ρ_{22} with oscillation frequency. One can then isolate the interference term, as was done in the preceding section for a quantum beat signal, by measuring the difference in occupation probability of state $|i\rangle$

$$S_i(\Delta_1, \Delta_2) = \rho_{ii}(\Delta_1) - \rho_{ii}(\Delta_2) \qquad (i = 1, 2), \qquad (5.116)$$

for two appropriately chosen values of the relative phase Δ. One loses, of course, in overall intensity, but this need not be a problem for a sufficiently large initial flux of atoms. The pairs $\Delta = (\pi, 0)$ and $\Delta = (\pi/2, -\pi/2)$, as first shown by Ramsey, are particularly useful, and lead in the present case to the signals

$$S_1(\pi, 0) = 4e^{-\gamma_1 t} e^{G(T+T_i)} \mathrm{Re}\left[I_{11}^2 I_{12}^* I_{21}^* e^{i\Omega(T+T_i)}\right], \qquad (5.117)$$

$$S_1(\pi/2, -\pi/2) = 4e^{-\gamma_1 t} e^{G(T+T_i)} \mathrm{Im}\left[I_{11}^2 I_{12}^* I_{21}^* e^{i\Omega(T+T_i)}\right], \qquad (5.118)$$

with comparable expressions for S_2 derivable from (5.113).

In Fig. 5.11 the variation with frequency of $S_1(\pi, 0)$ and $\rho_{11}(\Delta = 0)$ are superposed in the case of two coupled unstable states (hydrogen $3S$ and $3P$). From the scale of the figure, it is seen that the interference term is approxi-

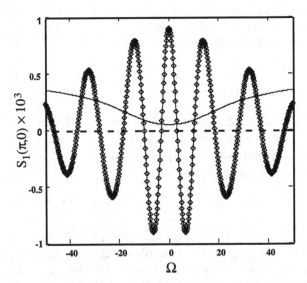

Fig. 5.11. Isolation of the quantum interference effect produced by two separated coherently oscillating fields with adjustable relative phase Δ. The oscillatory curve is the difference in occupation probabilities $\rho_{11}(\Delta = \pi) - \rho_{11}(\Delta = 0)$ of a long-lived state ($3S$ coupled to $3P$) as a function of the frequency interval $\Omega = \omega - \omega_0$. The interference is not visible in the frequency variation of the individual occupation probabilities (*thin solid line*). Theoretical parameters are: $V_{12} = 2\pi \times 6$ MHz, $c_1(0) = 1$, $c_2(0) = 0$, $\gamma_1 = 6.25 \times 10^6$ s^{-1}, $\gamma_2 = 172.4 \times 10^6$ s^{-1}; interaction time $T_i = 40$ ns, transit time $T = 50$ ns. (Adapted from Silverman [128])

mately one hundred times smaller than ρ_{11} in the vicinity of resonance, and therefore the contrast of the interference pattern is greatly enhanced in the difference signal $S_1(\pi, 0)$. In the absence of the counter-rotating component of the oscillating field, the interference pattern of $S_i(\pi, 0)$ $(i = 1, 2)$ is symmetric about the resonance frequency. The narrowing of the central portion of the line shape (compared with ρ_{ii}), and therefore the advantage to spectroscopy, are readily apparent. This narrowing becomes more pronounced, as already explained, for longer transit times T between interaction regions, as illustrated in Fig. 5.12.

Figure 5.12 also shows the results of choosing the alternative pair of phases, $\Delta = (\pi/2, -\pi/2)$. The line shape, resembling that of a dispersion curve, is antisymmetric about the resonance frequency $(\Omega = 0)$ at which point $S(\pi/2, -\pi/2)$ is theoretically null and the slope of the curve is steepest. This, too, is spectroscopically useful, for it can be more advantageous to determine the zero-crossing of a resonance line than to locate the maximum point in a region of near-zero slope. A line shape of the dispersion type is very sensitive to small shifts in the resonance frequency.

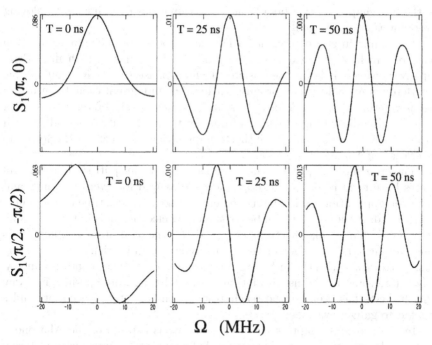

Fig. 5.12. Illustration of separated oscillating field interference patterns for different field-free transit times T and for two choices of pairs of relative phases $\Delta = (\pi, 0)$ and $\Delta = (\pi/2, -\pi/2)$. The greater T, the more rapid is the oscillatory structure, and therefore the narrower is the line shape near resonance. The initial amplitudes and decay rates are the same as for Fig. 5.11; $V_{12} = 2\pi \times 16$ MHz; $T_i = 60$ ns. (Adapted from Silverman [128])

5.6 Ion Interferometry and Tests of Gauge Invariance

Upon learning of the diffraction of helium atoms and hydrogen molecules from a crystal, I.I. Rabi allegedly remarked: "They will be diffracting grand pianos next!" While this is unlikely, to say the least, the coherent splitting and subsequent interference of beams of atoms, once also a remote possibility, has been achieved and is an actively pursued area of physics research [145]. Since atoms are not charged (like electrons) or penetrate matter (like neutrons), the construction of atom interferometers required first the development of new techniques for simulating necessary optical elements. Recent developments in optics and atomic physics have made possible various types of atom interferometers such as those based on (i) wavefront splitting at nanometer-sized mechanical structures [146], (ii) amplitude division at gratings constructed from standing waves of light [147], and (iii) pulsed laser-induced transitions within an atomic 'fountain' [148], i.e., a beam of atoms that first rises vertically through an interaction region and subsequently descends slowly through the same region under the action of gravity. The creation in the laboratory of Bose–Einstein condensates of atoms (which have a significant role to play in the last chapter of this book) has led to numerous experiments employing atom interferometry [149].

An atomic interferometer has a number of potential advantages, not the least of which is that atoms can be produced in beams of high flux. Principally, however, it is on the basis of its high phase sensitivity that atomic interferometers hold great promise as inertial and gravitational sensors in novel experiments to test relativity, search for new physical interactions (such as an intermediate range deviation from Newtonian gravity or so-called 'fifth force' [150]), or measure with hitherto unprecedented precision the local acceleration of gravity.

The optical techniques by which neutral atom beam splitters, mirrors, and lenses have been made are generally suitable as well for *charged* atoms, and the wide applications of ion interferometers at some future time can be safely anticipated. Although ions can be controlled by electric fields in the same way that point charges like electrons can (and that neutral atoms cannot), it is the internal structure of ions that in part makes them particularly interesting systems to study with the techniques employed in neutral atom interferometry. The interferometry of ions with internal state labeling would permit, in a way that hitherto has been beyond reach, tests of fundamental physical principles related to gauge invariance.

In the previous chapters we discussed various aspects of the Aharonov–Bohm (AB) effect, a quantum mechanical phenomenon concerning the interaction of a charged particle with the electromagnetic vector (or scalar) potential under conditions where static electric and magnetic fields are ideally null. Although once the subject of long theoretical and experimental controversy, the AB effect has since been established as a seminal part of quantum theory required by the gauge invariance of the equations of motion. The AB effect

illustrates the nonlocal influence of topology in physics, as well as the fundamental primacy of potentials (which relate to energy) over fields (which relate
to forces) [151]. However, as far as I am aware, *all* successful AB-effect experiments performed to date, whether based on particle beams or mesoscopic
metal rings (to be discussed in the next chapter), employed the same type
of structureless charged particle, the electron. The reason for this is obvious:
coherent electrons are readily available. Nevertheless, novel and conceptually
significant modifications of the AB effect can be studied by using composite
charged particles with an internal structure, such as ionized atoms [152].

One consequence of the usually assumed minimal coupling between particle charge and potential field is that the AB phase shift for single-electron
wave packets is independent of the electron energy–momentum–spin state, and
depends only on the magnetic flux enclosed by the particle path. Presuming
the same coupling holds within a composite structure of bound charges – and
this is one of the interesting issues worth testing – it can nevertheless follow
that the AB phase shift in ions may be observed on *selected excited states*.
We will examine this possibility momentarily. Furthermore, under appropriate
circumstances, the AB-phase shift can be manifested in the resonance fluorescence of the ions, but not, as traditionally detected, in the particle count rate
itself. This optical manifestation of the AB effect is still attributable to the
direct coupling of real particles with the vector potential, and is to be distinguished from predicted AB effects of the photon arising from the coupling of
virtual electron–positron pairs with an external potential [153].

As an illustration of the new kinds of experiments made possible by ion
interferometry, consider the interferometer configuration of Fig. 5.13 through
which passes a beam of ions of charge q. Each ion is assumed to be a three-
level system with ground state $|g\rangle$ and two close-lying excited states $|e_1\rangle$, $|e_2\rangle$
with respective energies ω_1, ω_2 (as usual in units of $\hbar$) and common decay
rate Γ. Although the figure shows an interferometer of the Mach–Zehnder
type, the beam splitters and mirrors need not be massive structures; they are
merely interaction regions for changing particle momentum or internal state,
and can in practice be regions of electromagnetic radiation such as employed
in atomic interferometers with internal state labeling [154]. The ions, excited
by laser to lower state $|e_1\rangle$, for example, are split equally at interaction region
(beam splitter) BS_1 into components that follow paths I or II through the
interferometer around a region of confined magnetic flux Φ.

Along the horizontal segment of each path the ions pass through a resonantly oscillating radiofrequency or microwave field that can induce transitions between the two excited states at the Bohr frequency $\Omega_0 = \omega_2 - \omega_1$.
The two transition fields V_1, V_2 oscillate coherently with some adjustable,
but well-defined, relative phase δ, as analyzed in the preceding section. After
traversing the oscillating fields, the ion beam recombines at interaction region
(beam splitter) BS_2 and passes on to detector D_1 or D_2. We assume the excited state lifetime $1/\Gamma$ to be sufficiently long (as in the case of metastable or
Rydberg states) for the ions to traverse the interferometer.

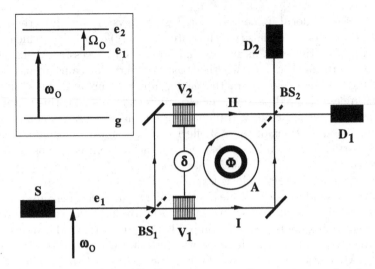

Fig. 5.13. Variations of the AB effect on a composite quantum system by means of ion interferometry. A beam of ions, excited into state e_1 by a laser, is coherently divided into components passing through one or the other of two coherently oscillating fields V_1 and V_2 (with relative phase δ) before recombining and exiting the interferometer. The effect of confined magnetic flux Φ may appear in either the excited ion count or in the ion resonance fluorescence, depending upon the transitions induced by the oscillating fields

Although, to avoid a linguistic awkwardness, the above description refers to ion 'beams' taking paths I and II, one must again bear in mind that, strictly speaking, it is single-ion wave packets that split and propagate through the interferometer. Moreover, in contrast to the standard Ramsey field configuration where ideally an entire atom or molecule passes sequentially through *both* fields with 100% probability, it is now the case that each component of the ion wave function passes through only *one* of the fields. Under the conditions of the experiment, the observer does not know through which oscillating field a given ion passes. There is thus the potential for interference to occur because of the geometrical (including topological) effects of spatial separation, as well as the temporal (or dynamic) effects of internal state transitions.

After passing interaction region BS_2, the amplitudes of the two components take the general forms

$$\text{Path I} \qquad \left[\alpha_{11}e^{i\chi_I^{(1)}}|e_1\rangle + \alpha_{12}e^{i\chi_I^{(2)}}|e_2\rangle\right]e^{i\gamma_I}\,, \qquad (5.119)$$

$$\text{Path II} \qquad \left[\alpha_{21}e^{i\chi_{II}^{(1)}}|e_1\rangle + \alpha_{22}e^{i\chi_{II}^{(2)}}|e_2\rangle\right]e^{i\gamma_{II}}\,, \qquad (5.120)$$

where $\chi_J^{(i)}$ is the phase shift of excited state i along path J with possible contributions from excitation, reflection and transmission, and space-time propagation in free space and within the transition regions V_J. The effect of the confined magnetic flux is contained in the nonintegrable phase shift

$$\gamma_J = \frac{q}{\hbar c} \int\limits_{\text{path } J} \boldsymbol{A} \cdot d\boldsymbol{s} \; , \tag{5.121}$$

where $\boldsymbol{A}(\boldsymbol{s})$ is the local vector potential field at some point $\boldsymbol{s}$ along path J.

The real-valued amplitudes α_{ij} ($i, j = 1, 2$) likewise express the net result of excitation, decay, diffraction, induced transition, and space-time propagation, but they are independent of the vector potential field (or magnetic flux). Their exact expressions can be constructed from the previously derived functions (I_{ij}) of the preceding section, but will not be needed to illustrate the points of principal interest in this section. Upon recombination, the net amplitude at each detector, for example D_1, takes the form

$$\psi(D_1) \sim \left[\alpha_{11} e^{i\chi_{\mathrm{I}}^{(1)}} e^{i\gamma_{\mathrm{I}}} + \alpha_{21} e^{i\chi_{\mathrm{II}}^{(1)}} e^{i\gamma_{\mathrm{II}}}\right] |e_1\rangle \tag{5.122}$$

$$+ \left[\alpha_{12} e^{i\chi_{\mathrm{I}}^{(2)}} e^{i\gamma_{\mathrm{I}}} + \alpha_{22} e^{i\chi_{\mathrm{II}}^{(2)}} e^{i\gamma_{\mathrm{II}}}\right] |e_2\rangle \; ,$$

where to avoid unnecessary complexity in the mathematical formalism, the additional effects of space-time propagation and decay over the path from BS_2 to D_1 have simply been absorbed in α_{ij} and $\chi_J^{(i)}$.

The arrival of excited ions can be observed in two experimentally distinct ways. One can, in principal, count particles, in which case the detector D_1 is sensitive only to excited states, as, for example, by means of a resonant ionization procedure. On the other hand, one can count the photons in the decay radiation, in which case D_1 represents a photomultiplier and associated electronics. The probability of direct particle detection at D_1 is given by

$$P(D_1) = |\psi(D_1)|^2 \tag{5.123}$$

for a suitably normalized wave function ψ. By contrast, the fluorescent signal with polarization $\boldsymbol{\varepsilon}$ deriving from electric dipole ($\boldsymbol{\mu}_{\mathrm{E}}$) transitions to some final state – here taken for simplicity to be the ground state $|g\rangle$ – is obtained from

$$S(D_1) = \mathrm{Tr}\left[\rho O_{\mathrm{det}}(\boldsymbol{\varepsilon})\right] \; , \tag{5.124}$$

where ρ is the density operator of the system

$$\rho = |\psi(D_1)\rangle\langle\psi(D_1)| \tag{5.125}$$

and $O_{\mathrm{det}}(\boldsymbol{\varepsilon})$ is the detection operator

$$O_{\mathrm{det}}(\boldsymbol{\varepsilon}) = (\boldsymbol{\mu}_{\mathrm{E}} \cdot \boldsymbol{\varepsilon})|g\rangle\langle g|(\boldsymbol{\mu}_{\mathrm{E}} \cdot \boldsymbol{\varepsilon})^\dagger \; , \tag{5.126}$$

as described in the chapter on quantum beats.

There are a variety of outcomes depending upon the effects of the oscillating fields V_1, V_2, but we shall consider two. Suppose that the transition regions convert incident states $|e_1\rangle$ into pure states $|e_2\rangle$. Then $\alpha_{21} = \alpha_{11} = 0$, and the probability of detecting an excited ion directly is

$$P(\mathrm{D}_1) = \alpha_{12}^2 + \alpha_{22}^2 + 2\alpha_{12}\alpha_{22}\cos\left(\Delta_2 + 2\pi\frac{\Phi}{\Phi_0}\right), \tag{5.127}$$

where

$$\Delta_i = \chi_{\mathrm{II}}^{(i)} - \chi_{\mathrm{I}}^{(i)} \qquad (i = 1, 2) \tag{5.128}$$

is the phase shift between component waves of state $|e_i\rangle$ due to a difference in optical path length through the interferometer. The ratio of magnetic flux enclosed by the interferometer to the fluxon is

$$\frac{\Phi}{\Phi_0} = \frac{q}{hc} = \oint_{\text{path (II–I)}} \boldsymbol{A} \cdot \mathrm{d}\boldsymbol{s} = \gamma_{\mathrm{II}} - \gamma_{\mathrm{I}}. \tag{5.129}$$

The optical signal deriving from the radiative decay of ions is in this case simply proportional to the particle detection probability

$$S(\mathrm{D}_1) = |\mu_{2g}|^2 P(\mathrm{D}_1), \tag{5.130}$$

where

$$\mu_{ig} = \langle e_i|\boldsymbol{\mu}_{\mathrm{E}} \cdot \boldsymbol{\varepsilon}|g\rangle \qquad (i = 1, 2) \tag{5.131}$$

is the electric dipole matrix element between states $|e_i\rangle$ and $|g\rangle$. It is to be noted that the phase Δ_2 which enters (5.127) and (5.130) is independent of observation time, since the phase factor $e^{-i\omega_2 t}$ appears in the amplitudes for both paths. Thus, (5.127) and (5.130) show a stationary manifestation of the AB effect in both the particle and photon counts. (A similar conclusion follows if pure $|e_1\rangle$ states emerge from both transition regions V_J.) A conceptually significant feature of this experiment – in which the electric charge, but *not* the detected bound state, has followed a closed contour about the magnetic flux – is that it tests the state independence of minimal charge coupling and the consistency in formulation of hierarchical quantum equations of motion (a point to be discussed further below).

Suppose, however, that transition region V_1 produces pure states $|e_1\rangle$ and V_2 produces pure states $|e_2\rangle$. Then $\alpha_{21} = \alpha_{12} = 0$, and it follows from (5.122) that the probability of particle detection and the optical signal are respectively

$$P(\mathrm{D}_1) = \alpha_{11}^2 + \alpha_{22}^2, \tag{5.132}$$

$$S(\mathrm{D}_1) = \alpha_{11}^2|\mu_{1g}|^2 + \alpha_{22}^2|\mu_{2g}|^2 + 2\alpha_{11}\alpha_{22}|\mu_{1g}||\mu_{2g}|\cos\left(\Delta_{12} + 2\pi\frac{\Phi}{\Phi_0}\right), \tag{5.133}$$

where

$$\Delta_{ij} = \chi_{\mathrm{II}}^{(j)} - \chi_{\mathrm{I}}^{(i)}. \tag{5.134}$$

In this case, the particle count rate shows no quantum interference at all – which is to be expected since the excited state *spatial* paths are known. On the other hand, the AB effect is now imprinted in the optical signal, which is permissible since the exact *temporal* path from ground state to excited state and back to ground state is not known (provided the bandwidth of the detector is wider than the total spectral emission of the excited manifold). The phase Δ_{12} (in which is absorbed any supplementary contribution from the possibly complex-valued electric dipole matrix elements) contains the observation time in the term $(\omega_2 - \omega_1)t = \Omega_0 t$. Thus, the optical signal manifests quantum beats at the Bohr frequency Ω_0 with an initial phase linearly dependent upon the enclosed magnetic flux.

The essential contribution of the two coherently oscillating transition regions is to ensure production of adjustable coherent superpositions of excited states. In an alternative configuration illustrated in Fig. 5.14, the separate beam-splitting and transition regions are replaced by two coherently oscillating counter-propagating sets of laser fields, with two co-propagating waves in each set to give, as before, a total of four interaction regions. A beam of two-level ions – one ground level and one pertinent excited level – enters the interferometer in the ground state. The atomic recoil (analogous to the Kapitza–Dirac effect[4]) produced by resonant absorption and stimulated emission of photons serves to split the ion wave packet coherently in each interaction region into two components depending on the ion state. Emerging from the final laser field are two separated beams, one of ground-state ions and the other of excited-state ions. An excited state ion, however, could have been generated by a transition (with corresponding deflection) induced by the fourth laser field in the lower ground-state beam (path I), or could have originated by interaction of a ground-state ion in the upper beam (path II) with the third laser field. Under the given experimental conditions, one can

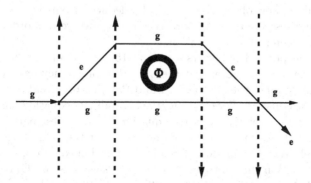

Fig. 5.14. Ion AB experiment employing an interferometer consisting of two pairs of coherently oscillating laser fields. Ions are separated according to their internal state by the momentum exchanges attendant to resonant photon absorption and emission

---

[4] The Kapitza–Dirac effect is the scattering of a particle beam from a periodic lattice of light created by a standing wave intensity pattern. See, for example, [155].

not know by which process an emerging ion was put into the excited state. Likewise for the output beam of ground-state ions. The probability that an excited- or ground-state ion emerges should therefore manifest a quantum interference with a phase dependent on the enclosed magnetic flux – provided that the process of state separation does not destroy the phase memory of the system. Such a system has in fact already been demonstrated in the construction of a neutral atom Sagnac interferometer[5] in which rotation, rather than magnetic flux, influences the relative phase. (The analogy between rotation and magnetism will be discussed in more detail in the following chapter.)

The above caveat raises an interesting general point regarding separability and coherence. It is well known that a Stern–Gerlach apparatus with a static inhomogeneous magnetic field dephases a superposition of spin substates to the extent that the substates are separated spatially [157]. The argument leading to this conclusion, however, does not apply to dynamic separation methods where time-varying fields do not literally separate previously existing states, but rather generate states in situ by inducing transitions.

Aside from the practical applications of atom interferometry – many of which apply as well to ions – there is a particular conceptual interest in examining the Aharonov–Bohm effect in a charged system with composite structure. In quantum mechanics the structure and interactions of atoms are routinely described – depending upon the question at hand – by different mathematical formulations of the equations of motion, e.g., those of the Dirac, Pauli, and Schrödinger theories. These three formulations are of course intimately related, the second derivable from the first in the approximation of small electron velocity, and the third derivable from the second upon neglect of spin. Nevertheless, each theory must be supplemented by physical assumptions prescribing how observables are to be deduced from the wave function. Furthermore, these theories did not arise in this hierarchical order, but were developed independently.

It has been pointed out years ago [158, 159], in a context different from that under examination here, that there is an inconsistency between the customary formulation of the Schrödinger and Pauli theories and results derived from the Dirac equation. For example, the Schrödinger equation has long been regarded as the nonrelativistic wave equation for a spinless particle, a seemingly reasonable interpretation since one arrives at the Schrödinger equation from a nonrelativistic reduction of the Klein–Gordon equation, as well as the Dirac, equation. On the contrary, according to Hestenes and Gurtler, the requirement that the Schrödinger theory be identical to the Pauli theory in the

---

[5] This interferometer is described in [154]. In a Sagnac interferometer two light (or particle) beams counter-propagate around the same closed path of a rotating interferometer. As a result of the Doppler effect, the recombined beam manifests an optical frequency shift proportional to the rotational angular frequency. It is of interest to note that there is a close analogy beween the Sagnac effect and the Aharonov–Bohm effect with isomorphic connection $qA/c = m\Omega \times r$ between angular velocity Ω and vector potential A. See, for example, [156].

absence of a magnetic field necessitates a different interpretation, namely that the Schrödinger equation describe – not a particle without spin – but a particle in a fixed eigenstate of spin. The difference is important and leads to different expressions for the charge current density and a re-interpretation of how the Schrödinger and Pauli currents are associated with momentum and energy. In the words of the authors [158, p. 574]:

> To put it bluntly, everyone to date has been using the wrong expression for charge current density in the Schrödinger theory. Of course there is no way that this error could be revealed directly by experiment, because the only direct experimental means of testing for the existence of a magnetization current is by introducing a magnetic field. But in that case everyone knows enough to discard the Schrödinger theory and use the Pauli or Dirac theories.

The interaction of ions with a vector potential field provides another significant circumstance in which the inconsistency between these various formalisms can not only potentially arise, but be tested. Succinctly stated, it is the replacement – required for the maintenance of gauge invariance by minimal coupling – of momentum $\boldsymbol{p}$ in the field-free Hamiltonian of a system by the operator $\boldsymbol{p} - (q/c)\boldsymbol{A}$ that ultimately leads to the unitary relation [see (1.21) and (1.22)]

$$\psi(\boldsymbol{x}) = \psi_0(\boldsymbol{x}) \exp\left(i\frac{q}{\hbar c} \int^{\boldsymbol{x}} \boldsymbol{A} \cdot d\boldsymbol{s}\right) , \tag{5.135}$$

from which arises the predicted AB phase shift in a space of appropriate topology. Here ψ_0 is a field-free solution of the Dirac equation – or of a *spinless* nonrelativistic wave equation – and ψ is the corresponding solution in the presence of a time-independent vector potential. However, the same application of minimal coupling to a nonrelativistic reduction of the Dirac equation with spin–orbit interaction (such as would apply to the internal structure of ions) leads to anomalous gauge-dependent, spin-dependent terms that, if truly present in the Hamiltonian, could conceivably alter the nature of the AB effect in ions through spin-assisted transitions.

There are theoretical reasons for believing these terms to be unwarranted, that correct implementation of gauge invariance must begin at the level of the Dirac equation, and that a proper nonrelativistic reduction of the Dirac equation does not lead in any gauge to nonvanishing spin-dependent terms under the conditions of the AB effect (where $\boldsymbol{A}$ is ordinarily assumed to be time-independent to avoid classical effects attributable to Faraday induction). Whether or not this reduction leads to gauge-invariant, spin-dependent effects that might show up under other circumstances is not, to my knowledge, a settled issue [160].

Nevertheless, for all its theoretical maturity, physics is ultimately an empirical science. Thus, even apart from the particular issue of hierarchy and gauge invariance discussed above, the uncertainty of an unexpected interaction between a vector potential field and the ionic internal degrees of freedom exists until the question is laid to rest by experiment.

Appendix 5A   Oscillatory Field Solution to the Two-State Schrödinger Equation

The oscillatory field (OF) solution to (5.45) can be expressed in the form

$$
\Psi(t) = \begin{pmatrix} \exp\left[-(\gamma_1/2 + i\omega_1)t\right] & 0 \\ 0 & \exp\left[-(\gamma_2/2 + i\omega_2)t\right] \end{pmatrix} \begin{pmatrix} J_{11} & J_{12} \\ J_{21} & J_{22} \end{pmatrix} \Psi(0) ,
$$

(5A.1)

where

$$
J_{11} = \frac{e^{(G-i\Omega)t/2}}{D} \left\{ (1 + \kappa_A^2 e^{2i\omega t})\left[(1 - \kappa_A K)^2 e^{-\mu t/2} + (\kappa_A + K)^2 e^{\mu t/2}\right] \right.
$$

$$
\left. -2(1 - e^{2i\omega t})\kappa_A(1 - \kappa_A K)(\kappa_A + K)\sinh\frac{\mu t}{2} \right\} ,
$$

(5A.2)

$$
J_{12} = \frac{-2e^{(G-i\Omega)t/2}}{D} \left\{ (1 + \kappa_A^2 e^{2i\omega t})(1 - \kappa_A K)(\kappa_A + K)\sinh\frac{\mu t}{2} \right.
$$

(5A.3)

$$
\left. -\frac{1}{2}(1 - e^{2i\omega t})\kappa_A\left[(1 - \kappa_A K)^2 e^{\mu t/2} + (\kappa_A + K)^2 e^{-\mu t/2}\right] \right\} ,
$$

$$
J_{21} = \frac{-2e^{-(G-i\Omega)t/2}}{D} \left\{ (1 + \kappa_A^2 e^{-2i\omega t})(1 - \kappa_A K)(\kappa_A + K)\sinh\frac{\mu t}{2} \right.
$$

$$
\left. +\frac{1}{2}(1 - e^{-2i\omega t})\kappa_A\left[(1 - \kappa_A K)^2 e^{-\mu t/2} + (\kappa_A + K)^2 e^{\mu t/2}\right] \right\} ,
$$

(5A.4)

$$
J_{22} = \frac{e^{-(G-i\Omega)t/2}}{D} \left\{ (1 + \kappa_A^2 e^{-2i\omega t})\left[(1 - \kappa_A K)^2 e^{\mu t/2} + (\kappa_A + K)^2 e^{-\mu t/2}\right] \right.
$$

$$
\left. +2(1 - e^{-2i\omega t})\kappa_A(1 - \kappa_A K)(\kappa_A + K)\sinh\frac{\mu t}{2} \right\} .
$$

(5A.5)

The denominator D is

$$D = (1 + \kappa_A^2)(1 + \Delta^2) , \qquad (5A.6)$$

with

$$\Delta^2 = \kappa_A^2 + K^2 + (\kappa_A K)^2 . \qquad (5A.7)$$

κ_A given by (5.62), is the off-diagonal element in the matrix C_A of (5.61) which diagonalizes the time-independent part of the Schrödinger equation in the anti-rotating frame. The function K, defined by

$$K = \frac{iV}{\varepsilon_A^+ + \lambda^- - i\omega} , \qquad (5A.8)$$

is the off-diagonal element in a matrix of the same form as in (5.61) which diagonalizes the time-independent part of the Schrödinger equation (5.64) transformed back into the rotating frame. $\varepsilon_A^\pm$ given in (5.65), are the eigenvalues obtained by diagonalization in the anti-rotating frame, and

$$\lambda^\pm = -\frac{1}{4}(\gamma_1 + \gamma_2) \pm i\mu \qquad (5A.9)$$

are the eigenvalues obtained by diagonalization in the rotating frame; μ, defined by (5.66), is interpretable as the precession frequency.

Upon approximating the radical $(1 + \vartheta^2)^{1/2}$ that appears in μ by 1, one obtains the rotating field solution, i.e., the approximate solution resulting from transformation of the original Schrödinger equation (5.45) directly into the rotating frame with subsequent neglect of all anti-resonant terms. In particular, it follows that $\lambda^\pm = \varepsilon_R^\pm$, and $K = \kappa_R$, and $\kappa_A = 0$, where the rotating-field eigenvalues are

$$\varepsilon^\pm = -\frac{1}{4}(\gamma_1 + \gamma_2) \pm i\nu , \qquad (5A.10)$$

with precession frequency ν defined by (5.57); the associated off-diagonal matrix element is

$$\kappa_R = \frac{\Theta}{\sqrt{1 + \Theta^2} + 1} , \qquad (5A.11)$$

with

$$\Theta = \frac{2iV_{12}}{G - i\Omega} , \qquad (5A.12)$$

to be compared with the corresponding equations (5.62) and (5.63).

Appendix 5B Generalized Rotating Field Theory and Optically-Induced Ground State Coherence in a 3-State Atom

The system of three states depicted in the left frame of Fig. 4.4 – one excited state and two degenerate ground states coupled by an optical perturbation of the form $V(t) = V \cos \omega t$ in the absence of a static external field – can be solved to good approximation by a generalization of the rotating field theory. A comprehensive explanation of this approach for a multilevel system of arbitrary number of coupled states is given in my book *Probing the Atom* [56], but the description can be considerably simplified for the present illustration of just three states with field-free eigenvalues (in units of $\hbar$) $\omega_1 = \omega_2 = 0$, $\omega_3 \equiv \omega_0$.

Substitution of the state vector $\Psi(t) = a_1|1\rangle + a_2|2\rangle + a_3|3\rangle$ into the Schrödinger equation with Hamiltonian (also in units of $\hbar$) $H = H_0 + V \cos \omega t$, in which $V_{12} = V_{21} = 0$ leads to the coupled equations

$$\frac{da_1}{dt} = -iV_{13}a_3 \cos \omega t \, ,$$

$$\frac{da_2}{dt} = -iV_{23}a_3 \cos \omega t \, , \tag{5B.1}$$

$$\frac{da_3}{dt} + i\omega_0 a_3 = (-iV_{31}a_1 - iV_{32}a_2) \cos \omega t \, .$$

It is assumed here that the elements of V are real, and therefore $V_{ij} = V_{ji}$. The strategy of the analysis is to remove the time-dependence from the set of equations so that the resulting transformed set can be integrated easily. To accomplish this use Euler's theorem to express the cosine in terms of exponentials, substitute into (5B.1) the redefined set of amplitudes

$$a_1 = e^{i\Omega t}b_1 \, ,$$

$$a_2 = e^{i\Omega t}b_2 \, , \tag{5B.2}$$

$$a_3 = e^{-i\omega_0 t}b_2 \, ,$$

where $\Omega = \omega - \omega_0$ is the displacement from resonance, and discard terms that oscillate at the antiresonant frequency $\omega + \omega_0$. The original set of equations then reduces to the time-independent set

$$\frac{db_1}{dt} = -i\Omega b_1 - i\frac{1}{2}V_{13}b_3 \, ,$$

$$\frac{db_2}{dt} = -i\Omega b_2 - i\frac{1}{2}V_{23}b_3 \, , \tag{5B.3}$$

$$\frac{db_3}{dt} = -i\frac{1}{2}V_{13}b_1 - i\frac{1}{2}V_{23}b_2 \, ,$$

which can be written in matrix form as

$$\frac{d\boldsymbol{b}}{dt} = -iM\boldsymbol{b} \, . \tag{5B.4}$$

To integrate (5B.4), one must transform M into diagonal form D

$$M = CDC^{-1} \, , \tag{5B.5}$$

where C is a matrix whose columns are the eigenvectors of M. The integration of (5B.4) then follows immediately

$$b(t) = Ce^{-iDt}C^{-1}b(0) , \qquad (5B.6)$$

where

$$(e^{-iDt})_{ij} = e^{-iD_{ii}t}\delta_{ij} ,$$

and the initial amplitudes are given by $b(0)$. For the system in question, I assume that both ground states are initially incoherently populated, i.e.,

$$b(0) = \frac{1}{\sqrt{2}} \begin{pmatrix} 1 \\ e^{i\phi} \\ 0 \end{pmatrix} , \qquad (5B.7)$$

where ϕ is a distributed phase.

Solution of the eigenvalue problem for M yields the eigenvalues

$$\varepsilon_0 = \Omega , \qquad \varepsilon_\pm = \frac{1}{2}\left(\Omega \pm \sqrt{\Omega^2 + V_{13}^2 + V_{23}^2}\right) \equiv \frac{1}{2}(\Omega \pm \Delta) \qquad (5B.8)$$

and corresponding matrices

$$C = \frac{1}{2} \begin{pmatrix} V_{23} & V_{13} & -V_{13} \\ -V_{13} & V_{23} & -V_{23} \\ 0 & \Delta - \Omega & \Delta + \Omega \end{pmatrix} , \qquad (5B.9)$$

$$C^{-1} = \begin{pmatrix} \dfrac{2V_{23}}{V_{13}^2 + V_{23}^2} & \dfrac{-2V_{13}}{V_{13}^2 + V_{23}^2} & 0 \\ \dfrac{V_{13}}{V_{13}^2 + V_{23}^2}\left(1 + \dfrac{\Omega}{\Delta}\right) & \dfrac{V_{23}}{V_{13}^2 + V_{23}^2}\left(1 + \dfrac{\Omega}{\Delta}\right) & \dfrac{1}{\Delta} \\ \dfrac{-V_{13}}{V_{13}^2 + V_{23}^2}\left(1 - \dfrac{\Omega}{\Delta}\right) & \dfrac{-V_{23}}{V_{13}^2 + V_{23}^2}\left(1 - \dfrac{\Omega}{\Delta}\right) & \dfrac{1}{\Delta} \end{pmatrix} . \qquad (5B.10)$$

Returning to the 'a' amplitudes by substituting the inverse transformation of (5B.2) into (5B.6) and (5B.7), and performing the matrix multiplication leads to a solution of the form

$$a_i(t) = C_{i1}C_{1i}^{-1} + C_{i2}C_{2i}^{-1}e^{i(\Omega - \Delta)t/2} + C_{i3}C_{3i}^{-1}e^{i(\Omega + \Delta)t/2} , \qquad (5B.11)$$

for $i = 1, 2, 3$, from which the elements

$$\rho_{ij}(t) = \overline{a_i a_j^*} \qquad (5B.12)$$

of the density matrix are constructed.

With regard to the free-induction decay experiment confirming the prediction of saturation regeneration of quantum beats, discussed in the preceding chapter, the density matrix element of interest is the time-independent part of $\rho_{12}(t) = \overline{a_1 a_2^*}$. Only those terms contribute, therefore, that result from multiplying a phase factor in (5B.11) by its complex conjugate. The final result is then

$$\rho_{12}^0 = \frac{V_{13} V_{23}/4}{\Omega^2 + V_{13}^2 + V_{23}^2} . \tag{5B.13}$$

It bears emphasizing that the solution (5B.6) and (5B.7), from which the entire density matrix of the system follows, is not a first-order perturbation or weak-pumping solution. The only approximation that has been made is to drop antiresonant terms (generalized rotating field approximation). The procedure outlined here can be applied as well to a system with unstable excited states and finite ground-state lifetime resulting from optical pumping, although the resulting equations will of course be more complicated.

Symmetries and Insights:
The Circulating Electron
in Electromagnetic Fields

6.1 Broken Symmetry of the Charged Planar Rotator

In the atomic model introduced by Niels Bohr in 1913 the electron was presumed to follow a circular orbit about the atomic nucleus. Although the nonrelativistic quantum theory of the hydrogen atom later confirmed the Bohr
atom energy spectrum (in absence of spin-related relativistic interactions), it
also showed that the concept of orbits was in general no longer tenable. The
bound electron could have any possible separation from the nucleus according
to a distribution function determined from the radial wave equation.

Nevertheless, the two-dimensional charged rotator (Fig. 6.1) – superficially
a Bohr atom without nucleus in which the electron is confined by unspecified
forces to a circular space of radius R – is an interesting quantum system to
consider. The orbiting particle does not, of course, follow a trajectory just
like that of the electron in the Bohr atom. A more appropriate, albeit ap-

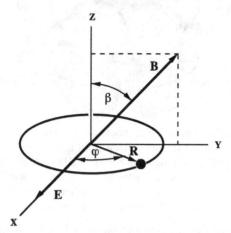

Fig. 6.1. Configuration of the electric (E) and magnetic (B) fields for the Stark
and Zeeman effects of the two-dimensional charged rotator circulating in a space of
radius R

proximate, real-world counterpart to this model system is a pi electron in a conjugated hydrocarbon ring,[1] as shown in Fig. 6.2. Indeed, the planar rotator (with concomitant application of the Pauli exclusion principle) yields a moderately successful prediction of the center of gravity of the rotational Raman spectrum of the hexagonal molecule benzene, the archetype of conjugated aromatic molecules [161].

The field-free planar rotator is a quantum system that, like the particle in a rectangular box with infinite walls, is trivially easy to analyze. The spectral degeneracy can be predicted simply on the basis of circular symmetry. Since the two possible classical motions of clockwise and counterclockwise rotation at the same angular frequency are energetically equivalent, all rotationally excited levels are doubly degenerate. The ground state, in which the angular momentum, and therefore the rotational energy, is zero, is nondegenerate. It is worth noting that there is no zero-point energy as, for example, in the case of the harmonic oscillator, since the electron is not bound in a potential well but occupies the entire space available to it. This space happens to be a one-dimensional space of constant curvature. From the form of the Hamiltonian for a particle with mass m confined to a ring of radius R

$$H_0 = \frac{L_z^2}{2mR^2} \, , \tag{6.1}$$

it follows immediately that the state vectors can be labeled by an integer quantum number μ (with $\mu = 0, \pm1, \pm2$, etc.) designating the eigenvalues $\mu\hbar$ of

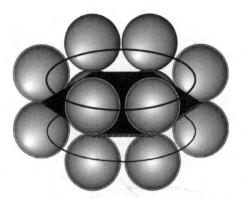

Fig. 6.2. The benzene molecule C_6H_6. Each vertex represents a carbon atom; the hydrogen atoms (one bonded to each carbon) are not shown. Sigma bonds, strong linkages of carbon $2s$ electrons directed between adjacent atoms in the molecular plane hold the framework together. Weaker pi bonds, linkages of carbon $2p$ electrons oriented normal to the plane, provide a ring-like network (indicated schematically by the rings) over which the pi electrons are delocalized

---

[1] The diagrammatic representation of a 'conjugated' molecule like benzene (C_6H_6) shows alternating single and double bonds, but, as a result of the pi electron delocalization, any pair of adjacent carbon atoms in the ring is indistinguishable from any other pair.

L_z, the component of angular momentum along the rotation axis (and, in fact, the only component of angular momentum). The energy eigenvalues are then

$$E_\mu = E_{-\mu} = \frac{\hbar^2 \mu^2}{2mR^2} \,, \tag{6.2}$$

with associated normalized wave functions

$$\langle \phi | \mu \rangle \equiv \psi_\mu(\phi) = \frac{e^{i\mu\phi}}{\sqrt{2\pi}} \tag{6.3}$$

that satisfy the boundary condition

$$\psi_\mu(\phi + 2\pi) = \psi_\mu(\phi) \tag{6.4}$$

required for single-valued functions. The azimuthal angle ϕ is measured in a positive sense from some arbitrarily chosen axis (which is usually the x axis of a Cartesian coordinate system). The states represented by wave function (6.3) are also eigenstates of the parity operator Π which reflects the particle through the origin ($\phi \to \phi + \pi$) with eigenvalues $e^{i\mu\pi} = (-1)^\mu$. Thus, states of even or odd μ respectively have even or odd parity. It should also be noted, in anticipation of an important point to be discussed shortly, that any linear superposition of states of fixed $|\mu|$ is also a valid solution of the Schrödinger equation.

With the introduction of external electric or magnetic fields or electro-magnetic potentials, the problem of the charged planar rotator ceases to be a trivial one [162, 163], and indeed full, correct solutions of these systems were first published long after the development of quantum theory [164, 165]. Although the system may still seem simple, this simplicity is somewhat de-ceiving. As one of few quantum mechanical systems involving electromagnetic interactions amenable to exact solution, there is much of conceptual impor-tance that one can learn from studying it. For one thing, the Schrödinger equations for both the Stark and Zeeman effects lead to differential equations of a type not frequently encountered in quantum mechanics: one with peri-odic coefficients and a three-term recursion relation. Such an equation is not expressible in terms of the generators of a particular symmetry group [known as the SU(1,1) dynamical group [166]] as are those of a number of commonly encountered solvable quantum systems, as for example the Coulomb and har-monic oscillator potentials. More interesting still, the energy spectrum of the resulting Stark and Zeeman states exhibits an unusual broken degeneracy pat-tern. Indeed, in at least one standard reference, the author had not suspected that the degeneracy is broken at all [167].

The planar rotator in a vector potential field in the *absence* of all elec-tric and magnetic fields – in other words, the bound-state counterpart to the AB effect – is a particularly interesting model system for examining another problematic aspect of the nonclassical effects of potentials. At issue is not the flux dependence of a diffraction pattern or scattering cross section, but rather the angular momentum spectrum, rotational behavior, and statistics of particles in rotational motion about tubes of magnetic flux [168, 169]. Some have proposed that a charged particle orbiting a magnetic flux tube can have a flux-dependent – and therefore arbitrary – angular momentum, in which case the wave function would acquire a flux-dependent rotational phase factor. An ensemble of such systems would follow statistics that interpolate between boson and fermion statistics [170]. Others have maintained instead that the orbital angular momentum spectrum remains integer-valued, and the wave function of the rotated particle is unaffected by the flux [171]. A collection of composite rotator–flux systems would then obey the traditional quantum statistics determined by the spin of the particle. Thus, as with earlier controversies of the AB effect, these, too, raise questions of fundamental import that transcend in scope the particularities of the individual systems under study. These questions concern, for example, the role, spectrum, and observability of various gauge invariant and non-invariant dynamical quantities in quantum mechanics.

In the following sections we will examine the quantum behavior of a two-dimension rotator in the presence of electric and magnetic fields and the interaction of a rotator with a vector potential in an otherwise field-free space.

6.2 The Planar Rotator in an Electric Field

Let us start with Stark effect. Without loss of generality consider the electron (of charge $q = -e$) to be confined to a circle of radius R in the x–y plane centered on the z axis. Since only components of an electric field E within the orbital plane can influence the state of motion of the electron, we orient E along the x axis as shown in Fig. 6.1. The Hamiltonian characterizing the total rotational and electric energy of a two-dimensional rotator with electric dipole moment

$$\boldsymbol{\mu}_{\mathrm{E}} = q\boldsymbol{r} = qR(\cos\phi, \sin\phi, 0) \tag{6.5}$$

is then

$$H = H_0 + H_{\mathrm{E}} = \frac{L_z^2}{2mR^2} - qEx , \tag{6.6}$$

or, in a coordinate representation,

$$H = -\varepsilon \left(\frac{\partial^2}{\partial\phi^2} - \lambda\cos\phi \right) , \tag{6.7}$$

with energy parameter

$$\varepsilon = \frac{\hbar^2}{2mR^2} \tag{6.8}$$

and electric interaction parameter (which is positive and dimensionless)

$$\lambda = \frac{2meR^3E}{\hbar^2} . \tag{6.9}$$

One may be tempted on the basis of common textbook examples of perturbation theory to predict that the perturbed energy spectrum of the planar rotator would display a quadratic Stark effect in the nondegenerate ground state (i.e., energy increases as E^2), and a linear Stark effect in the doubly degenerate excited states (energy of one state increases and that of the other decreases as E). This description of the degeneracy breaking, however, is not correct. The matrix elements of H_E between the unperturbed states are readily calculated to be

$$\langle \mu | H_E | \mu' \rangle = \frac{1}{2} eER \left(\delta_{\mu,\mu'+1} + \delta_{\mu,\mu'-1} \right) . \tag{6.10}$$

As expected, the first-order contribution to the ground state vanishes since H_E has no diagonal matrix elements. It also follows from relation (6.10), perhaps surprisingly, that there can be no first-order splitting of the degenerate pairs $|\pm|\mu|\rangle$. For splitting to occur, as in the case of the degenerate $2S$, $2P(m = 0)$ states of hydrogen (in the absence of spin), the degenerate states must be coupled by a matrix element of the interaction. In the case of hydrogen, the element $\langle 2P(m = 0)|z|2S \rangle$ is not zero. In the case of the planar rotator, however, the matrix element $\langle -\mu | x | \mu \rangle$ vanishes unless $-\mu = \mu \pm 1$, leading to $\mu = \pm 1/2$. Since there are no half-integer eigenvalues in the spectrum of the orbital angular momentum L_z, breaking of the excited state degeneracy must occur at least quadratically in E.

Additional physical insight can be gained by considering the relationships between parity and reflection symmetry and the electric dipole moment. The ground states of both the planar rotator and the hydrogen atom are parity eigenstates. Hence, the expectation value of the electric dipole moment in these states vanishes identically because

$$\langle 0 | \boldsymbol{\mu}_E | 0 \rangle = \langle 0 \Pi | (\Pi \boldsymbol{\mu}_E \Pi) | \Pi 0 \rangle = \langle 0 | - \boldsymbol{\mu}_E | 0 \rangle = 0 , \tag{6.11}$$

where $\Pi^2 = \Pi$. However, the external electric field can induce a dipole moment in the nondegenerate ground state proportional to E, in which the proportionality constant is the polarizability α. The induced moment then couples to the field giving an electric energy proportional to E^2. Classically, this energy is given by $-\alpha E^2/2$.

For degenerate states a different mechanism is possible. Although two degenerate eigenstates of the Hamiltonian may each be a parity eigenstate, linear combinations of these states need not be. Thus, for example, the hydrogen $2S$ and $2P(m = 0)$ states have parity eigenvalues $+1, -1$ respectively, but the degenerate linear superpositions

$$\frac{1}{\sqrt{2}} \left[|2S\rangle \pm |2P(m = 0)\rangle \right]$$

are obviously not parity eigenvectors. They form field-free degenerate eigenstates with a nonvanishing mean electric dipole moment. The external field can couple directly to these dipole moments, and the energy acquired by their alignment is therefore linear in E.

For the two-dimensional rotator the comparable linear combinations

$$|\mu\pm\rangle = \frac{1}{\sqrt{2}}\Big(|\mu\rangle \pm |-\mu\rangle\Big)$$

are also not parity eigenstates. However, unlike the hydrogenic states, they do not represent zero-field states with a nonvanishing electric dipole moment. The reason for this is traceable to the reflection symmetry of the Hamiltonian (6.6) and (6.7). Under the operations Π_x and Π_y for reflection across the x and y axes, the angular coordinate ϕ undergoes the changes $(\phi \to -\phi)$ and $(\phi \to \pi - \phi)$, respectively, and

$$\Pi_x|\mu\pm\rangle = \pm|\mu\pm\rangle , \tag{6.12}$$

$$\Pi_y|\mu\pm\rangle = \pm e^{i\mu\pi}|\mu\pm\rangle . \tag{6.13}$$

It is then straightforward to show by steps analogous to those in (6.11) that

$$\langle\mu\pm|x|\mu\pm\rangle = \langle\mu\pm|x|\mu\mp\rangle = 0 . \tag{6.14}$$

Because the $|\mu\pm\rangle$ are linearly independent states, there is no possible linear combination of them that can lead to a nonvanishing electric dipole moment. Unlike the case of the hydrogen atom, therefore, the symmetry of the planar rotator precludes a linear Stark effect. It is worth remarking at this point that the relations summarized in (6.14) are precisely the mathematical conditions for the failure of first-order degenerate perturbation theory to lift the degeneracy of a pair of states.

Let us examine more quantitatively the application of perturbation theory to this system. The states are to be labeled as $|\eta_{\mu\pi}\rangle$, where $\eta^2 \to \mu^2$ for $\lambda \to 0$ (vanishing electric field) and $\pi = \pm$ is the eigenvalue under the reflection Π_x. The energy eigenvalues can be written in the form

$$E_{\mu\pi} = \varepsilon\eta_{\mu\pi}^2 , \tag{6.15}$$

and are therefore completely specified by the dimensionless functions $\eta_{\mu\pi}(\lambda)$.

Upon application of standard second-order perturbation theory [172], one obtains

$$\eta_{0+} = -\lambda^2/2 \tag{6.16}$$

for the ground-state. Only the states with $\mu = \pm 1$ contribute to the infinite summation over unperturbed eigenstates. In view of the preceding discussion, the ground-state polarizability of the two-dimensional rotator is $\alpha_0 = 2R^4/a_0$, where $a_0 = \hbar^2/me^2$ is the Bohr radius; a planar rotator the size of a Bohr atom has twice the polarizability.

Applying second-order *degenerate* perturbation theory to the excited states requires solving the secular equation of a 2×2 matrix where each of the elements is an infinite summation over the unperturbed eigenstates.

The two solutions of the resulting quadratic equation yield the second-order correction to the energy of the two degenerate states. For the $\mu = \pm 1$ states one obtains

$$\eta_{1+} = 1 + 5\lambda^2/12 \,, \tag{6.17}$$

$$\eta_{1-} = 1 - \lambda^2/12 \,, \tag{6.18}$$

which leads to a level separation of $\lambda^2/2$. For all higher states, $|\mu| > 1$, the energy to order E^2 is determined from the equation

$$\eta_{\mu\pm}^2 = \mu^2 + \frac{\lambda^2/2}{4\mu^2 - 1} \qquad (|\mu| > 1) \,, \tag{6.19}$$

which is a function of the square of μ. It may seem surprising that to second order in E the degeneracy of only the $\mu = \pm 1$ states is lifted. I will comment on this matter in greater detail shortly.

To remove the degeneracy of the states with $|\mu| > 1$ one can try degenerate perturbation theory in still higher orders, but the calculations rapidly become tedious. It is better, instead, to reconsider the entire Schrödinger equation and cast it in a form such that the properties of the exact solution may be obtained most readily. Substitution into the Hamiltonian (6.7) of

$$\phi = 2\theta \,,$$
$$a = 4\mu^2 \,, \tag{6.20}$$
$$q = 2\lambda$$

brings the Schrödinger equation into the form

$$\left[\frac{\mathrm{d}^2}{\mathrm{d}\theta^2} + (a - 2q\cos 2\theta) \right] \psi_\mu(\theta) = 0 \,, \tag{6.21}$$

recognizable to mathematicians as the canonical form of Mathieu's equation [173]. [The specific context should eliminate any confusion between use of q for electric charge or for the Mathieu parameter defined in (6.20).]

The Mathieu equation has its origin in investigations of vibrating systems with elliptical boundary conditions and is not uncommonly encountered in wave propagation problems in electromagnetism and acoustics, as well as electron propagation through crystal lattices [174]. Although the above linear differential equation with periodic coefficients is invariant under the transformation $\theta \to \theta + \pi$, the solutions need not be, a point of interest in view of the significance to physics of spontaneous symmetry breaking in quantum field theory. When the parameter a assumes certain characteristic values (related to the energy eigenvalues in the present case), the solution, known as Mathieu functions, are periodic with periods of either π or 2π; for arbitrary a the solutions are not periodic (and are not called Mathieu functions). Upon substitution of $z = \cos(2\theta)$, the equation assumes the so-called Lindemann form which resembles somewhat the hypergeometric equation [175]. It is quite different, however, having two regular singularities at 0 and 1 and an irregular singularity at ∞. This equation is not one of the factorization types expressible in terms of the generators of the SU(1,1) dynamical group.

The physical requirement that a quantum mechanical wave function be single-valued (a limitation that does not necessarily apply in a multi-connected space, as will be seen shortly) and continuous restricts our attention to the periodic solutions or Mathieu functions. These solutions can be classified into four types on the basis of their periodicity and reflection symmetry under Π_x. They are traditionally designated as follows (where $k = 0, 1, 2, \ldots$):

- $\mathrm{ce}_{2k}(\theta, q)$: even solutions of period π that reduce to $\cos(2k\theta)$ as $q \to 0$; associated eigenvalues are $a = a_{2k}$.
- $\mathrm{ce}_{2k+1}(\theta, q)$: even solutions of period 2π that reduce to $\cos\left[(2k+1)\theta\right]$ as $q \to 0$; associated eigenvalues are $a = a_{2k+1}$.
- $\mathrm{se}_{2k+1}(\theta, q)$: odd solutions of period π that reduce to $\sin\left[(2k+1)\theta\right]$ as $q \to 0$; associated eigenvalues are $a = b_{2k+1}$.
- $\mathrm{se}_{2k+2}(\theta, q)$: odd solutions of period 2π that reduce to $\sin\left[(2k+2)\theta\right]$ as $q \to 0$; associated eigenvalues are $a = b_{2k+2}$.

The symbols ce and se derive from E.T. Whittaker's designations cosine-elliptic and sine-elliptic, respectively.

Each of the above solutions can be represented by an infinite Fourier series of either $\cos(j\theta)$ or $\sin(j\theta)$ – depending on the reflection symmetry – where the integer j is of the same form $(2k,\ 2k+1,\ 2k+2)$ as that which characterizes the solution. Thus, for example, the even-index Mathieu functions can be represented as

$$\mathrm{ce}_{2K}(\theta, q) = \sum_{k=0}^{\infty} A_{2k}^{2K} \cos(2k\theta) \,, \tag{6.22}$$

$$\mathrm{se}_{2K+2}(\theta, q) = \sum_{k=0}^{\infty} B_{2k+2}^{2K+2} \sin\left[(2k+2)\theta\right] \,. \tag{6.23}$$

The determination of the Fourier coefficients and eigenvalues can be accomplished to arbitrary accuracy through more powerful mathematical techniques than perturbation theory. Two such methods employ continued fractions and the solution of Hill's determinantal equation. There is no general closed-form expression for the eigenvalues, as is the case with the hypergeometric equation and its related forms. This is a consequence of the fact that the Mathieu equation leads to an irreducible three-term, rather than two-term, recursion relation.

With regard to the planar rotator it is important to recognize that the physically significant angular variable is ϕ, not θ. Thus, the criterion of wave function continuity, $\psi(\phi + 2\pi) = \psi(\phi)$, implies that $\psi(\theta + \pi) = \psi(\theta)$, in which case the admissible solutions are only those Mathieu functions with periodicity π, i.e., $\mathrm{ce}_{2k}(\theta, q)$ and $\mathrm{se}_{2k+2}(\theta, q)$. From relations (6.15), (6.20), and the form of the solutions (6.22) and (6.23), it is clear that the eigenfunctions of the Hamiltonian and the associated energy eigenvalues can be written as

$$\psi_{\mu+}(\phi) = \mathrm{ce}_{2\mu}\left(\frac{1}{2}\phi, \lambda\right), \qquad E_{\mu+} = \frac{1}{4}\varepsilon a_{2\mu} \qquad (\mu = 0, 1, 2, 3, \ldots), \quad (6.24)$$

$$\psi_{\mu-}(\phi) = \mathrm{se}_{2\mu}\left(\frac{1}{2}\phi, \lambda\right), \qquad E_{\mu-} = \frac{1}{4}\varepsilon b_{2\mu} \qquad (\mu = 0, 1, 2, 3, \ldots). \quad (6.25)$$

In the limit of vanishing electric field ($\lambda \rightarrow 0$), the above solutions reduce to $\cos(\mu\phi)$ and $\sin(\mu\phi)$ respectively, i.e., to the coordinate representation of the basis $|\mu\pm\rangle$. This is an important point, for it shows that even though the original basis $|\pm|\mu|\rangle$ is uncoupled by H_E, the choice of which linear superpositions of degenerate states are required for the analytical continuity of a perturbation series is not arbitrary. We shall return to this point again.

From the characteristic values of Mathieu functions (summarized to order q^6 in Table 6.1) one can verify the eigenvalues η_0, $\eta_{1\pm}$ previously obtained by perturbation theory and readily derive the energies of the next two excited states

$$\eta_{2+}^2 = 4^2 + \frac{1}{30}\lambda^2 + \frac{433}{216\,000}\lambda^4 + \cdots, \qquad (6.26)$$

$$\eta_{2-}^2 = 4^2 + \frac{1}{30}\lambda^2 - \frac{317}{216\,000}\lambda^4 + \cdots. \qquad (6.27)$$

Table 6.1. Characteristic values of Mathieu functions (to order q^6)

$$a_0 = -\frac{1}{2}q^2 + \frac{7}{128}q^4 - \frac{29}{2304}q^6$$

$$a_1 = 1 + q - \frac{1}{8}q^2 - \frac{1}{64}q^3 - \frac{1}{1536}q^4 + \frac{11}{36\,864}q^5 + \frac{49}{589\,824}q^6$$

$$b_1 = (\text{substitute } -q \text{ for } q)$$

$$a_2 = 4 + \frac{5}{12}q^2 - \frac{763}{13\,824}q^4 + \frac{1\,002\,401}{79\,626\,240}q^6$$

$$b_2 = 4 - \frac{1}{12}q^2 + \frac{5}{13\,824}q^4 - \frac{289}{79\,626\,240}q^6$$

$$a_3 = 9 + \frac{1}{16}q^2 + \frac{1}{64}q^3 + \frac{13}{20\,480}q^4 - \frac{5}{16\,384}q^5 - \frac{1961}{23\,592\,960}q^6$$

$$b_3 = (\text{substitute } -q \text{ for } q)$$

$$a_4 = 16 + \frac{1}{30}q^2 + \frac{433}{864\,000}q^4 - \frac{5\,701}{2\,721\,600\,000}q^6$$

$$b_4 = 16 + \frac{1}{30}q^2 - \frac{317}{864\,000}q^4 + \frac{10\,049}{2\,721\,600\,000}q^6$$

$$a_5 = 25 + \frac{1}{48}q^2 + \frac{11}{774\,144}q^4 + \frac{1}{147\,456}q^5 + \frac{37}{891\,813\,888}q^6$$

$$b_5 = (\text{substitute } -q \text{ for } q)$$

$$a_6 = 36 + \frac{1}{70}q^2 + \frac{187}{43\,904\,000}q^4 + \frac{6\,743\,617}{92\,935\,987\,200\,000}q^6$$

$$b_6 = 36 + \frac{1}{70}q^2 + \frac{187}{43\,904\,000}q^4 - \frac{5\,861\,633}{92\,935\,987\,200\,000}q^6$$

The unusual broken degeneracy pattern evident in the expressions for the $\mu = 0, 1, 2$ manifolds does in fact extend to all excited levels of the planar rotator: each field-free degenerate pair of states of fixed μ is broken only in order $E^{2\mu}$. Thus, degenerate perturbation theory in sixth order would be needed to break the degeneracy within the $\mu = 3$ level. Upon lifting of the degeneracy, it is the states of odd reflection parity that lie lower in energy.

From Fig. 6.3, which shows the characteristic curves of the Mathieu functions, i.e., the variation of a_μ and b_μ with q, it is seen that the greater the value of $|\mu|$, the greater is the field strength required to effect a noticeable separation between the field-free degenerate states. For sufficiently large values of E, the interaction with the electric field dominates the Hamiltonian, and one would then expect the energy to become negative. This expectation is borne out in Fig. 6.3. In the high-q region of the diagram the eigenvalue expansions are no longer valid, and one must utilize the asymptotic properties of the Mathieu functions

$$a_{2k}(q) = b_{2k+1}(q) \sim -2q + (8k+2)q^{1/2} , \tag{6.28}$$

$$a_{2k+1}(q) = b_{2k+2}(q) \sim -2q + (8k+6)q^{1/2} , \tag{6.29}$$

for $k = 0, 1, 2, \ldots$, and q large and positive, to determine the energy spectrum. The resulting energy eigenvalues take the form

$$E_{\mu\pm} = -eRE + \left(\mu \pm \frac{1}{2}\right)\hbar\left(\frac{eE}{mR}\right)^{1/2} , \tag{6.30}$$

which is amenable to a simple physical interpretation. The first term, which is negative and dominates the expression for large E, represents the energy of

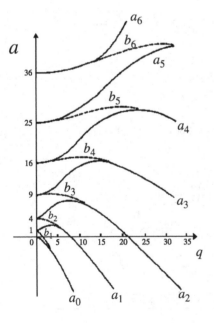

Fig. 6.3. Variation of the Mathieu functions a_μ and b_μ as a function of the parameter q which, for the two-dimensional rotator, depends on the electric or magnetic field strength. (Adapted from Silverman [162])

a classical electric dipole $\mu_E = eR$ aligned parallel to the field. The second term characterizes a harmonic oscillator spectrum with equidistant level separations of $\hbar\omega$ where

$$\omega = \left(\frac{eE}{mR}\right)^{1/2} = \left(\frac{\mu_E E}{I}\right)^{1/2} \tag{6.31}$$

is the oscillation angular frequency of a dipole μ_E with moment of inertia $I = mR^2$. The reason for this behavior will become apparent when we examine the eigenfunctions.

Before concluding this discussion of the energy spectrum of the planar rotator in an electric field, I would like to clarify an important general point concerning degenerate perturbation theory and the symmetry properties of the Hamiltonian. In particular, consider the seemingly paradoxical question: Why does one use degenerate perturbation theory when treating the interactions of degenerate states? It would seem that the answer is obvious. Nondegenerate perturbation theory fails, as is signaled by the appearance of terms with vanishing energy denominators. Degenerate perturbation theory eliminates these terms through construction of linear combinations of unperturbed states for which the matrix elements in the numerator likewise vanish. Thus, singular terms do not appear.

Suppose, however, that the degenerate basis with which one begins is already uncoupled by the interaction; this is the case with the planar rotator since $\langle -\mu|H_E|\mu\rangle = 0$. Under these circumstances, where singularities do not occur, it may seem – and the pedagogical literature has encouraged this view – that it is permissible to use nondegenerate perturbation theory. This procedure cannot be correct, for the application of nondegenerate perturbation theory to the two-dimensional rotator leads to energy eigenvalues that are functions of μ^2 and therefore remain doubly degenerate to all orders. The error lies in the failure to recognize the *primary objective* of degenerate perturbation theory which is not merely to eliminate singular terms, but to select unambiguously the appropriate linear superposition of degenerate states from among an infinite number of possibilities. To start with any other combination leads to a discontinuous change of states in the limit of vanishing interaction parameter, so that the perturbative expansions are not valid.

But how is one to know whether or not a particular interaction term–let us call it H' – removes the original degeneracy in some order? This knowledge is presumably the outcome of the calculation – yet which calculation one performs (degenerate vs. nondegenerate perturbation theory) seems to depend on it. For the case of the planar rotator the fact that the eigenvalues of degenerate perturbation theory agree with the characteristic values of the Mathieu functions is proof that the degeneracy is indeed lifted and that one must construct an appropriate initial basis $\left(\cos(\mu\phi), \sin(\mu\phi)\right)$ even though the matrix elements between the original basis $(e^{i\mu\phi}, e^{-i\mu\phi})$ vanish. What must one do, however, when an exact solution is not known? Would it actually have been necessary to employ eighth order perturbation theory to discover that the degeneracy is lifted between $\mu = 4$ states?

It is at this point where the theory of groups comes to the rescue. One must examine the symmetries of the unperturbed Hamiltonian H_0 and the perturbation H' in order to ascertain whether the original degenerate basis, which spans an irreducible representation of the symmetry group H_0, remains an irreducible representation of the (possibly lower) symmetry group of $H_0 + H'$.

In the case of the planar rotator the Hamiltonian $H_0 = L_z^2/2mR^2$ is invariant under the elements of the continuous group SO(2), the group of rotations in the plane. It is also invariant under the discrete symmetry group comprising the identity I, the reflections Π_x and Π_y, and the inversion Π; this group is isomorphic to the four-group ($Vierergruppe$) V. The total symmetry of H_0 is the two-dimensional rotation–reflection group, a mixed continuous group that can be parameterized by a continuous parameter ϕ (the rotation angle) and a discrete parameter $d = \pm 1$ (a determinant). One matrix representation of this group is

$$\{\phi, d\} = \begin{pmatrix} \cos\phi & \sin\phi \\ -d\sin\phi & d\cos\phi \end{pmatrix} , \tag{6.32}$$

where

$$\{\phi, d\} \cdot \{\phi', d'\} = \{d'\phi + \phi', dd'\} \tag{6.33}$$

gives the group composition function [176]. This group is clearly non-Abelian (i.e., the group operations do not commute) although the subgroups SO(2) and V are Abelian. The basis $e^{\pm i\mu\phi}$ spans a two-dimensional representation for $\mu \neq 0$.

Upon addition of an electric field $\boldsymbol{E}$ perpendicular to the rotation axis the total Hamiltonian $H = H_0 + H'$ (with $H' = -\boldsymbol{\mu}_E \cdot \boldsymbol{E}$) is invariant under a much smaller symmetry group comprising only the elements I and Π_x. This group is isomorphic to the so-called symmetric or permutation group S_2 which is Abelian and therefore can have only one-dimensional representations. There are two such representations designated respectively Γ^S (which is symmetric under Π_x) and Γ^A (which is antisymmetric under Π_x). A simple character analysis shows that if $\Gamma_{\mu d}$ is an irreducible representation of the rotation–reflection group, then $\Gamma_{\mu d} \rightarrow \Gamma^S + \Gamma^A$ upon the lowering of the symmetry to S_2 by addition of the electric field. Thus, the degenerate basis splits under the action of $\boldsymbol{E}$; the exact solutions $ce_{2\mu}(\phi/2, \lambda)$ and $se_{2\mu}(\phi/2, \lambda)$ display the group-theoretically predicted symmetries.

Group theoretical arguments cannot in general give the scale of an interaction, i.e., they cannot yield the magnitude of the splitting. The power of group theory is that it can give the degeneracy pattern which, moreover, is valid to all orders of perturbation theory. An important example is the three-dimensional rigid rotator in an electric field along the rotation axis [177]. Group theory shows that the degeneracy of states with different angular momentum quantum numbers l and $|\mu|$ is lifted, but, for given l, there is a residual degeneracy of states differing only in the sign of μ; this degeneracy is never lifted as long as the axial symmetry is not broken.

Mention should also be made of so-called accidental degeneracies [178], i.e., the occurrences of degenerate states belonging to different irreducible representations of the symmetry group of the Hamiltonian. The hydrogen atom, or more precisely the spinless Coulomb problem, is a well-known example; the degeneracy among the n^2 states of fixed principal quantum number n and different orbital quantum numbers l cannot be explained on the basis of the most obvious symmetry group, SO(3), the three-dimensional rotation group. In such cases one can usually find a larger invariance group that does account for the degeneracies. For hydrogen this group is SO(4), the four-dimensional rotation group. Accidental degeneracy, while characteristic of a specific dynamical law, is not necessarily an unexplainable degeneracy.

In short, then, if by group theoretical or other means it is clear that an interaction will lift the degeneracy of an initially degenerate set of states, nondegenerate perturbation theory cannot be used even though the states of the basis may not be directly coupled by the interaction.

Let us consider next the wave functions and corresponding electron distributions of the planar rotator. The exact solutions (to within a normalization constant) are the Mathieu functions given by (6.24) and (6.25) which are representable in the following Fourier decompositions to order q^2

$$ce_k(\phi, q) = \cos k\phi - \frac{1}{4}q \left[\frac{\cos(k+2)\phi}{k+1} - \frac{\cos(k-2)\phi}{k-1} \right] \tag{6.34}$$
$$+ \frac{1}{32}q^2 \left[\frac{\cos(k+4)\phi}{(k+1)(k+2)} + \frac{\cos(k-4)\phi}{(k-1)(k-2)} \right] ,$$

$$se_k(\phi, q) = \sin \phi - \frac{1}{4}q \left[\frac{\sin(k+2)\phi}{k+1} - \frac{\sin(k-2)\phi}{k-1} \right] \tag{6.35}$$
$$+ \frac{1}{32}q^2 \left[\frac{\sin(k+4)\phi}{(k+1)(k+2)} + \frac{\sin(k-4)\phi}{(k-1)(k-2)} \right] ,$$

valid for $q^2/[2(k^2 - 1)] \ll k^2$ with $k > 0$. To first order in q (or λ) these expansions correspond to the results of first-order perturbation theory for which, as emphasized previously, analytical continuity at $\lambda = 0$ requires the use of the appropriate zero-field states $|\mu\pm\rangle$.

The ground-state wave function, obtained from first-order perturbation theory, is

$$\psi_{0+} = \frac{1}{\sqrt{2\pi}}(1 - \lambda \cos \phi) \tag{6.36}$$

and leads to the probability distribution (to order λ)

$$P_{0+}(\phi) = |\psi_{0+}(\phi)|^2 = \frac{1}{2\pi}(1 - 2\lambda \cos \phi) . \tag{6.37}$$

Thus, the electron density is seen to be greatest at $\phi = \pi$ and least at $\phi = 0$. This can be understood on the basis of a classical model; a static electric dipole will align itself along the field so as to minimize its energy. The wave functions of the excited states, deducible from relations (6.34) and (6.35), lead again in first order to the probability densities

$$P_{\mu+}(\phi) = \frac{1}{\pi} \left[\cos^2(\mu\phi) \left(1 + \frac{2\lambda}{4\mu^2 - 1} \cos\phi \right) + \frac{2\mu\lambda}{4\mu^2 - 1} \sin(2\mu\phi) \sin\phi \right],$$
(6.38)

$$P_{\mu-}(\phi) = \frac{1}{\pi} \left[\sin^2(\mu\phi) \left(1 + \frac{2\lambda}{4\mu^2 - 1} \cos\phi \right) - \frac{2\mu\lambda}{4\mu^2 - 1} \sin(2\mu\phi) \sin\phi \right].$$
(6.39)

Recall that in perturbation theory wave functions to order λ^{k-1} determine the energy eigenvalues to order λ^k. Since degeneracy breaking in the present case does not occur until order $\lambda^{2\mu}$ for degenerate states $|\pm\mu\rangle$, the wave functions characterizing nondegenerate states must be of order $\lambda^{2\mu-1}$. To order λ^1 therefore, (6.38) and (6.39) represent a pair of degenerate states.

The physical interpretation of the above excited-state probability densities which characterize a rotating electric dipole is not as straightforward as the interpretation of the ground-state distribution characterizing a stationary dipole. For small λ the dominating feature of the probability densities is the $\cos^2(\mu\phi)$ and $\sin^2(\mu\phi)$ dependence in the states of even and odd reflection parity, respectively. The average probability distribution of the two states, however,

$$\overline{P}_\mu = \frac{1}{2}(P_{\mu+} + P_{\mu-}) = \frac{1}{2\pi} \left(1 + \frac{2\lambda}{4\mu^2 - 1} \cos\phi \right)$$
(6.40)

yields a simpler expression that is amenable to interpretation according to a classical model. Equation (6.40) predicts a maximum electron density at $\phi = 0$ and a minimum density at $\phi = \pi$ in direct opposition to the ground-state distribution. One can account for this by arguing that the classical rotating dipole passes the minimum potential region around $\phi = \pi$ with a higher speed than that with which it passes the maximum potential region around $\phi = 0$. Hence the dipole spends less time in the vicinity of $\phi = \pi$ than it does in the region of $\phi = 0$. It is of interest to note that by incorrectly applying nondegenerate perturbation theory to the planar rotator one obtains (6.40) directly as the excited-state electron density.

For high field strengths the perturbation expansions in λ are no longer valid, and one must utilize asymptotic expressions for the Mathieu functions which, for sufficiently large and positive q, can be written in terms of elliptic functions or Hermite polynomials. These expressions are not particularly illuminating and will not be discussed here further. The most important attribute of the asymptotic Mathieu functions, however, is that they are significant only in the vicinity of $\theta = \pm\pi/2$, dropping off rapidly to very small values near $\theta = 0$ and π as q increases. Thus the strong-field wave functions of both reflection symmetries become more tightly confined about $\phi = \pi$ as the field strength increases. The system is again interpretable in terms of a classical dipole aligned closely along the field direction. The dipole is not stationary, however, but undergoes small-amplitude oscillations about the equilibrium position. This is readily confirmed by expanding the Hamiltonian (6.7) about the point $\phi = \pi$ to obtain the Hamiltonian of a harmonic oscillator. In this way the occurrence of an oscillator spectrum in (6.30) is accounted for.

6.3 The Planar Rotator in a Magnetic Field

The interaction of the planar rotator with a magnetic field (Zeeman effect) provides some instructive similarities and contrasts to the previously discussed behavior in an electric field. As before, symmetry considerations can reveal significant features of the system before any detailed analytic solution of the equation of motion is attempted. Let the rotator (of radius R) again be in the x–y plane with the magnetic field $\boldsymbol{B}$ in the y–z plane at angle β to the rotation axis z as shown in Fig. 6.1. The position vector of the electron is completely determined by the rotation angle ϕ with respect to the x axis.

The total Hamiltonian of the system is then expressible in the form (derived in Appendix 6A)

$$ H = H_0 + H_{\mathrm{M}} = \frac{L_z^2}{2mR^2} + \omega_{\mathrm{L}} L_z \cos\beta + \frac{1}{2} m\omega_{\mathrm{L}}^2 R^2 \left(\cos^2\phi + \sin^2\phi \cos^2\beta \right) , $$

$$ (6.41) $$

where

$$ \omega_{\mathrm{L}} = \frac{eB}{2mc} \tag{6.42} $$

is the Larmor frequency (a positive number for an electron). The term in H_{M} linear in ω_{L} and independent of the electron coordinate is the paramagnetic term; the diamagnetic contribution to H_{M} is quadratic in ω_{L} and proportional to the component of the magnetic field normal to the electron position vector. As seen from (6.41), there are two inequivalent basic orientations of the magnetic field that can influence the state of motion of the electron within the plane of rotation:

- $\boldsymbol{B}$ along the rotation axis,
- $\boldsymbol{B}$ normal to the rotation axis.

In the case of the Stark effect there is only one orientation of the electric field; an electric field along the rotation axis merely results in a constant potential that can be set equal to zero by locating the electron in the plane $z = 0$.

The case of $\boldsymbol{B}$ parallel to the rotation axis, however, is trivially solvable, for the energy eigenvectors are simply the eigenstates $|\mu\rangle$ of L_z. The eigenvalues are then

$$ E_\mu = \varepsilon\mu^2 + \mu\hbar\omega_{\mathrm{L}} + \frac{1}{2} m\omega_{\mathrm{L}}^2 R^2 , \tag{6.43} $$

where ε has been defined in (6.8). Thus, an axially-oriented magnetic field lifts the double degeneracy of the excited states, whereas an axial electric field has no effect. It is perhaps not superfluous to point out the reason for this. An electric field (a polar vector) represents a preferential *direction*. By contrast, a magnetic field (an axial vector) is a chiral structure with a preferential *sense* (clockwise or counterclockwise) or *handedness* (left or right). These characteristics, of which much more will be said in Chap. 7, stem from the origin of the two fields: an electric field is produced by charges, whereas a magnetic field arises from currents. The states $|\pm|\mu|\rangle$ have probability current densities that circulate parallel and antiparallel, respectively, to the current producing the magnetic field and therefore represent two inequivalent orientations of a magnetic dipole in the magnetic field.

The case of $\boldsymbol{B}$ lying within the plane of rotation ($\beta = \pi/2$) is not trivially solvable and will be the focus of our attention in this section. In the analysis of the rotator in an electric field, it was shown that the Stark solutions span the symmetric and antisymmetric one-dimensional irreducible representations of a group isomorphic to the symmetric group S_2. An examination of the Hamiltonian including the rotational kinetic energy and diamagnetic potential energy – there is no paramagnetic term (term linear in the Larmor frequency) for the chosen field orientation – shows that it is invariant under a group of operations comprising the identity, reflection across an axis in the rotation plane normal to the field, and inversion through the origin. This group is isomorphic to the (Abelian) four-group V. Since the degenerate field-free basis spans a two-dimensional irreducible representation of the rotation-reflection group, it must split under the magnetic interaction into nondegenerate states that span the irreducible representations of V. A simple character analysis shows that states of even or odd $|\mu|$, respectively, span representations of even or odd inversion parity. However, the two states of the same inversion parity that belong to a rotational level of given $|\mu|$ differ in their reflection parities. One would therefore expect the Zeeman effect to give rise to stationary states of four possible symmetry types in contrast to the two possible symmetry types of the Stark effect. The higher symmetry also leads to marked differences in the energy spectra between the two cases, particularly in the high-field domain. Let us now examine this system in more detail.

In a coordinate representation the Hamiltonian (6.41) without paramagnetic term can be written in the form

$$H = -\varepsilon \left(\frac{\partial^2}{\partial \phi^2} - \xi^2 \cos \phi \right) , \tag{6.44}$$

where the magnetic interaction parameter ξ is

$$\xi = \frac{m \omega_{\mathrm{L}} R^2}{\hbar} . \tag{6.45}$$

With energy eigenvalues again expressed in the form of relation (6.15) (but with the appropriate quantum labels to be specified), Hamiltonian (6.44) leads to a Schrödinger equation which can be cast once more into the canonical form of Mathieu's equation

$$\left[\frac{\mathrm{d}^2}{\mathrm{d}\phi^2} + (a - 2q \cos 2\phi) \right] \psi_\eta(\phi) = 0 , \tag{6.46}$$

where

$$a = \mu^2 - \frac{1}{2}\xi^2 \qquad (6.47)$$

and

$$q = \frac{1}{4}\xi^2 . \qquad (6.48)$$

Note that the physically significant angular coordinate ϕ appears, and not $\theta = \phi/2$ as in the case of the Stark effect, (6.21). As a consequence, the boundary condition $\psi(\phi + 2\pi) = \psi(\phi)$ imposed by the continuity of the wave function can be satisfied by Mathieu functions of both 2π and π periodicity.

The exact solutions (up to a normalization constant) are therefore the four types of Mathieu functions described earlier, and the wave functions and energy eigenvalues can be written as

$$\psi_{\mu+}^{\pi}(\phi) = ce_\mu(\phi, \xi) , \qquad E_{\mu+}^{\pi} = \varepsilon\left[a_\mu(\xi) + \frac{\xi^2}{2}\right] \qquad (\mu = 0, 2, 4, \ldots) , \quad (6.49)$$

$$\psi_{\mu-}^{\pi}(\phi) = se_\mu(\phi, \xi) , \qquad E_{\mu-}^{\pi} = \varepsilon\left[b_\mu(\xi) + \frac{\xi^2}{2}\right] \qquad (\mu = 2, 4, 6, \ldots) , \quad (6.50)$$

$$\psi_{\mu+}^{2\pi}(\phi) = ce_\mu(\phi, \xi) , \qquad E_{\mu+}^{2\pi} = \varepsilon\left[a_\mu(\xi) + \frac{\xi^2}{2}\right] \qquad (\mu = 1, 3, 5, \ldots) , \quad (6.51)$$

$$\psi_{\mu-}^{2\pi}(\phi) = se_\mu(\phi, \xi) , \qquad E_{\mu-}^{2\pi} = \varepsilon\left[b_\mu(\xi) + \frac{\xi^2}{2}\right] \qquad (\mu = 1, 3, 5, \ldots) , \quad (6.52)$$

where, in addition to eigenvalues of L_z, the states are labeled by their periodicity and reflection symmetry with respect to Π_x.

For a weak magnetic field the eigenvalues can be expanded in a perturbation series in the parameter ξ^2. From Table 6.1, the energy eigenvalues of the ground state and first six excited states, truncated at the lowest order in ξ^2 that breaks the degeneracy, are as follows:

$$\left(\eta_{0+}^{\pi}\right)^2 = \frac{1}{2}\xi^2 , \qquad (6.53)$$

$$\left(\eta_{1+}^{2\pi}\right)^2 = 1 + \frac{3}{4}\xi^2 , \qquad (6.54)$$

$$\left(\eta_{1-}^{2\pi}\right)^2 = 1 + \frac{1}{4}\xi^2 , \qquad (6.55)$$

$$\left(\eta_{2+}^{\pi}\right)^2 = 4 + \frac{1}{2}\xi^2 + \frac{5}{192}\xi^4 , \qquad (6.56)$$

$$\left(\eta_{2-}^{\pi}\right)^2 = 4 + \frac{1}{2}\xi^2 - \frac{1}{192}\xi^4 , \qquad (6.57)$$

$$\left(\eta_{3+}^{2\pi}\right)^2 = 9 + \frac{1}{2}\xi^2 + \frac{1}{256}\xi^4 + \frac{1}{4096}\xi^6 , \qquad (6.58)$$

$$\left(\eta_{3-}^{2\pi}\right)^2 = 9 + \frac{1}{2}\xi^2 + \frac{1}{256}\xi^4 - \frac{1}{4096}\xi^6 . \qquad (6.59)$$

We see from the above relations that the double degeneracy of zero-field states of specified $|\mu|$ is broken by a magnetic field (in the rotator plane) in order ξ^{μ} in contrast to the Stark effect in which degeneracy breaking occurs in order $\lambda^{2\mu}$.

When the magnetic field is sufficiently great that the eigenvalue expansions are no longer valid, one must again turn to the asymptotic properties of the Mathieu functions. From equations (6.28), (6.29) and (6.47), (6.48) we find that the quadratic dependence of the energy on ξ (or B) drops out, and the eigenvalue spectrum again assumes a form highly suggestive of that of a harmonic oscillator

$$E_{\mu+}^{2\pi} = \left(\mu + \frac{1}{2}\right)\hbar\omega_{\mathrm{L}} \qquad (\mu = 0, 2, 4, \ldots) , \qquad (6.60)$$

$$E_{\mu-}^{2\pi} = \left(\mu - \frac{1}{2}\right)\hbar\omega_{\mathrm{L}} \qquad (\mu = 2, 4, 6, \ldots) , \qquad (6.61)$$

$$E_{\mu\pm}^{\pi} = \left(\mu \pm \frac{1}{2}\right)\hbar\omega_{\mathrm{L}} \qquad (\mu = 1, 3, 5, \ldots) . \qquad (6.62)$$

An interesting and significant distinction between the high-field Zeeman and Stark spectra is that states split by an electric field E remain nondegenerate for all values of E, but a very strong magnetic field actually recreates a double degeneracy as evident in Fig. 6.3 and analytically verified in relations (6.60)–(6.62). A given degenerate level consists of two states of opposite reflection symmetry and periodicity. In the next higher level the periodicity labels are reversed. Thus, for example, the lowest level consists of the states ce_0^{π} and $se_1^{2\pi}$; the first excited level consists of the states $ce_1^{2\pi}$ and se_2^{π}. Every two successive levels exhaust all four types of Mathieu functions.

These high-field states give rise to a quasi-Landau energy spectrum. The true Landau energies for a free electron orbiting in a plane perpendicular to a static uniform magnetic field are

$$E_n = \left(n + \frac{1}{2}\right)\hbar\omega_{\mathrm{C}} \qquad (n = 0, 1, 2, \ldots) , \qquad (6.63)$$

where

$$\omega_{\mathrm{C}} = 2\omega_{\mathrm{L}} \qquad (6.64)$$

is the cyclotron frequency. Differences between the spectra of (6.60)–(6.62) and (6.63) and (6.64) are to be expected since the two physical systems are not exactly comparable, differing in both the orientation of the magnetic field (normal versus parallel to the rotation axis) and the accessible space (a circular perimeter versus the entire x–y plane). The resemblance of the Landau or quasi-Landau spectrum to the harmonic oscillator spectrum is again not merely fortuitous, but a consequence of the fact that the diamagnetic contribution to the Hamiltonian (6.41) is of the form of a harmonic oscillator potential.

Although the exact solutions given above allow one to deduce all the properties of the planar rotator in a magnetic field, several important physical concepts can be further clarified by examining the system from the perspectives of both perturbation theory and group theory, as was done with the Stark effect. In the absence of a paramagnetic term, the matrix elements of H_{M} in the field-free basis $|\mu\rangle$ are

$$\langle\mu|H_{\mathrm{M}}|\mu'\rangle = \frac{1}{2}\varepsilon\xi^2\left[\delta_{\mu,\mu'} + \frac{1}{2}(\delta_{\mu,\mu'+2} + \delta_{\mu,\mu'-2})\right], \qquad (6.65)$$

from which it is clear that the diagonal elements of H_{M} are nonvanishing in contrast to the diagonal elements of H_{E}. This indicates that the states $|\mu\rangle$ already constitute a field-free basis with nonvanishing magnetic dipole moment. A comparable electric dipole moment is precluded by the symmetry of the Hamiltonian (6.6). The off-diagonal elements $(H_{\mathrm{M}})_{-\mu,\mu}$ are in general zero. There is an exception, however, for $\mu = \pm 1$ where, for each choice of sign, one term in (6.65) survives. As a consequence of this coupling, the double degeneracy of the first excited level is split by first-order degenerate perturbation theory in contrast to the Stark effect. A splitting linear in H_{M}, however, is still quadratic in the magnetic field. Thus, the symmetry of the planar rotator is such that, despite the degeneracy of its excited states, no splitting linear in either the electric or magnetic field occurs for a weak field normal to the rotation axis. The degeneracy of the states with $|\mu| > 1$ can be removed by successively higher orders of perturbation theory, and these calculations must yield, of course, the same expansions as obtained from the characteristic values of the Mathieu functions.

Having begun the discussion of the magnetic interaction of the planar rotator with a brief remark on symmetry, let us conclude by noting once more the power of group theory to provide a more fundamental way of understanding the correlations of state symmetries with rotational excitation other than by solving the Schrödinger equation and discovering these correlations in the Mathieu function solutions. In particular, I address the questions why is it possible for a strong magnetic field to recreate degeneracy among states, whereas the degeneracy lifted by an electric field remains broken independent of the field strength, and what leads to the correlations in periodicity and reflection symmetry of the strong-field states that become degenerate. The answers lie in the symmetry of the Hamiltonian.

The rotational kinetic energy H_0, as discussed before, is invariant under the two-dimensional rotation-reflection group, and the zero-field basis $e^{\pm i\mu\phi}$ spans a two-dimensional irreducible representation $\Gamma_{\mu d}$ of this group. The total Hamiltonian (6.41), as pointed out earlier, is invariant under a smaller group of symmetry operations, the Abelian four-group or V, comprising the elements I, Π_x, Π_y and Π. It is clear, therefore, that the degenerate basis must split in a magnetic field although, since V is a higher symmetry group than S_2 and has twice as many irreducible representations, it is perhaps not so obvious which zero-field states go into which irreducible representations.

Using the character table of the four-group and the orthogonality properties of group characters, one can verify that the degeneracy breaking must occur as follows

$$\Gamma_{\mu d} \longrightarrow \begin{cases} \Gamma_1 + \Gamma_2 , & \mu \text{ even}, \\ \Gamma_3 + \Gamma_4 , & \mu \text{ odd}, \end{cases} \qquad (6.66)$$

where the irreducible representations are defined and correlated with the Mathieu functions in Table 6.2. The above decomposition implies that states of even or odd rotational quantum number μ are respectively split into states of even or odd inversion parity. Since $\Pi_x \Pi_y = \Pi$, even (odd) inversion parity comes about only if the reflection parities of the states are respectively the same (opposite). Translated into the symmetry properties of the Mathieu functions, the above group theoretical results require that field-free states of even or odd μ split into pairs ce_μ, se_μ with respectively even or odd μ. These split pairs remain split independent of the strength of the magnetic field.

However, there is *no* group theoretical restriction against a basis state of Γ_1 or Γ_2 becoming degenerate with a basis state of Γ_3 or Γ_4 as the field strength increases. This follows from the fact that these two pairs of representations arise from different irreducible representations of the rotation-reflection group. Thus, *should degeneracy occur*, group theory requires that the degenerate states occur in the pairs $(ce_{2\mu}, se_{2\mu+1})$ and $(ce_{2\mu+1}, se_{2\mu+2})$, where, for both cases, $\mu = 0$, 1, 2, etc.

Group theory can not say whether a degeneracy in this case will occur – and, in fact, an *exact* degeneracy in the high-field spectrum does not really occur. The asymptotic relation $a_\mu \approx b_{\mu+1}$ between the characteristic values of the Mathieu functions holds ideally only in the limit $q \to \infty$; for finite q the difference $b_{\mu+1} - a_\mu$ falls off exponentially with the square root of q, a difference that is effectively negligible for the high values of q at which the asymptotic formulas are valid.

We will not consider the properties of the Zeeman wave functions in the same detail as for the Stark effect because the discussions would be similar. From the Fourier decompositions given by relations (6.34) and (6.35) one can construct the perturbation series for each state. The basic distinction to keep in mind is that the angular variable which appears is now ϕ, the electron coordinate. This has an important consequence for the high-field asymptotic wave

Table 6.2. Character table of the four group (V)

V	I	Π_x	Π_y	Π	Basis $(n = 0, 1, 2, \ldots)$
Γ_1	1	1	1	1	ce_{2n}^{π}
Γ_2	1	-1	-1	1	se_{2n+2}^{π}
Γ_3	1	1	-1	-1	$ce_{2n+1}^{2\pi}$
Γ_4	1	-1	1	-1	$se_{2n+1}^{2\pi}$
Γ_{nd}	2	0	0	$(-2)^n$	$e^{in\phi}, e^{-in\phi}$

functions. As discussed in the previous section, the Mathieu functions for large q assumed their maximum values in the vicinity of $\theta = \pm\pi/2$ and dropped rapidly to small values at $\theta = 0$, π, which represented the same location. Since $\theta = \phi$ in the present case, the electron density is greatest about the two points where the magnetic field axis intersects the orbital circle. Expansion of the Hamiltonian about these points again results in the Hamiltonian of a harmonic oscillator. Interpreted classically, the rotator can no longer rotate freely, but executes small-amplitude oscillations about the direction either parallel or antiparallel to the field. This is consistent with the observation that the Mathieu functions are linear superpositions of components with both positive and negative magnetic dipole moments.

6.4 The Planar Rotator in a Vector Potential Field

We have seen that the presence of electric and magnetic fields alter the energy and wave functions of a charged particle confined to a circular orbit (or, more precisely, to a linear space of constant curvature $1/R$). They break the level degeneracy in interesting ways that can be elucidated by symmetry arguments and give rise to a range of motions that, depending on field strength, vary between free rotation with well defined angular momentum and harmonic oscillation about points of equilibrium. While there are aspects of the interaction of the two-dimensional rotator with electromagnetic fields that can be correctly determined only by quantum mechanics, the system nevertheless has classical counterparts in the behavior of electric and magnetic dipoles which, at least in the domain of applicability of the Bohr correspondence principle, yield quantitative results comparable to those arrived at quantum mechanically.

Consider next, however, the configuration of Fig. 6.4 for which – depending on the history of the system – there may or may not be a classical analogue.

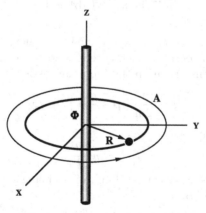

Fig. 6.4. Configuration of the two-dimensional charged rotator in a vector potential field A produced by magnetic flux Φ manifesting a bound-state AB effect

This is the bound-state counterpart of the AB effect treated earlier in the context of a charged particle beam. The particle of mass m and charge g is constrained to a circular space of radius R through which is now threaded, so to speak, a long tube of magnetic flux. We can imagine the flux confined to the interior of a long solenoid of radius $a \ll R$. Then, as in Chap. 1, the circulating electron is in an environment permeated by a vector potential field [specified in the Coulomb gauge by (1.27)] ideally free of electric and magnetic fields, excluding the fields of the particle, itself. Although there is no splitting of wave packets in this case, and hence no recombination and interference to speak of, there can still be a physical effect of the local vector potential (or remote magnetic field) on the state of motion of the orbiting particle. The nature of this effect is the issue we will examine.

Since there is a certain measure of arbitrariness in the specification of a gauge field like the magnetic vector potential, it is pertinent to comment briefly here on the matter of gauge invariance and observability of a dynamical variable. It is well known that to be an observable, a dynamical variable must be representable by a Hermitian operator with complete set of eigenstates. For a mechanical system coupled to a gauge field, a further criterion must be met so that physically meaningful quantities do not depend on the arbitrary choice of gauge. All theoretical expressions, e.g., quantum mechanical expectation values and transition matrix elements, representing measurable quantities must be invariant under a gauge transformation as summarized by relations (1.17)–(1.19). In the case of a charged particle orbiting a tube of magnetic flux, a gauge transformation cannot change the magnetic field $\boldsymbol{B}$ inside the tube if the vector potentials $\boldsymbol{A}$ and $\boldsymbol{A}'$ outside the tube satisfy Stokes' theorem

$$\oint_C \boldsymbol{A} \cdot \mathrm{d}\boldsymbol{s} = \oint_C \boldsymbol{A}' \cdot \mathrm{d}\boldsymbol{s} = \text{total magnetic flux} = \varPhi \,, \qquad (6.67)$$

where the contour circumscribes the solenoid. It then follows from (1.17) that

$$\oint_C \boldsymbol{\nabla}\varLambda \cdot \mathrm{d}\boldsymbol{s} = \oint_C \mathrm{d}\varLambda = 0 \,, \qquad (6.68)$$

and the gauge function $\varLambda(\boldsymbol{r}, t)$ is single-valued, a mathematical property of importance in previous defenses of the theoretical existence of the AB effect [179].

Under a gauge transformation the (nonrelativistic) Hamiltonian

$$H = \frac{1}{2m} \left(\boldsymbol{p} - \frac{q\boldsymbol{A}}{c} \right)^2 + q\phi \qquad (6.69)$$

and Schrödinger equation

$$H\varPsi = \mathrm{i}\hbar \frac{\partial \varPsi}{\partial t} \qquad (6.70)$$

are form-invariant, the transformed Hamiltonian being given by relation (1.25) with $U = \mathrm{e}^{\mathrm{i}q\varLambda/\hbar c}$ as follows:

$$H' = \mathrm{e}^{\mathrm{i}q\varLambda/\hbar c} H \mathrm{e}^{-\mathrm{i}q\varLambda/\hbar c} - \frac{q}{\hbar c} \frac{\partial \varLambda}{\partial t} = \frac{1}{2m} \left(\boldsymbol{p} - \frac{q\boldsymbol{A}'}{c} \right)^2 + q\phi' \,. \qquad (6.71)$$

Thus, the time-evolution of the physical system is unaffected by a gauge transformation, i.e., Ψ' evolves under H' in the same way that Ψ evolves under H.

The eigenvalue spectrum of H, however, is not invariant under a gauge transformation. If $\Psi_E = \Psi_E^0 e^{-iEt/\hbar}$ is a stationary state of H' with eigenvalue E, then $\Psi'_E = e^{iq\Lambda/\hbar c}\Psi_E$ is also an eigenstate, but not necessarily a *stationary* state, of H' with eigenvalue $E' = E - (q/\hbar c)\partial\Lambda/\partial t$. Thus, E' is both gauge- and time-dependent unless Λ is independent of time. If the latter condition holds, the Hamiltonians H' and H are simply related by a unitary transformation and necessarily have the same eigenvalue spectrum.

Since the eigenvalues of H are not gauge-invariant, does this imply that energy is not an observable? The answer is clearly 'no', for the reason that in quantum mechanics, as in classical mechanics, the Hamiltonian need not represent the energy of the system (a known, but not widely appreciated, point). A suitable operator that *does* represent the energy of the system should have a time-independent scalar potential $\phi(x)$. Under an arbitrary gauge transformation, the true energy operator W, whose eigenvalues are obtained from

$$W\Psi = E\Psi , \tag{6.72}$$

and characterize the allowed energies of the system, becomes [180]

$$W = H' + \frac{q}{\hbar c}\frac{\partial\Lambda}{\partial t} = e^{iq\Lambda/\hbar c}He^{-iq\Lambda/\hbar c} . \tag{6.73}$$

It is evident that the eigenvalue spectrum and expectation values of W are independent of gauge and correspond to observable quantities. W, of course, does not ordinarily generate the time evolution of the system; that is the role of the Hamiltonian.

The Hamiltonian (6.69) of the planar rotator in the vector potential field (1.27) of a tube of magnetic flux (with scalar potential $\phi = 0$) can be succinctly written in the form

$$H = \varepsilon K_z^2 = \varepsilon(L_z - \alpha)^2 , \tag{6.74}$$

where K_z is the component of the kinetic angular momentum

$$\boldsymbol{K} = \boldsymbol{r} \times \left(\boldsymbol{p} - \frac{q\boldsymbol{A}}{c}\right) \tag{6.75}$$

normal to the rotator plane (the only component) and

$$\alpha = \frac{\Phi}{\Phi_0} \tag{6.76}$$

is the ratio of enclosed magnetic flux to the fluxon Φ_0 [see (1.30)].

Since the distinctions between canonical and kinetic dynamical variables have at times been the source of some confusion, it is worth noting here the properties and roles of $\boldsymbol{K}$ and the canonical orbital angular momentum $\boldsymbol{L}$. The components of $\boldsymbol{L}$, which satisfy the commutation relations

$$[L_i, L_j] = i\hbar \sum_k \varepsilon_{ijk} L_k , \tag{6.77}$$

where ε_{ijk} is again the Levi-Civita permutation symbol (or the completely antisymmetric tensor), are the generators of an infinitesimal rotational transformation of angle θ about direction $\boldsymbol{n}$ according to

$$\langle \boldsymbol{r} | \left(1 + \frac{\mathrm{i}}{\hbar} \boldsymbol{n} \cdot \boldsymbol{L} \mathrm{d}\theta \right) = \langle \boldsymbol{r} + (\boldsymbol{n} \times \boldsymbol{r}) \mathrm{d}\theta | \, . \tag{6.78}$$

The matrix elements of $\boldsymbol{L}$, however, are not invariant under a gauge transformation, but transform as follows

$$\langle \boldsymbol{L} \rangle_{\Psi'} = \langle \boldsymbol{L} \rangle_{\Psi} + \frac{q}{\hbar c} \langle \boldsymbol{r} \times \boldsymbol{\nabla} \Lambda \rangle_{\Psi} \, , \tag{6.79}$$

where the bracket represents either a transition moment or expectation value. $\boldsymbol{L}$, therefore, ceases in general to be an observable in a mechanical system coupled to an electromagnetic vector potential. By contrast, the dynamical variable $\boldsymbol{K}$ represents the mechanical angular momentum of a particle (corresponding, for example, to mvR for a classical particle in uniform circular motion with speed v); it has gauge-invariant matrix elements and satisfies a different set of commutation relations [181]

$$[K_i, K_j] = \mathrm{i}\hbar \sum_k \varepsilon_{ijk} \left\{ K_k + \frac{e}{\hbar c} x_k [\boldsymbol{x} \cdot (\boldsymbol{\nabla} \times \boldsymbol{A})] \right\} \, . \tag{6.80}$$

For a particle in the field-free region outside the solenoid (where $\boldsymbol{B} = \boldsymbol{\nabla} \times \boldsymbol{A} = 0$) the commutation relations of $\boldsymbol{K}$ reduce to those of $\boldsymbol{L}$.

Consider a system for which the magnetic flux Φ is constant in time, and therefore does not produce a local electric field at the particle by means of Faraday's law of induction. Thus, with α independent of time, the Schrödinger equation

$$\varepsilon \left(-\frac{\mathrm{i}}{\hbar} \frac{\partial}{\partial \phi} - \alpha \right)^2 \Psi = \mathrm{i}\hbar \frac{\partial \Psi}{\partial t} \tag{6.81}$$

readily admits of *two* stationary-state solutions with the following wave functions, energy, and angular momentum eigenvalues:

- Solution I

$$\Psi_\mu^{\mathrm{I}}(\phi, t) = \frac{1}{\sqrt{2\pi}} \mathrm{e}^{\mathrm{i}(\mu+\alpha)\phi} \mathrm{e}^{-\mathrm{i}E_\mu^0 t/\hbar} \qquad (\mu = 0, \pm 1, \pm 2, \ldots) \, , \tag{6.82}$$

$$\text{energy} \qquad E_\mu^0 = \mu^2 \varepsilon \, , \tag{6.83}$$

$$\text{canonical angular momentum} \qquad l_\mu = (\mu + \alpha)\hbar \, , \tag{6.84}$$

$$\text{kinetic angular momentum} \qquad k_\mu = \mu\hbar \, . \tag{6.85}$$

- Solution II

$$\Psi_\mu^{\mathrm{II}}(\phi, t) = \frac{1}{\sqrt{2\pi}} \mathrm{e}^{\mathrm{i}\mu\phi} \mathrm{e}^{-\mathrm{i}E_\mu t/\hbar} \qquad (\mu = 0, \pm 1, \pm 2, \ldots) \, , \tag{6.86}$$

$$\text{energy} \qquad E_\mu = (\mu - \alpha)^2 \varepsilon \, , \tag{6.87}$$

$$\text{canonical angular momentum} \qquad l_\mu = \mu\hbar \, , \tag{6.88}$$

$$\text{kinetic angular momentum} \qquad k_\mu = (\mu - \alpha)\hbar \, . \tag{6.89}$$

A cursory examination of the solutions shows that if the magnetic flux were quantized in units of the fluxon, then α would be integer-valued and the two solutions entirely equivalent; i.e., to each state of Solution I labeled by μ would correspond a state of Solution II with identical eigenvalues although different label. However, the flux through a current ring (like the planar rotator) is not ordinarily quantized in which case α can take on a continuum of values, and the two sets of solutions are not equivalent. Solution I has the energy spectrum of the field-free planar rotator, an integer kinetic angular momentum spectrum, and a flux-dependent canonical angular momentum spectrum. For non-integral values of α, the wave function is multiple-valued. Solution II, by contrast, is characterized by an integer-valued canonical angular momentum spectrum, a flux-dependent kinetic angular momentum spectrum, and a single-valued wave function. The property of single-valuedness was invoked earlier [see (6.4)] to obtain the wave functions and energy spectrum of the field-free rotator; without single-valuedness, the angular momentum (canonical and kinetic are in this case equivalent) eigenvalue spectrum would not be integer-valued. Does this mean one should discard Solution I as unphysical? On the other hand, Solution I (6.82) has the form of (1.21) which gives rise to the AB effect. What is one to make of this?

The physical content of these two sets of solutions may be understood by examining them as special cases of the general solution to the rotator in the presence of a time-dependent flux. This solution is given by

$$\psi(t) = \exp\left[-\frac{i}{\hbar}\int^t H(t')dt'\right]\psi(0) = \exp\left\{-i\varepsilon\int^t \left[L_z - \alpha(t')\right]^2 dt'\right\}\psi(0) ,$$
(6.90)

where the first equality is the general solution to the time-dependent Schrödinger equation whenever the Hamiltonian at two different times commutes, i.e., $[H(t), H(t')] = 0$. If, at $t = 0$, the potential-free rotator is in an eigenstate of angular momentum with quantum number μ, then, by relation (1.21), the state of the rotator in the presence of a static vector potential $\boldsymbol{A}_0$ with sole tangential component $A_{0\phi}(r) = \Phi(0)/2\pi r$ is represented by the wave function

$$\psi_\mu(0) = \frac{1}{\sqrt{2\pi}}e^{i(\mu+\alpha_0)\phi} ,$$
(6.91)

where evaluation of the phase integral in (1.21) has led to

$$\frac{e}{\hbar c}\int^{\boldsymbol{x}} \boldsymbol{A}_0\cdot d\boldsymbol{s} = \frac{e}{\hbar c}\int^{\phi} \frac{\Phi(0)}{2\pi R}d\phi' = \frac{\Phi(0)}{\Phi_0}\phi = \alpha_0\phi .$$
(6.92)

Substitution of relation (6.91) into (6.90) then yields the solution

$$\psi_\mu(\phi, t) = \frac{1}{\sqrt{2\pi}}e^{i(\mu+\alpha_0)\phi}\exp\left\{-i\varepsilon\int^t \left[\mu + \alpha_0 - \alpha(t')\right]^2 dt'\right\} .$$
(6.93)

Consider two cases for the time dependence of $\alpha(t)$.

In the first case the flux $\Phi(0)$ is null for $t < 0$ and is subsequently brought to the constant value Φ at $t = 0_+$. Then $\alpha_0 = 0$ for $t \leq 0$, and $\alpha(t) = \alpha = $ constant for $t > 0$. By Faraday's law of induction the initiation of the magnetic flux generates an electric field, whose tangential (and only) component is given by

$$E_\phi(t) = -\frac{1}{2\pi Rc}\frac{\partial \Phi}{\partial t} , \tag{6.94}$$

which exerts a torque on the rotating particle

$$\tau_z = qRE_\phi = -\frac{q}{2\pi}\frac{\partial \Phi}{\partial t} \tag{6.95}$$

along the axis of rotation. The torque changes the initial angular momentum $L_z = \mu\hbar$ by the amount

$$\Delta L_z = \int_0^\infty \tau_z dt = -\frac{q\Phi}{2\pi c} = -\alpha\hbar , \tag{6.96}$$

thereby giving the state a final mechanical angular momentum

$$K_z = L_z + \Delta L_z = (\mu - \alpha)\hbar . \tag{6.97}$$

In a similar way the induced electric field also does work on the particle at a rate

$$\frac{dW}{dt} = \omega_{rot}\tau_z = \frac{L_z}{mR^2}\tau_z = -2\varepsilon(\mu - \alpha)\frac{\partial \alpha}{\partial t} , \tag{6.98}$$

leading to a change in energy by the amount

$$\Delta E = \int \frac{dW}{dt} dt = -2\varepsilon\alpha\left(\mu - \frac{1}{2}\alpha\right) . \tag{6.99}$$

The final energy is then

$$E_\mu = E_\mu^0 + \Delta E = (\mu - \alpha)^2\varepsilon . \tag{6.100}$$

The properties of Solution II can therefore be thought to derive from the local interaction of the particle with an induced electric field over the period of initiation of the magnetic flux (which, in fact, can be implemented at an entirely arbitrary rate).

In the second case we allow the flux through the circular space of the rotator to remain at its initial value throughout the existence of the rotator system. Then $\alpha_0 = \alpha(t) = \alpha = $ constant, and it is seen that the resulting wave function is that of Solution I. As there is no induced electric field to exert torques and do work on the charged particle, one would expect – as indeed is the case – that there is no effect of the flux on the characteristic energy and mechanical angular momentum of the system. Nevertheless, there is a purely quantum mechanical influence, a bound-state AB effect, reflected in the flux-dependent shift of the canonical angular momentum eigenvalue spectrum.

It is interesting to note the general significance of past history on the present state of a quantum system as illustrated by the example of the two-dimensional rotator in a vector potential field. Although both solutions represent a rotator threaded by a 'constant' magnetic flux – constant, that is, from the perspective of one examining the system long after its formation – the manner of formation has played a seminal role. The properties of the orbiting particle, which at some point in its past experienced the initiation of the magnetic flux, are different from an apparently identical particle which began its orbital motion about an already existing tube of flux. There is nothing in this perspective that violates quantum mechanical principles although the present situation is not frequently encountered. After all, is not a $2S$ state the same for all hydrogen atoms whether it was the end result of an excitation from the $1S$ ground state or a decay transition from a $3P$ excited state? In most cases the answer is undoubtedly 'yes', but there are circumstances, such as just illustrated, where the mode of origination of a quantum system leaves a distinct legacy.

What about the admissibility of multiple-valued wave functions? This has been a long-debated and somewhat thorny issue. Some have argued that single-valuedness is a necessary criterion for solutions to be physically meaningful [179, 182]; others have taken the position that multiple-valued wave functions should not be rejected a priori [183, 184]. Whereas the imposition of single-valuedness as a criterion for an electromagnetic or gravitational field to be physically meaningful is equivalent to requiring that the classical forces acting on a particle be uniquely specified, the same criterion applied to a quantum field, such as the quantum mechanical wave function, is not so transparent and readily motivated; it is a *bilinear* product of wave functions (like the probability density), and not the wave function itself, that is directly related to experimentally observable quantities.

In a simply connected space, such as that of the planar rotator in the absence of magnetic flux, specification of single-valuedness ensures against spurious solutions introduced by the arbitrary choice of a particularly convenient coordinate system with the polar axis through the loop. An alternative choice of polar axis outside the loop leads to single-valued solutions only. The mathematical equivalence of the two descriptions reflects the topological circumstance that there is no unambiguous meaning to the 'inside' of a closed loop in a simply-connected space. Topologically, any such loop can be deformed continuously to a point. In the case of a non-simply connected space, however, such as that of the planar rotator threaded by a tube of magnetic flux, there is a significant topological distinction between loops that thread once, twice, thrice, etc., about a hole in the space. Two loops that wind about the excluded region of space a different number of times cannot be continuously deformed into one another because of the hole. Mathematically, they fall into different so-called homotopy classes.

The argument for rejection of multiple-valued wave functions has also been made on the grounds that they fail to regenerate the expected single-valued solutions upon adiabatic extinction of the magnetic flux. Such an argument would not be relevant, however, to a composite system for which the particle experiences an invariable flux beyond the control of the experimenter.

Finally, it has often been asserted that the two-dimensional system of a particle orbiting an infinitely long tube of flux is in reality just an idealization of a three-dimensional system with high, but finite, potential barrier and finite return flux. Then, there would again no longer be a distinction between paths that wind about a designated origin and paths that do not (for all such paths could be continuously deformed into one another), and wave functions should once more be single-valued. This point of view entirely sidesteps the original problem by converting it to a three-dimensional one. The cogency of the premise will not be discussed here further except to say that the two-dimensional system as introduced is a conceptually valid one, and numerous theoretical and experimental studies have shown that the physics of two dimensions usually differs qualitatively from the physics of three.

6.5 Fermions, Bosons, and Things In-Between

If real systems analogous to the planar rotator threaded by a temporally invariant magnetic flux should exist in nature, these systems would manifest curious physical properties indeed. As discussed earlier, the wave function of a particle rotated, let us say by an angle θ, undergoes a unitary transformation generated by the canonical angular momentum operator

$$\psi(\phi + \theta) = e^{-iL_z\theta/\hbar}\psi(\phi) . \tag{6.101}$$

Thus, an eigenfunction of L_z with the flux-dependent eigenvalue spectrum (6.84) subjected to an integral number of complete rotations – equivalent to the identity operation in a simply-connected space – takes the form

$$\psi(\phi + 2\pi n) = e^{-i2\pi n\Phi/\Phi_0}\psi(\phi) . \tag{6.102}$$

In the absence of magnetic flux, or more generally for a magnetic flux quantized in integral units of the fluxon, the initial and rotated wave functions are identical, and the system behaves like a scalar under rotation. For $n\Phi/\Phi_0$ equal to an odd-integer multiple of $1/2$, however, the phase factor equals -1, and the wave function rotates like a spinor. Spinors ordinarily provide a mathematical description of fermions, particles (like the electron or neutron) with half odd integer values of intrinsic angular momentum or spin. In the present circumstances, the actual spin of the system is irrelevant, and therefore even a spinless charged boson could, according to (6.102) exhibit the rotation properties of a fermion [168,169]. Of course, the system must not be thought of as the particle alone, but rather as the composite bound system of a particle and magnetic flux tube. If the magnetic flux is arbitrary, then the phase factor in (6.102) is neither $+1$ nor -1, but is in general a complex number.

The connection between the angular momentum (and therefore the rotational properties) of individual particles and the quantum statistical behavior of many such identical particles has never been particularly transparent as quantum analyses go, and, in fact, actually lies outside quantum mechanics proper, having its origin in arcane propositions relating to the microscopic causality of relativistic quantum fields [185,186]. It is indeed amazing, as some authors have pointed out, that "such a deep connection, essential to the stability of matter in circumstances apparently very remote from the relativistic domain, does require these concepts. We do not know any alternative basis for it" [186]. The end result, however, can be simply stated and is widely known – namely, all matter should fall into one of two categories of particles: bosons and fermions. The wave function of a multi-boson system is unaltered by the exchange of location of any two particles, one consequence of which is that there is no upper limit to the number of bosons that can occupy a given quantum state. In contrast, the wave function of a multi-fermion system undergoes a change of sign for each pair of particles exchanged, and not more than one fermion within the system can occupy a specified quantum state. From a group theoretical perspective the wave function of n bosons or n fermions is respectively the fully symmetric or fully antisymmetric irreducible representation of the permutation group S_n, the other representations of S_n apparently not being used by Nature in this regard.[2]

However, theoretical examination of composites like the planar rotator bound to a tube of magnetic flux indicates that the exchange of two identical composites multiplies the wave function of the total system by a complex phase factor like that in (6.102). Since the phase in this factor can have any value, the composite of particle and flux has been termed an 'anyon'. The study of anyons raises the fascinating question of whether Fermi–Dirac and Bose–Einstein statistics are actually special cases of a more general quantum statistics and, indeed, a general theory of the quantum statistics of two-dimensional systems has been developed [188].

Do anyons – or anything closely resembling them in the three-dimensional world – actually exist? The answer is quite possibly an affirmative one, although definitive proof has yet to be provided. Nevertheless, their existence has been postulated in theories of diverse phenomena like the fractional quantum Hall effect (where the effective charge carriers appear to come in fractions of a single electron charge) and high-temperature superconductivity [189].

Two objections are perhaps likely to arise at this point concerning the direct observability, even in principle, of the flux-dependent phase factor. First, how would one be able to know whether the bound particle has rotated around the flux tube or not? The canonical angular momentum and the angular coordinate ϕ (or, more precisely, harmonic functions of ϕ) are effectively conjugate variables which, by the uncertainty principle cannot be known simultaneously.

---

[2] The total number of classes $r(n)$ of the permutation group S_n is equal to the number of ways to partition n into a set of positive integers that sum to n. This is given by the coefficient of x_n in the formal power series expansion of the Euler generating function. See [187].

Since the states of the planar rotator are characterized by sharp (albeit flux-dependent) values of angular momentum, the wave function is delocalized over the entire available space like a standing wave, rather than like a particle. Second, the phase factor is a global factor independent of the state of the system, and therefore vanishes when one calculates expectation values or transition probabilities. How can it then enter the mathematical expression for some observable quantity?

Although it is impossible to ascertain the angle of rotation of a particle in one of the planar rotator eigenstates, one can always conceive of a particle in a state described by a linear superposition of such eigenstates spanning a range of quantum numbers μ. The greater the uncertainty in angular momentum, the more localized will be the wave packet representing the state of the particle. Moreover, since each superposed eigenstate contributes the same flux-dependent phase factor under rotation, the rotationally transformed wave packet will have exactly the same form as expression (6.102). With regard to the second question, it should be noted that under the conditions of a quantum interference experiment in which rotational pathways about two magnetic flux tubes are available to the particle, the phase in (6.102) would cease to be global, the experimental outcome then being sensitive to the *relative* phase of the two probability amplitudes.

To test the predicted influence of the isolated magnetic flux, it is not necessary to limit one's considerations to a strictly periodic system like the 2D

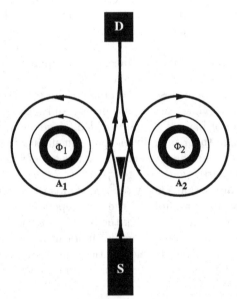

Fig. 6.5. Split-beam electron interference experiment manifesting the influence of winding number in the AB effect. Depending on the value of the magnetic flux, a spinless charged particle can behave under rotation like a fermion. (Adapted from Silverman [168])

rotator. Consider instead a configuration such as that of Fig. 6.5 in which a collimated beam of charged particles (with or without spin) is split into two coherent beams and made to circulate in orbits of equal radius and opposite sense about similar solenoids generating fluxes Φ_1, Φ_2 respectively [168]. As before, use of the classical terminology of 'splitting a beam', should not disguise the intrinsically quantum nature of the experiment; what is intended is that one particle at a time propagate through the apparatus – just as in the interference experiments with an electron microscope beam. It is therefore the two components of a *single-particle* wave packet that propagate around the two solenoids. If ever probed, one would find, of course, that there is but one particle and that it had taken one circular path or the other, but not both. To make particles circulate in the prescribed manner one could in principle employ uniform, time-independent background magnetic fields of equal strength and opposite orientation; these background fields would then contribute no net relative phase to the AB phase shift.

Upon recombination of the coherently split beam after n revolutions about one or the other solenoid, the wave function (for parallel tubes of flux) takes the form

$$\psi(\phi + 2\pi n) = \frac{1}{\sqrt{2}}\left[e^{-i2\pi n\Phi_1/\Phi_0}\psi_1(\phi) + e^{-i2\pi n\Phi_2/\Phi_0}\psi_2(\phi)\right] . \qquad (6.103)$$

For equal amplitudes ψ_1 and ψ_2 the state (6.103) leads to a forward beam intensity

$$I(2\pi n) \propto |\psi(\phi + 2\pi n)|^2 \propto \cos^2\left(2\pi n\frac{\Phi_1 + \Phi_2}{\Phi_0}\right) I_0 , \qquad (6.104)$$

where I_0 is the incident beam intensity. Thus, the predicted phase factor can be made to have experimental consequences in a split-beam quantum interference configuration. Although there is no direct contact between the particles of the beam and the magnetic fields within the two solenoids, the forward beam intensity reveals the number of times a particle has circulated around one of the solenoids (although, of course, not which one).

The quantum number n in (6.102) and ensuing relations is related to the topological concept of winding number which plays a fundamental role in the study of the connectivity of spaces. Paths through a space can be defined in such a way that they form a group known as the homotopy group [190]. One might think that there would ordinarily be an infinite number of different paths, but from the perspective of group theory this is not necessarily so. There are as many distinct homotopic classes of paths – paths that cannot be continuously deformed into one another (i.e., without cutting and pasting) – as there are distinct group operations. In a three-dimensional space without holes, representable by the unit sphere, there are only two classes of paths. All paths that begin at the origin and cut the surface of the sphere an even number of times can be deformed to a single point at the origin. Conversely, all paths that cut the surface an odd number of times can be deformed into a simple closed path between the origin and a single point on the surface. Thus, there are only two homotopy classes for a three-dimensional sphere: even and odd. Surprisingly, the smaller two-dimensional space of an annulus is homotopically the richer. Paths that have the same end points and the same winding number can be deformed into one another and therefore belong to the same class; they are homotopically equivalent. The homotopy group of an annulus contains an infinite number of classes.

In the two-slit AB experiment of Chap. 1, only the classical paths available to the electron – those that pass above and below the solenoid – were included in the superposition of probability amplitudes [see (1.28)]. Is it conceivable, however, that every once in a while an electron in the beam makes one or more full loops about the solenoid before propagating to the detector? The existence of such nonclassical paths of higher winding number can be justified by theory [191], but the amplitudes for these processes must be very weak, for no evidence of their contribution has yet been discerned in the resulting interference patterns.

6.6 Quantum Interference in a Metal Ring

Although electron beam experiments in vacuum have so far exhibited no distinguishable quantum effects attributable to particle winding, such effects *have* been seen in a two-dimensional system that at first thought would appear most unlikely to display any quantum interference at all. Experimental investigations of very small, normal (i.e., not superconducting) metal rings have revealed surprising quantum behavior in total contrast to that anticipated from the classical theory of metals (or even from a number of incautious applications of quantum theory) [192].

Figure 6.6 shows a representative sample, a gold ring of approximately 1 μm in diameter with wire thickness of a few hundredths of a micron. The ring, which contains about 10^8 atoms, is said to be of mesoscopic size, i.e., a scale between atomic and macroscopic dimensions. (Cells of the human body are roughly 5–20 μm in size.) The tiny ring is part of a circuit; electrons are introduced at one terminal, flow through the ring, and exit at the diametrically opposite terminal.

Except for topology, there is seemingly little in common between a metal ring with its lattice of some millions of positive ions and ambient sea of bound and conduction electrons, and the single electron circulating in a space devoid of other matter. Indeed, the latter system does not have a resistance. By contrast, a conduction electron in a normal disordered metal ring does not propagate freely, but diffuses through the metal in the manner of a random walk, undergoing collisions with impurity atoms and lattice defects. The mean free path of such an electron, typically on the order of 10 nm (0.01 μm), is about 100 times smaller than the sample length. As a consequence of the frequent collisions, it is intuitively reasonable to suppose that the phase of the wave function of the electron at the exit terminal should be essentially unrelated to the phase at the entrance. In other words, there should be no electron coherence across the ring, so that the probability amplitudes for passage between entrance and exit by clockwise or counterclockwise paths about the hole could not interfere. Thus, one would not expect to observe the AB effect in a mesoscopic metal ring subjected to a magnetic field through the central hole.

Contrary to such expectations, the metal ring *does* exhibit the AB effect, as revealed by oscillations (with period $\Phi_0 = hc/e$) in the resistance of the ring as a function of the magnetic field, as also shown in Fig. 6.6. Strictly speaking, the conditions for the AB effect do not hold in these experiments, for the magnetic field penetrates the entire ring and not solely the central hole. The unbound electrons are therefore subject to the Lorentz force, but this is *not* the source of the oscillations (although field penetration of the ring has important consequences as will be discussed shortly). The main point

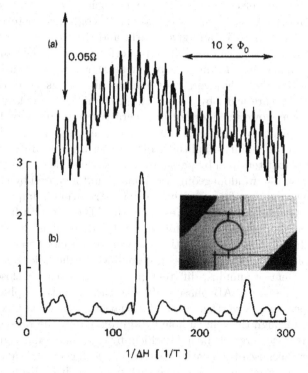

Fig. 6.6. Electron interference in a mesoscopic normal metal ring shown in the *insert*. The resistance as a function of magnetic field (*curve a*) displays oscillations with period Φ_0 attributable to the AB effect for electron trajectories between the entrance and exit terminals. The Fourier transform (*curve b*) also reveals an oscillation period of $\Phi_0/2$ for cyclic electron trajectories beginning and ending at the point of entry. (Courtesy of S. Washburn)

at present which requires explanation is the surprising survival of electron coherence. How can a particle that has undergone hundreds or thousands of scatterings exhibit self-interference?

The solution to the enigma is that *elastic* collisions, which predominate in a reasonably pure metal at low temperature (below 1 K), do not cause loss of phase memory; the electron coherence is destroyed only by inelastic collisions. Consequently, the coherence length of the electron is not to be identified with the total mean free path, but rather only with the mean distance between inelastic scatterings or other phase-randomizing events, which can attain values of a few microns – i.e., a size larger than the ring. Until the experiments on mesoscopic rings, prevailing opinion held that *all* scattering would destroy the electron coherence. This belief is now known to be mistaken.

The idea of coherence, which ordinarily refers to an ensemble of similarly prepared systems, must be interpreted a little differently in the present case, since the phase of the exiting electron wave function is a function of the random path through the metal. One, therefore, cannot prepare the quantum states of conduction electrons in the ring in such a way as to control the phase of exiting electrons. Nevertheless, as elastic collisions are reversible, the phase is well determined for any given path, and therefore all electrons that follow the identical path will exhibit self-interference with the same relative phase. The conductance of a mesoscopic metal ring can vary markedly from one sample to another, the average over many such rings yielding the bulk value deducible theoretically by a quantum mechanical ensemble average. An ensemble average, however, does not in general apply to an individual ring.

As a consequence of mesoscopic coherence, there is a nonvanishing interference between probability amplitudes of a carrier that has diffused from the entrance to the exit terminal by passing (once) to one side or the other of the hole. In addition to a random geometrical phase shift acquired by propagation and scattering (and therefore sensitive to the distribution of impurities in the metal), the interference term acquires, as in the AB effect with free particles, a supplementary magnetic phase shift $(2\pi\Phi/\Phi_0)$ determined by the amount of magnetic flux enclosed by the path. However, because the magnetic field in the mesoscopic ring experiments was not confined to the hole, but uniformly permeated the entire annulus, different pairs of paths can enclose different values of flux. Thus, the AB phase shift, like the geometrical phase shift, is also path dependent. Nevertheless, the fluctuation in enclosed flux is smaller than the flux through the central hole by approximately the ratio of the area of the annulus to the area of the hole, which, by experimental design, is a very small number. Because of the path sensitivity of both geometric and magnetic phase shifts, the variation of resistance with magnetic field displays an aperiodic random background upon which is superposed – as the signature of the AB effect – an oscillatory fine structure with period Φ_0.

The periodicity of the oscillations can be accurately established by electronically determining the Fourier transform of the magneto-resistance data. With judicious choice of ring geometry, however, the Fourier transform shows an additional peak at a magnetic field corresponding to an oscillation period $\Phi_0/2$ as shown again Fig. 6.6. It is perhaps no surprise, in view of the preceding discussion on winding, that an oscillation with phase angle $\Phi/(\Phi_0/2) = 2\Phi/\Phi_0$ is attributable to interference involving two paths – one clockwise, the other counterclockwise – making one complete revolution about the ring. (Since electrons following a $360°$ path return to their point of injection and do not leave the ring, they therefore contribute to a diminution in conductance or increase in resistance.) The two counter-revolving paths need not be identical, in which case they enclose different values of flux, and the contribution of all such pairs of paths within the ring is again a statistical one. Theoretically, the total electrical resistance of the ring as a function of magnetic field B can be expressed in the form

$$R(B) = R_{\mathrm{b}} + \sum_{n=1} R_n \cos\left(\alpha_n + 2\pi n \frac{\Phi}{\Phi_0}\right) , \qquad (6.105)$$

where R_{b} is the aperiodic background comprising the classical resistance (including a magneto-resistance term proportional to B^2) and the zero-frequency part of the AB effect, and R_n and α_n (for integer $n = 1, 2, \ldots$) are random functions of B approximately inversely proportional to the winding number n and varying over a domain of B proportional to the area of the annulus.

There is one circumstance, however, where the AB effect in a mesoscopic disordered metal is *not* a statistical one, by which I mean it is totally insensitive to the distribution of impurities along path. This is the case where the magnetic field is confined to the central hole, and contributing paths through the metal are precisely time-reversed images of one another. In other words, if one path through the metal involves electron scattering at impurity sites $1, 2, 3, \ldots, N$, then the other path involves the same scattering events in reverse order, $N, \ldots, 3, 2, 1$. Such pairs of time-reversed trajectories can *only* occur for complete revolutions about the ring, and therefore give rise to magneto-resistance oscillations with maximum period $\Phi_0/2$.

Ironically, it was just this condition that prevailed in the first observation of the AB effect in a normal metal structure – with the configuration of a cylinder, not a two-dimensional ring [193]. The sample was a quartz fiber about 1 cm in length and 1 μm in diameter upon which a metal film had been deposited. In stark contrast to the features of the AB effect in a ring described above, the variation of sample resistance with magnetic field yielded a smoothly oscillatory function with period $\Phi_0/2$ only. There was no contribution at the fundamental period Φ_0. Here is an example in which the AB effect derives exclusively from the occurrence of winding.

One might have expected that mesoscopic cylinders and rings – which, after all, are topologically equivalent – should exhibit similar magneto-resistance behavior. Why, then, were no AB oscillations observed at the fundamental period Φ_0 in the cylinder? A cylinder may be thought of as a parallel stack of mesoscopic rings. If the length of the stack is much greater than the coherence length of a ring, the AB oscillations with period Φ_0 from different rings (i.e., from segments of the cylinder separated by more than the coherence length) are uncorrelated and average out, leaving only a contribution from the special subset of time-reversed paths. In essence, the three-dimensional cylindrical geometry executes an ensemble average of mesoscopic rings [194].

Nevertheless, whether in a single ring or cylindrical stack, the contribution of time-reversed paths to the AB effect is rapidly extinguished when magnetic flux penetrates the metal to an extent on the order of Φ_0. Different pairs of paths then enclose different values of flux, and the resulting oscillations are no longer in phase.

Appendix 6A   Magnetic Hamiltonian of the Two-Dimensional Rotator

The Hamiltonian of a particle with charge q and mass m in a static magnetic field $\boldsymbol{B}$ is

$$H_{\mathrm{M}} = \frac{\boldsymbol{P}^2}{2m} = \frac{1}{2m}\left(\boldsymbol{p} - \frac{q\boldsymbol{A}}{c}\right)^2 , \tag{6A.1}$$

in which $\boldsymbol{P}$ is the kinetic linear momentum, $\boldsymbol{p}$ is the canonical linear momentum, and the vector potential $\boldsymbol{A}$ can be written in the form

$$\boldsymbol{A} = -\frac{1}{2}\boldsymbol{r} \times \boldsymbol{B} , \tag{6A.2}$$

whose curl is identically $\boldsymbol{B}$. Expanding the square in (6A.1) with substitution of (6A.2) leads to the expression

$$H_{\mathrm{M}} = \frac{\boldsymbol{p}^2}{2m} + \boldsymbol{\omega}_{\mathrm{L}}\cdot\boldsymbol{L} + \frac{1}{2}m\omega_{\mathrm{L}}^2 r^2 \sin^2\theta , \tag{6A.3}$$

in which

$$\boldsymbol{\omega}_{\mathrm{L}} = -\frac{q\boldsymbol{B}}{2mc} \tag{6A.4}$$

is the Larmor angular velocity and θ is the angle between the particle coordinate vector $\boldsymbol{r}$ (of constant magnitude R) and the magnetic field. For a negatively charged particle ($q = -e$), $\boldsymbol{\omega}_{\mathrm{L}}$ and $\boldsymbol{B}$ are parallel.

In the example of the two-dimensional electron rotator depicted in Fig. 6.1, the Cartesian components of the coordinate vector and magnetic field are

$$\boldsymbol{r} = R(\cos\phi, \sin\phi, 0) , \tag{6A.5}$$

$$\boldsymbol{B} = B(0, \sin\beta, \cos\beta) , \tag{6A.6}$$

from which it follows that $\cos\theta = \sin\phi\sin\beta$. Moreover, the differential operator $\boldsymbol{p}^2 = -\hbar^2\nabla^2$, where ∇^2 is the two-dimensional Laplacian, is simply related to the square of the angular momentum operator $L_z^2 = -\nabla^2/\hbar^2 R^2$ by $\boldsymbol{p}^2 = L_z^2/R^2$. Reduction of (6A.3), with account taken of the preceding relations, leads directly to the expression

$$H_{\mathrm{M}} = \frac{L_z^2}{2mR^2} + \omega_{\mathrm{L}} L_z \cos\beta + \frac{1}{2} m\omega_{\mathrm{L}}^2 R^2 (\cos^2\phi + \sin^2\phi\cos^2\beta) \qquad (6A.7)$$

of (6.41).

Chiral Asymmetry:
The Quantum Physics of Handedness

7.1 Optical Activity of Mirror-Image Molecules

By the end of the second decade of the 19th century – long before the discovery of X-rays and the invention of the electron microscope – it still would have been possible, using nothing more than a beam of light and two polarizing crystals (e.g., calcite), to determine that the intrinsic structure of at least some molecules was three-dimensional (assuming one believed in molecules then). The revealing phenomenon of 'optical activity' – a rotation of the plane of polarization of the light upon transmission through the sample as shown in Fig. 7.1 – does not occur for all molecules, but only for those which cannot be superimposed on their mirror image. Such a structure, subsequently termed 'dissymmetric' by Louis Pasteur but referred to as 'chiral' (derived from the Greek for 'hand') in current terminology – must necessarily be three-dimensional because a flat object and its mirror image could always be made to superimpose. Reflection, as illustrated in Fig. 7.2, reverses the chirality or handedness of an object, thereby interchanging, for example, the right and left winding of a helix or screw.

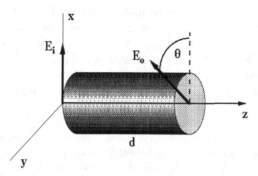

Fig. 7.1. Optical rotation of linearly polarized light by a chiral sample. Upon emerging from a sample of length d, the polarization vector E_o has been turned through an angle θ relative to the incident polarization vector E_i

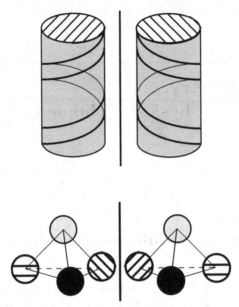

Fig. 7.2. Mirror reflection reverses the handedness of chiral objects. a chiral object and its mirror image cannot be superposed. The tetrahedron with four different vertex objects is representative of an asymmetric carbon atom, a common chiral building block in chemistry. Some materials, like crystalline quartz, are constructed from *achiral* molecules arranged in a helical structure

For a molecule to exist in two nonsuperposable mirror-image forms (called enantiomers), it must have no center or planes of symmetry, nor any rotation-reflection axes.[1] A dissymetric shape is not necessarily devoid of *all* symmetry, however; it may possess pure rotation axes like the C_2 or two-fold rotation axis of a helix. One of the most common manifestations of natural molecular chirality is that associated with tetrahedrally valent carbon atoms bonded to four different substituents (Fig. 7.2). Materials composed of distinct enantiomeric forms of the same molecule have identical bulk chemical and physical properties such as mass density, fusion and vaporization points, and rates of reaction (when chemically combining with *nonchiral* reactants). They also rotate the polarization of a linearly polarized light beam to the same extent per unit length of material, but in *opposite* senses. It is by their optical activity that one can most readily distinguish mirror-image molecules.

---

[1] Optically active substances can also be constructed from mirror-inequivalent arrangements of achiral molecules. Crystalline quartz is one such example; repeating units of silicon dioxide wind in helical fashion (with left or right circulations) about the optic axis. Unlike substances composed of intrinsically chiral molecules, however, the chirality, and therefore the optical activity, vanish when these enantiomorphic forms are melted or dissolved in solution. Thus, fused quartz exhibits no optical activity.

In certain cases, however, one could distinguish enantiomeric forms of a substance without any optical apparatus at all, but merely by their aroma. How can the human nose serve as a chiral detector? Herein is one indication of perhaps the most outstanding unsolved problem of the life sciences: the origin of biomolecular homochirality. All life forms – from the lowliest virus to the human body – are built from only one of the two possible enantiomeric forms of its chiral constituents and not from both forms in more-or-less equal measure as perhaps one might have expected. An organism may be composed, for example, of right-handed sugar molecules or left-handed amino acids. Indeed, the capacity (if not necessity) to synthesize and consume homochiral molecules is the very hallmark of the living state.[2]

Light, too, comes in two basic enantiomeric forms although they are not usually designated as such. These are the left and right circular polarizations. From the perspective of classical physical optics – as first explained by Augustin Fresnel in the 1820s – the phenomenon of optical rotation is attributable to circular birefringence, i.e., the difference in indices of refraction (n_L, n_R) for left and right circularly polarized light. A circularly polarized wave of angular frequency ω and wave number $k_{L,R} = n_{L,R}\omega/c$ can be regarded as a superposition of two orthogonal linearly polarized waves oscillating with a relative phase of $\pm\pi/2$ radians as represented by the polarization states

$$|L\rangle = \frac{1}{\sqrt{2}}(\boldsymbol{x} + e^{i\pi/2}\boldsymbol{y}) = \frac{1}{\sqrt{2}}(\boldsymbol{x} + i\boldsymbol{y}) , \qquad (7.1)$$

$$|R\rangle = \frac{1}{\sqrt{2}}(\boldsymbol{x} + e^{-i\pi/2}\boldsymbol{y}) = \frac{1}{\sqrt{2}}(\boldsymbol{x} - i\boldsymbol{y}) . \qquad (7.2)$$

Here $|L\rangle$ and $|R\rangle$ are complex-valued unit vectors representing left and right circular polarizations respectively for a wave propagating along the z axis. The unit vectors $\boldsymbol{x}$ and $\boldsymbol{y}$ (together with $\boldsymbol{z}$) are directed along the corresponding axes of an orthogonal right-handed coordinate system, i.e., $\boldsymbol{x} \times \boldsymbol{y} \cdot \boldsymbol{z} = 1$. The use of Dirac notation in (7.1) and (7.2) signifies that, from a quantum perspective, the same expressions serve as the basis states of circularly polarized photons. To obtain the physical traveling wave at location z and time t, which in classical optics is necessarily a real-valued function, one multiplies the expressions in (7.1) and (7.2) by a phase factor of the form $e^{i(kz-\omega t)}$ and then takes the real part. This leads to expressions

$$\boldsymbol{L}(z,t) = \frac{1}{\sqrt{2}}\left[\boldsymbol{x}\cos(k_L z - \omega t) + \boldsymbol{y}\sin(k_L z - \omega t)\right] , \qquad (7.3)$$

$$\boldsymbol{R}(z,t) = \frac{1}{\sqrt{2}}\left[\boldsymbol{x}\cos(k_R z - \omega t) - \boldsymbol{y}\sin(k_R z - \omega t)\right] , \qquad (7.4)$$

in which the feature of 'rotation' is plainly evident.

According to standard optical convention, the electric field of a left circularly polarized (LCP) wave rotates towards the left side of an observer facing

---

[2] Thorough discussions of the problem of biomolecular homochirality may be found in [195], and in the special issue of Biosystems [196].

the light source-with corresponding definition of right circular polarization (RCP) [7].[3] We will not need to distinguish by separate symbols the complex and real polarization vectors. The complex form is generally the more convenient to use, since the differential effects of a medium on the two states of polarization can be accounted for simply by appropriate phase factors of the form $e^{i\phi}$ where, as we will see shortly, ϕ can be complex-valued if light absorption (or emission) occurs.

Consider, therefore, an x-polarized wave incident upon a sample of transparent optically active active material (see, again, Fig. 7.1). Representing the incident wave by its electric field $\boldsymbol{E}_{\text{in}}$, and inverting the relations (7.1) and (7.2), one can write

$$\boldsymbol{E}_{\text{in}} = \boldsymbol{x} = \frac{1}{\sqrt{2}}(\boldsymbol{L} + \boldsymbol{R}) . \tag{7.5}$$

Inside the chiral medium, the left- and right-circularly polarized components of the incident wave advance at a phase velocity determined by their respective refractive indices $n_{\text{L,R}}$. The recombined wave emerging from the sample after a geometric path length d then has the form

$$\boldsymbol{E}_{\text{out}} = \frac{1}{\sqrt{2}}\left(\boldsymbol{L}e^{in_{\text{L}}\omega d/c} + \boldsymbol{R}e^{in_{\text{R}}\omega d/c}\right) = \frac{1}{\sqrt{2}}(\boldsymbol{x}\cos\theta - \boldsymbol{y}\sin\theta)e^{in\omega d/c} \tag{7.6}$$

of a linearly polarized wave rotated by the angle

$$\theta = \frac{(n_{\text{L}} - n_{\text{R}})\omega d}{2c} \tag{7.7}$$

in a medium of mean refractive index

$$n = \frac{n_{\text{L}} + n_{\text{R}}}{2} . \tag{7.8}$$

If $n_{\text{L}} > n_{\text{R}}$ the rotation angle is positive and the polarization vector is rotated towards the right side of an observer facing the source. In other words, the rotation is in the sense of the wave with the larger phase velocity. The global phase factor in (7.6) is of no consequence unless the wave is subsequently superposed with a reference wave.

Optical rotation of linearly polarized light is but one example of optical activity, the assortment of optical responses that can distinguish chiral forms. Circular dichroism is another in which linearly polarized light propagating through a nontransparent chiral substance is converted to elliptically polarized light. Elliptical polarization is a linear superposition of linear and circular polarizations; the motion of the electric vector projected onto a plane perpendicular to the direction of light propagation traces out an ellipse. In this case the phenomenon is attributable to the difference in absorption of left and right circular polarizations. Both optical rotation and circular dichro-

---

[3] It is a common error to think that the electric vector of circularly polarized light traces out a circle in time. Since the light wave is advancing as the electric vector is rotating, the actual locus of points traced out would resemble something like the twisted ribbon on a barber's pole.

ism can be treated together by assigning different complex-valued refractive indices of the form $\tilde{n} = n + i\kappa$ for left and right circular polarizations. The absorption coefficient κ then leads to a diminution in amplitude with distance: $e^{i\tilde{n}d/c} = e^{ind/c}e^{-\kappa d/c}$. In addition to the manifestations of optical activity in transmitted light, there are other effects, more difficult to observe, but of conceptual and practical importance, associated with reflected light [197][4] or scattered light [199]. This includes, for example, the difference in reflection of incident LCP and RCP light, and the conversion by reflection of an incident linearly polarized wave into an elliptically polarized wave.

Optics, alone, does not account for the origin of the circular birefringence of chiral media. For this, one must have a microscopic model of the way in which a chiral molecule interacts with electromagnetic radiation. Within the framework of classical physics, a simplified heuristic explanation of optical rotation may be had by examining a system of conducting helices irradiated by linearly polarized electromagnetic waves. Consider first a single helix. The electric field of the incident wave drives charges back and forth along the helical pathway thereby inducing oscillating electric and magnetic dipole moments which radiate secondary (or scattered) waves with their own characteristic dipole patterns. The electric dipole arises because of charge separation along the helix; the magnetic dipole is a consequence of the projected circular motion of the charges in a plane perpendicular to the helical axis. Depending upon the handedness of the helix, the induced moments will be either parallel or antiparallel to one another, and the net electric field of the scattered radiation from both dipoles can therefore take on two different orientations relative to the electric field of the incident wave. The plane of vibration of the resultant transmitted wave, a superposition of the incident and scattered waves, will consequently depend on the sense of the helix. Although, in a system of randomly oriented helical molecules, the extent of optical rotation will depend on the projection of the incident electric field along each helix, the sense of rotation will always be the same for helices of the same handedness.

From the perspective of quantum theory, optical rotation is the outcome of a quantum interference process occurring in the elastic scattering of incident photons by the induced electric and magnetic dipoles of a chiral molecule. There are several interesting features that distinguish this process from other examples of quantum interference we have considered previously. First, limiting this discussion to nonabsorbing molecules, one can remark that the light scattering involves virtual, as opposed to real, processes. A molecule, initially in its ground state, undergoes a nonresonant cyclic transition to an excited state and back again to the ground state by one of several indistinguishable interaction pathways. The apparent violation in energy conservation for each

---

[4] A long-standing controversy over the phenomenological description of optical activity resolved theoretically in this paper is described in detail in [114], Chap. 4. Reflective measures of chirality and the experimental procedures for observing them are described in [198].

one-way transition is, of course, undetectable, for the final state of the system conserves energy, and any attempt to probe the system during scattering would perturb it and destroy the sought-for interference effect. Second, the various interaction pathways correspond to different time sequences of absorption and emission by the electric and magnetic dipoles. However, since electric and magnetic dipole interactions are connecting the *same* two states (ground and excited), the states of the chiral molecule cannot be eigenstates of well-defined parity, and the process would appear to violate parity conservation. This is a consequence of the *molecular* dissymmetry. (The constituent atoms have a center of symmetry.) Yet structural optical activity is, after all, the result of an electromagnetic process-and the laws of electrodynamics, both classical and quantum, are invariant under parity transformations.

The invariance of electromagnetism under both parity (or coordinate inversion $x \to -x$) and time reversal ($t \to -t$) can be demonstrated directly from Maxwell's equations. Under inversion of coordinates the electric and magnetic fields transform as follows: $E(x) \to -E(-x)$, $B(x) \to B(-x)$. Likewise, reversing time leads to the transformation: $E(t) \to E(-t)$, $B(t) \to -B(-t)$. Since all fields of a particular kind, irrespective of the charge or current configuration producing them, must behave in the same way under symmetry transformations, one can understand the above transformations by examining the simple systems of a charged parallel-plate capacitor and a current-carrying solenoid. Under inversion, the plates of the capacitor exchange positions, and the orientation of the electric field is thereby reversed. Since the charges are stationary, they (and the electric field) are unaffected by time reversal. Similarly, coordinate inversion has no effect on either the handedness of the solenoid or the direction of charge flow, in which case the magnetic field is unaffected. Under time reversal, however, charge flows through the solenoid in the opposite direction, and the orientation of the magnetic field is therefore reversed. The net effect of these transformations is to leave Maxwell's equations invariant.

In the following section we will examine more closely optical activity as a quantum interference process and the issue of symmetry. It will be seen that optical activity deriving from chiral molecular structure does not violate the conservation laws of electrodynamics. On the other hand, optical activity in unbound atoms would. Atomic optical rotation and other manifestations of optical activity in atoms have in fact been observed and can not be accounted for by electromagnetic processes alone, but arise from the weak nuclear interactions [200].

7.2 Quantum Interference and Parity Conservation

The interaction of an electromagnetic wave with a chiral molecule or crystal is in general a complicated process to treat. However, to illustrate simply how just one facet of optical activity, namely optical rotation, arises as a quantum interference process, we adopt a model [201] which discards all but the most essential features of the interaction and reduces the problem to the familiar form of a two-level system. In this unadorned quantum electrodynamic (QED) model, the direction of propagation, wave number, total energy, and occupation number of an incident photon are unchanged; the influence of the medium is manifested exclusively in its effects on the state of the photon's polarization. Since the photon is undeviated in its passage through the sample, the process is known as forward scattering.

The quantum optical attribute corresponding to circular polarization is technically designated 'helicity', the projection $\boldsymbol{S} \cdot \boldsymbol{k}/|\boldsymbol{k}|$ of the photon angular momentum $\boldsymbol{S}$ (in units of $\hbar$) onto the linear momentum $\boldsymbol{k}$ (also in units of $\hbar$). Since the photon is a massless spin-1 boson, this projection may have the two values ± 1, which corresponds, respectively to LCP and RCP states. Suppose the incident light propagates along the z axis and is polarized along the x axis. One can represent this state quantum mechanically by the spinor

$$\begin{pmatrix} 1 \\ 0 \end{pmatrix} ,$$

where the corresponding spinor

$$\begin{pmatrix} 0 \\ 1 \end{pmatrix} ,$$

characterizes a y-polarized photon. The complete state of a forward scattered photon (within the framework of our assumptions above) then takes the form

$$\begin{pmatrix} \phi_1 \\ \phi_2 \end{pmatrix} ,$$

where the complex-valued amplitudes ϕ_i ($i = 1, 2$) are to be determined by solution of the Schrödinger equation with appropriate Hamiltonian. The general form, however, is one which we have encountered in Chap. 5

$$H\Psi = \begin{pmatrix} \Omega & \overline{\omega} \\ \overline{\omega}^* & \Omega \end{pmatrix} \begin{pmatrix} \phi_1 \\ \phi_2 \end{pmatrix} = \mathrm{i}\frac{\partial\Psi}{\partial t} . \tag{7.9}$$

All elements of the Hamiltonian matrix are again expressed in angular frequency units. The diagonal elements characterize the energy of the light in the chiral medium which is the same for either state of linear polarization. The off-diagonal elements characterize the interaction (to be specified shortly) which breaks the degeneracy of the polarization states and induces transitions between them. Since no absorption or dissipation occurs in the processes under consideration, the Hamiltonian is Hermitian, and the off-diagonal elements in general satisfy the relation $h_{ij} = h_{ji}^*$ as indicated explicitly in the matrix form of H in (7.9). The interaction $\overline{\omega}$ can be written in the form

$$\overline{\omega} = M\mathrm{e}^{\mathrm{i}\alpha} , \tag{7.10}$$

with real-valued modulus M and phase α.

The chiral molecules and light together comprise a single, closed system in which total energy is conserved and the Hamiltonian is independent of time. Using the relations of Chap. 5 [in particular (5.7), (5.10) and (5.11)] with the correspondences

$$h_0 = \Omega , \quad h_1 = \mathrm{Re}(\bar{\omega}) , \quad h_3 = \mathrm{Im}(\bar{\omega}) , \quad h_3 = 0 , \tag{7.11}$$

one can integrate (7.9) immediately to obtain the polarization state

$$\begin{pmatrix} \phi_1 \\ \phi_2 \end{pmatrix} = e^{-i\Omega t} \begin{pmatrix} \cos Mt & -ie^{i\alpha} \sin Mt \\ -ie^{-i\alpha} \sin Mt & \cos Mt \end{pmatrix} \begin{pmatrix} \phi_1^0 \\ \phi_2^0 \end{pmatrix} . \tag{7.12}$$

If the incident light is x-polarized, the above state reduces to

$$\begin{pmatrix} \phi_1 \\ \phi_2 \end{pmatrix} = e^{-i\Omega t} \begin{pmatrix} \cos Mt \\ -ie^{-i\alpha} \sin Mt \end{pmatrix} , \tag{7.13}$$

which represents the geometrical polarization vector

$$e = x \cos \frac{Mnd}{c} - y e^{i(\pi/2 - \alpha)} \sin \frac{Mnd}{c} , \tag{7.14}$$

where, as usual, the global phase factor has been suppressed, and the time for a photon with phase velocity c/n to propagate a length d is $t = nd/c$. (There is no distinction to be made in this idealized model between phase velocity and group velocity because the photons are assumed to be in monochromatic plane wave states.)

It will be demonstrated soon that the matrix element $\bar{\omega}$ is a pure imaginary number, with a phase $\alpha = \pm\pi/2$ depending on the light frequency and the sign of a particular product of electric and magnetic dipole matrix elements referred to as the 'rotational strength'. If the rotational strength is positive and the frequency is below that of the nearest electronic resonance, then $\alpha = \pi/2$, and therefore $\bar{\omega} = iM$. The matrix element $\bar{\omega}$ changes sign as the light frequency passes through each resonance. Ordinarily, molecular electronic transitions fall in the ultraviolet portion of the spectrum, and one observes optical rotation by means of lower-frequency visible light. Substitution of $\alpha = \pi/2$ in (7.14) leads to a polarization vector identical in form to that in relation (7.6) with the optical rotation angle now given by

$$\theta = \frac{Mnd}{c} . \tag{7.15}$$

Although (7.6) [or (7.14)] and (7.13) characterize the same physical state, their interpretations are quite different. In the classical picture of optical rotation the electric vector of the light wave undergoes a continuous rotation as the beam passes through the chiral medium. At any moment the polarization is a well-determined quantity. According to QED, however, the exact polarization of the photon is uncertain until a measurement is performed. A measurement at time t would reveal a state of x-polarization with a probability

$$P_x(t) = \cos^2(Mt) \tag{7.16}$$

and a state of y-polarization with corresponding probability

$$P_y(t) = \sin^2(Mt) .\tag{7.17}$$

Rotating the analyzer (aligned initially along x) through the angle θ of (7.15) would reveal a polarization state of (7.14) with 100% probability.

Comparing (7.7) and (7.15) allows one to express the circular birefringence in terms of the QED matrix element

$$n_{\text{L}} - n_{\text{R}} = \frac{2Mn}{\omega} .\tag{7.18}$$

Since the two refractive indices for circular polarization must reduce to the mean index n in the absence of a chiral interaction ($\varpi = 0$), and interchange under a change of sign of ϖ, it follows that

$$n_{\text{L,R}} = n\left(1 \pm \frac{M}{\omega}\right) ,\tag{7.19}$$

where the upper sign corresponds to left circular polarization.

The consistency of the above reasoning, based on a comparison of results from QED and classical optics, can be checked by returning to the Schrödinger equation (7.9) and diagonalizing the Hamiltonian to obtain the eigenvalues and eigenvectors of the photon in a chiral medium. For the present case of a frequency below resonance where $\alpha = \pi/2$, solution of this simple eigenvalue problem leads to the characteristic frequencies

$$\Omega_{\pm} = \Omega \pm M \tag{7.20}$$

and associated vectors (normalized to unit magnitude)

$$\phi_{\pm} = \frac{1}{\sqrt{2}}\begin{pmatrix} 1 \\ \mp i \end{pmatrix} .\tag{7.21}$$

Inspection of relations (7.1) and (7.2) and (7.21) shows that $\phi_{\pm}$ are states of right and left circular polarization, respectively. However, do the eigenvalues in (7.20) correlate correctly with the appropriate refractive indices? Applying the dispersion relation between frequency and wave number

$$\Omega = \frac{ck}{n} ,\tag{7.22}$$

to the eigenfrequencies in (7.20) leads to the expressions

$$\frac{ck}{n_{\pm}} = \frac{ck}{n} \pm M ,\tag{7.23}$$

from which the chiral refractive indices can be determined. Since the chiral interaction M is orders of magnitude smaller than an optical frequency (as will be demonstrated shortly), one can solve for $n_\pm$ in the approximation of $M \ll ck$ to find

$$n_\pm = n \left(1 \mp \frac{Mn}{ck}\right) = n \left(1 \mp \frac{M}{\Omega}\right) , \tag{7.24}$$

in agreement with (7.19) and (7.21) where n_+ is associated with right circular polarization and n_- with left circular polarization. According to our model, there is no energy gained or lost when a photon enters the chiral medium; the diagonal element Ω can then be identified with the vacuum angular frequency ω. The significance of the eigenvectors $\phi_\pm$ is that they propagate through the chiral medium *unchanged* in form and at well-defined phase speeds characterized by their respective indices of refraction.

We examine next the dynamical part of the problem leading to the interaction element $Me^{i\alpha}$. To simplify matters, consider first the interaction of a photon and a single chiral molecule with ground state $|0\rangle$ and spectrum of excited states $|n\rangle$. The Hamiltonian for the total system of radiation and molecule is then the sum

$$H = H_{\mathrm{rad}} + H_{\mathrm{mol}} + H_{\mathrm{int}} , \tag{7.25}$$

where the interaction term

$$H_{\mathrm{int}} = -\boldsymbol{\mu}_{\mathrm{E}} \cdot \boldsymbol{E} - \boldsymbol{\mu}_{\mathrm{M}} \cdot \boldsymbol{B} \tag{7.26}$$

expresses the coupling of the molecular electric dipole moment $\boldsymbol{\mu}_{\mathrm{E}}$ and magnetic dipole moment $\boldsymbol{\mu}_{\mathrm{M}}$ to the electric and magnetic fields of the photon. Explicit expressions for the other terms of the Hamiltonian are not needed; the first leads to the energy of the free radiation field (represented in our problem by the frequency ω) and the second to the molecular energies Ω_n which we shall assume to be known. As before, the Hamiltonians are in units of $\hbar$, and we will refer to energy and angular frequency interchangeably.

One might wonder how the interaction Hamiltonian of (7.26) relates to the standard Hamiltonian

$$H = \frac{1}{2m} \left(\boldsymbol{p} - \frac{q\boldsymbol{A}}{c}\right)^2 + q\phi$$

obtained by minimal coupling of a charged particle with linear momentum $\boldsymbol{p}$ and charge q to electromagnetic vector and scalar potentials $(\boldsymbol{A}, \phi)$ in an explicitly gauge-invariant way. This question is by no means a trivial one, and has led to repeated discussion in the physics literature even though the problem was effectively resolved long ago [202]. In brief, the electric and magnetic dipole interaction terms correspond to the terms $-q\boldsymbol{A} \cdot \boldsymbol{p}/mc + q^2 A^2/2mc^2$ in the expansion of the preceding Hamiltonian after implementation of an appropriate gauge transformation.

Within the framework of QED, the electric and magnetic fields in (7.26) are operators which act on photon states $|\boldsymbol{k}\sigma\rangle$ of momentum $\boldsymbol{k}$ and polarization label $\sigma = 1, 2$ designating the two orthonormal linearly polarized basis vectors $e^{(\sigma)}$ for each wave vector k. Strictly speaking, one should write $e^{(\sigma)}(k)$, but we will not do this to avoid encumbering the notation unnecessarily, since the direction of k – and therefore of $e^{(\sigma)}$ – remains unchanged throughout the scattering. The total system of radiation and matter can then be represented initially by a state vector of the form $|0; k_1\rangle$ for a molecule in its ground state and one photon of polarization $e^{(1)}$. From the explicit representation of the radiation fields in terms of creation $a(k, \sigma)^\dagger$ and annihilation $a(k, \sigma)$ operators [203]

$$E(r) = \sum_{k,\sigma} \left(\frac{2\pi\hbar\omega}{V}\right)^{1/2} ie^{(\sigma)}\left[a(k,\sigma)e^{ik\cdot r} - a(k,\sigma)^\dagger e^{-ik\cdot r}\right], \qquad (7.27)$$

$$B(r) = \sum_{k,\sigma} \left(\frac{2\pi\hbar\omega}{V}\right)^{1/2} \frac{ik \times e^{(\sigma)}}{k}\left[a(k,\sigma)e^{ik\cdot r} - a(k,\sigma)^\dagger e^{-ik\cdot r}\right], \qquad (7.28)$$

it is seen that E and B create or annihilate one photon at a time. In the above expressions V is the quantization volume – i.e., the volume within which the field modes are defined – which we identify here with the total volume of the sample containing chiral molecules. Although the frequency of the incident photon is assumed not to correspond to a resonance of the molecule, the Hamiltonian H_{int} induces transitions to intermediate states of the form $|n; 0\rangle$ in which the incoming photon has been absorbed and $|n; k_1, k_2\rangle$ in which an additional photon has been emitted, the molecule in both cases being in an excited state. These violations of energy conservation are immediately rectified by de-excitation of the molecule and subsequent emission or absorption of a photon to result in the final state $|0; k_2\rangle$. Thus, in the overall energy-conserving process of forward elastic scattering, the molecular state is unchanged and an incoming photon of polarization $e^{(1)}$ emerges with polarization $e^{(2)}$.

The probability amplitude for this process is given by perturbation theory to lowest nonvanishing order as

$$V_{12} = \sum_I \frac{\langle 0; k_2|H_{\text{int}}|I\rangle\langle I|H_{\text{int}}|0; k_1\rangle}{E_0 - E_I}, \qquad (7.29)$$

where the sum is over all possible intermediate states I. The energy (frequency) of the initial state is $E_0 = \Omega_0 + \omega$; the energy of an intermediate state I is either $E_I = \Omega_n$ or $E_I = \Omega_n + 2\omega$ depending on whether the transition was effected by absorption or emission of a photon. Substitution of H_{int} into (7.29) and evaluation of the resulting expression by means of (7.27) and (7.28), lead to a sum of four distinct matrix elements which contribute to the change in photon polarization. These are represented diagrammatically in Fig. 7.3 in which the time-ordered sequence of events in each diagram proceeds from bottom to top. At each vertex the interaction between molecule and light can be mediated by either an electric or magnetic dipole coupling in which the coupled field either annihilates the incoming photon or creates the outgoing photon. Two interactions (electric or magnetic) of the same kind do not appear in any of the diagrams, for these processes do not change photon polarization.

It may seem particularly strange that a molecule can radiate an outgoing photon from its *ground* state – i.e., *before* the arrival of the incoming photon – but such processes are permitted by QED and must in fact be included if the calculation is to be in accord with experiment. Indeed, the individual vertices of every diagram in the figure violate energy conservation; only *in toto* does a diagram represent a process in accord with physical law. The separate scattering processes of Fig. 7.3 are indistinguishable; all that can be observed is that a photon $e^{(1)}$ is incident on a sample of ground-state molecules, and a photon $e^{(2)}$ leaves the same sample. One can disregard, I suppose, the physical interpretation of the diagrams and consider the associated mathematics sim-

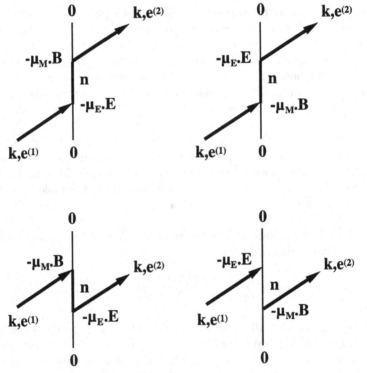

Fig. 7.3. Time-ordered diagrams of forward scattering processes contributing to optical rotation. *Vertical lines* represent the transition of a molecule out of, and back into, its ground state 0 via a virtual transition to intermediate excited state n. *Oblique lines* represent absorption of a photon with polarization $e^{(1)}$ and emission of a photon with polarization $e^{(2)}$. The diagrams are distinguished by the order of electric and magnetic dipole interactions and by the sequence of absorption and emission

ply as an exercise in the approximate solution of a differential equation. However, I think this would take away from physics much of the richness of its imagery, for the purpose of physics is not merely to compute, but to understand.

Evaluation of the matrix element (7.29) leads to the expression

$$V_{12} = \frac{4\pi\omega}{V} \sum_n \frac{1}{\omega^2 - \Omega_{n0}^2} \Big\{ i\omega \text{Im} \big[(\mu_E)_{0n;1}(\mu_M)_{n0;1} + (\mu_E)_{0n;2}(\mu_M)_{n0;2} \big] \quad (7.30)$$

$$- \Omega_{n0} \text{Re} \big[(\mu_E)_{0n;1}(\mu_M)_{n0;1} - (\mu_E)_{0n;2}(\mu_M)_{n0;2} \big] \Big\} ,$$

where the energy interval between ground and excited state is

$$\Omega_{n0} = \Omega_n - \Omega_0 \quad (7.31)$$

and the dipole matrix elements are defined by

$$(\mu_K)_{0n;\sigma} \equiv \langle 0 | \mu_K \cdot e^{(\sigma)} | n \rangle , \quad (7.32)$$

with $K = $ E, M and $\sigma = 1, 2$. Equation (7.30) pertains to a single molecule. In an actual experiment the sample contains a density of η molecules per unit of volume randomly oriented and uniformly distributed over the volume V. One must therefore average (7.30) over all orientations of the electric and magnetic dipole moments. Then, presuming that each molecule contributes individually to the overall optical rotation (i.e., that there is no cooperative interaction between molecules), one must multiply V_{12} by the number of molecules in the sample, ηV. This leads to the final expression for $Me^{i\alpha}$ in the Schrödinger equation (7.9)

$$\bar{\omega} \equiv Me^{i\alpha} = \frac{8\pi\eta\omega^2 e^{i\pi/2}}{3} \sum_n \frac{R_{n0}}{\omega^2 - \Omega_{n0}^2} , \quad (7.33)$$

where

$$R_{n0} \equiv \text{Im} \big[(\mu_E)_{0n} \cdot (\mu_M)_{n0} \big] \quad (7.34)$$

is designated the rotational strength for level n.

Although the exact evaluation of R_{n0}, and therefore M, for a real molecule is a difficult calculation, it is worthwhile to make a rough estimate to see the extent of contribution of the chiral interaction to the refractive indices of (7.19). We can approximate the electric dipole moment by ea_0, the product of electron charge and the Bohr radius of a ground-state hydrogen atom. Similarly, the magnetic dipole moment can represented by the Bohr magneton, $e\hbar/2mc$, where m is the electron mass. We then obtain $R_{n0} \sim 2.3 \times 10^{-38}$ in cgs units – or $\sim 2.2 \times 10^{-11}$ when divided by $\hbar$ – which is more or less an upper limit to actual values, since the induced electric and magnetic dipoles need not in general be parallel nor is the product of the matrix elements necessarily a pure imaginary number. Assuming, further, a sample density of that of water ($\eta \sim 3.3 \times 10^{22}$ molecules/cm^3), a red probe beam ($\lambda = 600$ nm, or $\omega = 3.14 \times 10^{15}$ s^{-1}), and a resonance in the ultraviolet ($\lambda = 100$ nm or $\Omega_{n0} = 6\omega$), we obtain $M \sim 1.7 \times 10^{11}$ s^{-1}. The chiral parameter in (7.19), is then $M/\omega \sim 5.3 \times 10^{-5}$. Though small, a chiral parameter of this magnitude is easily measurable and is actually one to two orders of magnitude larger than the corresponding parameters of many naturally occurring chiral molecules. Indeed, one of the recent significant achievements in chiral metrology was the measurement to one part in 10^7 of the difference with which chiral matter reflects LCP and RCP light [204].[5]

We return again to the question of whether or not optical activity – as represented, for example, by the matrix element (7.33) – violates the conservation of parity, since R_{n0} would be expected to vanish for states of well-defined parity. The fact that the eigenstates of each enantiomeric form of a molecule are not parity eigenstates does not mean, however, that optical activity is necessarily a parity nonconserving process. This question can be answered only by considering the entire system, matter plus radiation. Figure 7.4 illustrates the configuration of a linearly polarized wave incident from the right on a sample of chiral molecules which rotates the polarization clockwise for an observer facing the source. The mirror image of this process transforms the configuration to a linearly polarized wave incident from the left upon the other enantiomeric form of the original molecules; the polarization of the forward scattered light is rotated counterclockwise to an observer facing the source. But this is exactly what one expects to happen. The two enantiomeric forms of a chiral molecule rotate linearly polarized light in opposite senses, the sense of rotation being defined with respect to the direction of light propagation. (There is no other unique direction in the system.) Since both the original and mirror-image processes occur in nature, there is no violation of parity conservation. The mere fact that one can physically separate (or synthesize separately) the two forms of a chiral molecule and carry out optical experiments on only one of these forms does not constitute any violation of physical law.

Suppose, however, that instead of molecules in Fig. 7.4, the sample consisted of unbound atoms. Electrodynamically, an atom has a center of symmetry, since the electrons are bound to the nucleus by the isotropic Coulomb force. One would therefore not expect an atom to come in enantiomeric forms or to rotate the plane of incident linearly polarized light. Nevertheless, atoms have been shown both theoretically and experimentally to be optically active.

One significant outcome of the unification of electromagnetism and the weak nuclear interactions into a single 'electroweak' theory, was the prediction of weak neutral currents – in effect, a charge-preserving interaction between charged particles mediated by the exchange of the Z$_0$ boson, a neutral particle with mass one hundred times that of the proton mass. Because the range of an interaction is of the order of the Compton wavelength $\lambda_C = \hbar/Mc$ for

---

[5] The conceptual background to this experiment, as well as the experimental details, are discussed in greater detail in [7].

a mediating particle of mass M, the effects of weak neutral currents in atoms are largely confined to a region of radius $10^{-7}a_0$ about the atomic center. Only electronic S states substantially overlap the atomic nucleus. Thus, as a result of the weak interaction between atomic S electrons with nucleons in the nucleus, electronic S and P states are mixed to a very small extent

$$|S'\rangle \sim |S\rangle + \varepsilon|P\rangle \qquad (7.35)$$

(with $\varepsilon \sim 10^{-11}$ in hydrogen) and are no longer exact parity eigenstates [205]. The strength of the coupling grows rapidly with atomic number Z, since the orbital radius of an S electron decreases with Z, and the electron orbital velocity near the nucleus – upon which the weak interaction also depends – increases with Z.

There is an important distinction, however, between structural optical activity and optical activity attributable to weak neutral currents. Whereas the mirror-image process of the former leads, as illustrated in Fig. 7.4, to another process allowed in nature, the mirror-image process of the latter does not occur. The weak nuclear interactions are truly parity violating, and therefore atoms come in but one chiral form; the enantiomeric form does not exist.

Although the chirality of the weak interactions lies essentially in mathematical laws rather than in an explicitly visible geometric structure (as in the case of molecules), one can nevertheless construct dynamical quantities that reveal a sense of 'handed' motion [206]. Figure 7.5 illustrates the electron probability current density for the hydrogenic $2P_{1/2}$ state

$$\boldsymbol{J}(\boldsymbol{r}) = \mathrm{Re}\langle 2P'_{1/2}|\boldsymbol{p}/m|2P'_{1/2}\rangle , \qquad (7.36)$$

where $\boldsymbol{p}$ is the linear momentum operator, m is the electron mass, and the designation P' indicates that the actual state has a weak admixture of $2S_{1/2}$ similar to that expressed in (7.35). S states of other principal quantum number are present as well, but their contributions are comparatively negligible. Each streamline, or locus of points everywhere tangent to $\boldsymbol{J}$, manifests a helical

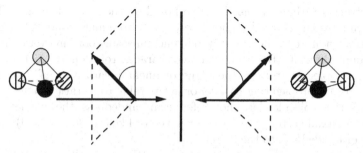

Fig. 7.4. Structural optical activity is an electromagnetic process that conserves parity. The mirror inversion (*left*) of the optical rotation of linearly polarized light transmitted through an enantiomerically pure sample (*right*) is also an allowable process – namely, a rotation in the opposite sense by a sample of opposite chirality

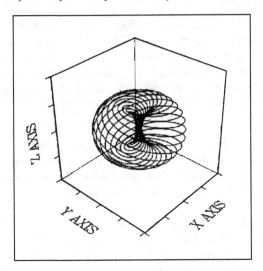

Fig. 7.5. Helicity of the hydrogen $2P_{1/2}$ probability density current resulting from weak neutral currents. Shown is the streamline of probability current density $\boldsymbol{J}$ with greatly exaggerated mixing coefficient $\varepsilon = 0.5$ and initial point taken to be $(x, y, z) = (6, 0, 0)$ in units of the Bohr radius. (Adapted from Hegstrom et al. [206])

structure as it winds in a preferred sense over a toroidal surface whose axis of rotation (z axis) is the quantization axis. The pitch of the helix, determined by the mixing amplitude ε, is greatly exaggerated in the figure for purposes of visibility. In the absence of weak neutral currents, however, ε would be zero, and the corresponding streamlines of a pure $2P_{1/2}$ state of sharp parity would generate circles about the axis of quantization.

7.3 Optical Activity of Rotating Matter

It has been stressed before that optical activity is displayed by chiral materials – i.e., structures than can not be superposed on their mirror image. The required chirality, however, need not always arise from matter alone, but can be a property intrinsic to the larger system encompassing both matter and fields. For example, consider the phenomenon of Faraday rotation in which a sample of *achiral* molecules rotates the plane of linear polarization of a transmitted light beam propagating parallel or antiparallel to a static magnetic field. Although the molecules have no preferential handedness, the magnetic field, which is an axial vector (and not a polar vector like the electric field) imparts a sense of handedness to the system.

Faraday rotation, like natural optical rotation, is parity conserving. Figure 7.6 illustrates the clockwise optical rotation of a light beam propagating to the left, parallel to the magnetic field. Under mirror reflection, the molecules are unaffected since they are presumed to be achiral, the magnetic field is

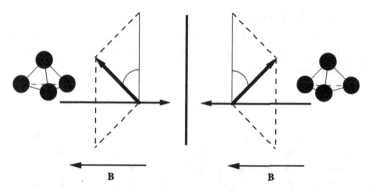

Fig. 7.6. The Faraday effect, like structural optical activity, is also an electromagnetic process that conserves parity. The mirror inversion (*left*) of the optical rotation of linearly polarized light propagating through an *achiral* sample parallel to a static magnetic field is an allowable process – namely, a rotation in the same sense relative to the magnetic field direction

unchanged (as discussed at the beginning of the previous section), and the light, now propagating to the right (antiparallel to the magnetic field), displays a counterclockwise optical rotation. This is exactly what one would expect. Faraday rotation occurs in a fixed sense with respect to the *magnetic field* and not with respect to the direction of light propagation as in the case of structural optical activity. The stark contrast between the Faraday effect and (field-free) optical rotation by intrinsically chiral molecules can be demonstrated by actually reflecting a transmitted light beam back through the sample with a mirror. Upon emerging from a sample of naturally chiral molecules, the net optical rotation will be *zero*. The rotation along the return path will have reversed the rotation along the forward path because the wave vector of the light is reversed. For both passages, however, the rotation will have occurred in a fixed *sense* (clockwise or counterclockwise) to an observer facing the source. On the other hand, if the system consists of achiral molecules in a static magnetic field, the optical rotation of the emerging light will be *twice* that of a one-way passage, since the orientation of the magnetic field has remained the same.

From the perspective of quantum theory, the presence of a static magnetic field B splits the degeneracy of magnetic substates of the molecules (Zeeman effect) and leads, by means of perturbation theory, to molecular polarizabilities that depend differently on the magnitude B for left and right circularly polarized light. This chiral asymmetry in the polarizabilities translates, through standard relations of electrodynamics, into chirally inequivalent indices of refraction, and hence, by (7.7) and (7.8) into an optical rotation. We will examine these connections shortly in a different and somewhat unusual context.

There is an insightful analogy known as Larmor's theorem in classical mechanics [207], whereby the motion of a charged particle (charge q, mass m)

in a static magnetic field $\boldsymbol{B}$ can be analyzed to a first-order approximation in $\boldsymbol{B}$ as if the particle were in a field-free environment in a rotating reference frame. The hypothetical angular velocity of mechanical rotation $\boldsymbol{\Omega}_\mathrm{L}$ is related to the magnetic field as follows

$$\boldsymbol{\Omega}_\mathrm{L} = -\frac{q\boldsymbol{B}}{2mc} \, , \tag{7.37}$$

where the magnitude of $\boldsymbol{\Omega}_\mathrm{L}$ is the Larmor frequency. The converse of this analogy – namely, that a system undergoing uniform rotation can be treated as if it were subjected to a static magnetic field in a stationary reference frame – has interesting and not widely realized implications for molecular structure and the manifestation of optical activity [208]. Indeed, in judging whether or not a system is chiral, one must take account of not only the material sample and any electromagnetic fields present, but the reference frame as well.

Let us designate the Hamiltonian and state vector of a quantum system in an inertial reference frame by H and $|\Psi\rangle$ and the corresponding quantities in a rotated reference frame by H' and $|\Psi'\rangle$. Upon rotation of the system through an angle θ about the unit vector $\boldsymbol{n}$, the two state vectors are related by a unitary transformation

$$|\Psi'\rangle = U|\Psi\rangle = \mathrm{e}^{-\mathrm{i}\boldsymbol{J}\cdot\boldsymbol{n}\theta}|\Psi\rangle \, , \tag{7.38}$$

where the generator of rotation $\boldsymbol{J}$ is the total angular momentum of the system (and not to be confused with the current density of the previous section). Substitution of (7.38) into the Schrödinger form of the equation of motion

$$H|\Psi\rangle = -\mathrm{i}\frac{\mathrm{d}}{\mathrm{d}t}|\Psi\rangle \tag{7.39}$$

yields an equation of identical form in $|\Psi'\rangle$, but with the transformed Hamiltonian

$$H' = UHU^{-1} + \mathrm{i}U\frac{\mathrm{d}U^{-1}}{\mathrm{d}t} \, . \tag{7.40}$$

Use of the explicit expression for U from (7.38) and assumption of rotational invariance in the inertial frame ($[H, \boldsymbol{J}] = 0$) results in the Hamiltonian

$$H' = H - \boldsymbol{\Omega} \cdot \boldsymbol{J} \, , \tag{7.41}$$

where $\boldsymbol{\Omega} = \boldsymbol{n}\mathrm{d}\theta/\mathrm{d}t$ is the angular velocity of rotation. For a nonrelativistic system in an inertial reference frame, the Hamiltonian can be expressed as the sum of two terms, one for the motion of the center of mass and the other governing the internal dynamics of the system. This same separation can be performed for a rotating system, and it will be hereafter assumed that H' determines the bound state energy spectrum and that $\boldsymbol{J}$ refers to the total relative angular momentum of the constituent particles. From (7.41) it follows immediately that the energy eigenvectors of the Schrödinger equation are the same in the rotating and inertial frames, but the eigenvalues of the initially degenerate magnetic substates now depend on the magnetic quantum number m_J (i.e., the eigenvalues of J_z, where the quantization axis is identified with the axis of rotation). Two substates of the same manifold differing in magnetic quantum number by $\Delta m_J = 1$ are separated by the energy (frequency) interval Ω. In keeping with the aforementioned analogy, the additional term in (7.41) has the form of the magnetic Hamiltonian $H_M = -(q/2mc)\mathbf{L} \cdot \mathbf{B}$ for a bound charged particle with orbital angular momentum $\mathbf{L}$. In other words, the isomorphic relation

$$\Omega \Longleftrightarrow \frac{q\mathbf{B}}{2mc} \qquad (7.42)$$

has the opposite sign to that of Larmor's theorem (7.37). This is to be expected, since in Larmor's theorem Ω_L was chosen to *cancel* (to first order) the effects of the extant magnetic field. It is also important to note that the quantum mechanical generator of rotations $\mathbf{J}$ is not necessarily identical with $\mathbf{L}$, but can include nonclassical contributions from electron and nuclear spin angular momenta.

The fact that rotation formally influences the energy eigenstates of a quantum system as if a magnetic field were present suggests that intrinsically achiral rotating atoms and molecules should display optical activity analogous to the Faraday effect. Suppose such a sample to be irradiated with LCP or RCP light of amplitude E_0 and frequency ω propagating parallel to the axis of rotation. The interaction between the molecule and light is governed principally by the electric dipole Hamiltonian

$$H_E = -\boldsymbol{\mu} \cdot \mathbf{E}_{L,R} = -\frac{1}{2}E_0\left(\mu_{\mp}e^{i\omega t} + \mu_{\pm}e^{-i\omega t}\right), \qquad (7.43)$$

where

$$\mu_{\pm} = \mu_x \pm i\mu_y , \qquad (7.44)$$

and the upper and lower signs correspond respectively to LCP and RCP. In contrast to the analysis of Sect. 7.2, the rotation-induced optical activity which we are now considering does not arise as an interference between electric and magnetic dipole interactions, and we can neglect here the contribution of the latter which is ordinarily weaker than the former by the ratio v/c, where v is the bound electron speed. (When the angular momentum $\mathbf{J}$ derives exclusively from electron and nuclear spins, however – as in the case of ground-state hydrogen hyperfine states – the magnetic dipole coupling plays an important role [209].)

Let us assume that the ground state $|0\rangle$ of the system is an S state. It then follows from first-order time-dependent perturbation theory that the perturbed state vectors $|\Psi_{L,R}\rangle$ contain contributions from virtual transitions to higher P states. Calculation of the electric dipole moments induced by LCP and RCP light

$$\langle\boldsymbol{\mu}\rangle_{L,R} = \text{Re}\langle\Psi_{L,R}|\boldsymbol{\mu}|\Psi_{L,R}\rangle = \alpha_{L,R}\mathbf{E}_{L,R} \qquad (7.45)$$

then leads to the chirally inequivalent polarizabilities

$$\alpha_{L,R} = \frac{1}{\hbar} \sum_n \frac{\Omega_{n0}\mu_{n0}^2}{\Omega_{n0}^2 - (\omega \pm \Omega)^2} , \tag{7.46}$$

where

$$\mu_{n0}^2 \equiv |\langle n|\mu_+|0\rangle|^2 = |\langle n|\mu_-|0\rangle|^2 , \tag{7.47}$$

and Ω_{n0}, given by (7.31), is the energy interval between the ground state and excited state n. Only P states with $m_L = 1$ enter expression (7.46), the contributions from states with $m_L = -1$ having already been included through prior use of the symmetry in relations (7.47).

Neglecting, for simplicity, the possible distinction between the electric field of the incident wave and the local electric field at a molecular site – an approximation that could, if necessary, be improved by means of the Lorentz-Lorenz formula [210] – we deduce the LCP and RCP dielectric constants and refractive indices from the relation

$$\varepsilon_{L,R} = 1 + 4\pi\eta\alpha_{L,R} = n_{L,R}^2 , \tag{7.48}$$

where η is again the number of molecules per unit of volume. From (7.48) it then follows that matter in rotation should exhibit a circular birefringence of the form

$$n_L - n_R \approx \frac{8\pi\eta}{\hbar} \sum_k \frac{\Omega_{n0}\mu_{n0}^2}{(\Omega_{n0}^2 - \omega^2)^2}\omega\Omega , \tag{7.49}$$

which is linearly proportional to the angular frequency of rotation Ω.

The predicted effect is quite small in comparison to the circular birefringence (7.18) of a naturally optically active medium. For the same conditions as before of a sample with the density of water, a resonance at 100 nm, and a red probe beam of 600 nm, (7.49) leads to a proportionality coefficient of $\sim 2.3 \times 10^{-18}$ s, where we have approximated the dipole matrix element by the product of electron charge and the Bohr radius. The consequences of this circular birefringence, however, are not beyond detection. Thus, for a rotation rate of 100 Hz (which is probably close to the upper limit of what is achievable in the laboratory) and a total path length of 10 m (obtained by multiple reflection of the light through the sample), one could expect an optical rotation of about 4×10^{-6} degrees. Note that multiple passage of the light is helpful here, because the phenomenon is analogous to the Faraday effect. Use of a nonresonant ultraviolet probe beam could enhance the signal by one to two orders of magnitude.

Established techniques such as photoelastic modulation and synchronous detection can detect Faraday rotations with a sensitivity of 10^{-4} degrees, while recently developed laser polarimeters, designed for the study of optical activity associated with parity violations in atoms, can detect optical rotations at the level of 10^{-6} degrees with an expectation of improvement by several orders of magnitude [211].

The phenomenon of rotational optical activity, analyzed above within the framework of quantum mechanics, is amenable, as well, to a classical mechanical interpretation [212] which affords a visual image of the effect of classical forces on the orbits of the bound electrons. According to this viewpoint, it is principally the Coriolis pseudo-force that acts differently on particles circulating in clockwise or counterclockwise orbits as driven by the electric field of incident circularly polarized light. This leads to chirally inequivalent orbital radii, and therefore to different polarizabilities and refractive indices.

It is instructive to examine this point of view more closely in the simple case of a model atomic system with single particle of mass m and charge q in an isotropic harmonic oscillator potential $U(r) = m\Omega_0^2 r^2/2$ with angular frequency of oscillation Ω_0. If the particle is subjected to an electromagnetic plane wave $\boldsymbol{E}(t)$ of angular frequency ω propagating parallel to the rotation axis of a frame rotating with angular velocity $\boldsymbol{\Omega}$, it experiences an effective force[6]

$$\boldsymbol{F}_{\text{eff}} = \boldsymbol{F} - 2m(\boldsymbol{\Omega} \times \boldsymbol{v}) - m\boldsymbol{\Omega} \times (\boldsymbol{\Omega} \times \boldsymbol{v}) \,, \tag{7.50}$$

where the 'true' force, determined in an inertial frame, is

$$\boldsymbol{F} = -\boldsymbol{\nabla} U(\boldsymbol{r}) + q\boldsymbol{E} \,. \tag{7.51}$$

The second and third terms in (7.50) will be recognized as the Coriolis and centrifugal pseudo-forces.

Solution of Newton's equation of motion in the rotating frame

$$m\frac{\mathrm{d}^2 \boldsymbol{r}}{\mathrm{d}t^2} = \boldsymbol{F}_{\text{eff}} \tag{7.52}$$

for the magnitudes of the steady-state coordinates r_{L}, r_{R} produced by LCP and RCP waves, respectively, leads to the polarizabilities

$$\alpha_{\text{L,R}} = \frac{q^2/m}{\Omega_0^2 - (\omega \pm \Omega)^2} \,, \tag{7.53}$$

and ultimately to the circular birefringence

$$n_{\text{L}} - n_{\text{R}} \approx \frac{8\pi\eta q^2}{m} \frac{\omega\Omega}{(\Omega_0^2 - \omega^2)^2} \,. \tag{7.54}$$

In both the quantum and classical analyses the plane of polarization is rotated clockwise for an observer facing the light source.

---

[6] We have again separated the center of mass motion and the internal or relative motion. Equation (7.50) is the force on the 'relative' particle in a two-particle system – i.e., the hypothetical particle whose mass is the reduced mass $m_1 m_2/(m_1 + m_2)$ and whose velocity is the relative velocity $\boldsymbol{v}_2 - \boldsymbol{v}_1$. If the mass of one particle is much greater than that of the other, then the motion of the bound particle with smaller mass is virtually the same as that of the 'relative' particle.

One can formally correlate the classical expressions in (7.53) and (7.54) with the quantum mechanical expressions of (7.46) and (7.49) by identifying the classical dipole moment qr with the matrix element μ_{n0} and the oscillation frequency Ω_0 with the resonance frequency Ω_{n0} (for a particular state n in the summation), and equating twice the potential energy $m\Omega_0^2 r^2$ to $\hbar\Omega_{n0}$.

From (7.42) it is seen that the magnetic analogue of the Coriolis force is the Lorentz force

$$\boldsymbol{F}_{\mathrm{M}} = \frac{q}{c}\boldsymbol{v} \times \boldsymbol{B}\,, \tag{7.55}$$

and one could account classically for the Faraday rotation by a formally identical mathematical analysis. There is an important distinction between the two forces, Lorentz and Coriolis, however, which illustrates the limitations of the analogy between magnetism and rotation. Note that the chiral polarizabilities (7.53) are independent of the sign of charge for both the Faraday effect and rotational optical activity, although the Coriolis and Lorentz forces *differ* in this respect. Consider first the magnetic case illustrated in the left hand side of Fig. 7.7. If the charge q is positive, it is driven by an incident LCP wave (traveling parallel to the magnetic field) to orbit counterclockwise as viewed by an observer facing the source. The Lorentz force then accelerates the particle radially *outward*. If the charge is negative, the LCP wave drives it clockwise, but the Lorentz force, whose direction is also reversed, again accelerates the particle radially *outward*. In either case the LCP wave leads to an orbital radius, and therefore polarizability and refractive index, larger than those in the absence of the magnetic field. Correspondingly, a RCP wave leads to a smaller refractive index, with the result that the birefringence (7.54) is positive.

Since the Coriolis force, unlike the Lorentz force, is not proportional to electric charge, it accelerates counter-circulating positive and negative parti-

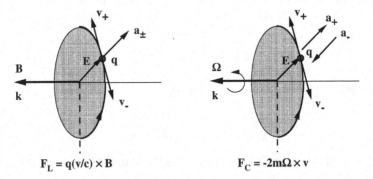

$$F_L = q(v/c) \times B \qquad\qquad F_C = -2m\Omega \times v$$

Fig. 7.7. Effects of Lorentz and Coriolis forces on a charged particle driven by an incident traveling wave E of left circular polarization. Irrespective of the sign of charge q, the Lorentz force of a magnetic field B parallel to the wave vector of the light is directed radially outward. The Coriolis force, attributable to a frame rotating with angular velocity Ω parallel to the wave vector of the light, is directed outward for a positive charge and inward for a negative charge. The instantaneous particle velocities and accelerations are $v_\pm$ and $a_\pm$ where $\pm$ specifies the sign of the charge

cles in opposite radial directions as shown in the right hand side of Fig. 7.7. From relation (7.42), it is seen that reversing the sign of the charge in a magnetic configuration actually corresponds to reversing the angular velocity of the rotating reference frame in the present circumstance. We are considering, however, a fixed sense of rotation. One might intuitively think, therefore, that by the foregoing argument the Coriolis force should lead to $n_L > n_R$ for a positive charge and $n_R > n_L$ for a negative charge – but this is not so. With account taken of all contributions to (7.50), it is found that the sign of q determines the relative orientation (parallel or antiparallel) of $E_{L,R}$ and the corresponding $r_{L,R}$, but not the magnitude of $r_{L,R}$.

Looked at from an inertial frame, the inequivalent action of LCP and RCP radiation is effectively attributable to the Doppler effect. A LCP wave of frequency ω propagating parallel to the axis of a frame rotating with angular frequency Ω is perceived to have the frequency $\omega + \Omega$ by an inertial observer. Similarly, the inertially measured frequency of the corresponding RCP wave would be $\omega - \Omega$. Thus, LCP and RCP waves of the same frequency in the rotating reference frame have different frequencies in an inertial reference frame in which case the chiral asymmetry expressed by (7.53) can be interpreted as a consequence of the frequency dispersion of the polarizability function. For a fixed sense of mechanical rotation, only the sense of rotation of the electric field of the incident light matters, and not the sign of the charge of the particle.

In general, theoretical analyses of terrestrial atomic and molecular systems ordinarily take for granted at the outset that the frame of reference is inertial, and actual experiments are usually executed under such conditions that this assumption is thought to be adequate. In point of fact, of course, the Earth is not an inertial frame, and the manifestations of the Earth's rotation on *macroscale* systems, both mechanical (e.g., Foucault pendulum) and electromagnetic (e.g., Sagnac effect, discussed in [212]), have been known for a long time. Neutron interferometry has made it possible to demonstrate the effect of the Earth's gravity and diurnal rotation on a quantum system [213]. Such experiments, which employ beams of free neutrons, raise interesting questions regarding the influence of the Earth's spin on the optical properties of bound-state systems. Would it be possible, for example, to observe optical activity induced in materials by the Earth's rotation where $\Omega/2\pi = 1.2 \times 10^{-5}$ Hz? The coresponding optical rotations would be smaller than those calculated for a rapidly rotating laboratory turntable by some seven orders of magnitude, and, until the present, were beyond the sensitivity of any technique known to the author. Nevertheless, the field of polarimetry has been advancing rapidly, and developments in ring-laser interferometry, in particular, give cause for a cautiously optimistic response [214].

A ring laser interferometer is a self-excited optical oscillator in which two counter-propagating beams can be made to interfere upon exiting through a mirror. If the ring is subject to a nonreciprocal effect on the two beams, the interference fringes will be shifted. One such cause, referred to as the Sagnac effect, would be a rotation of the rest frame of the interferometer. A second cause would be a circularly birefringent sample – either natural or rotationally induced – in one arm of the interferometer. It is instructive to examine the Sagnac effect briefly in order to appreciate the degree of sensitivity of ring laser interferometry.

Consider two light beams (angular frequency ω and wavelength λ) propagating in opposite senses around a circular ring interferometer of radius R. The circular geometry is not required, but is merely simpler to treat for purposes of illustration than a polygonal geometry. If the interferometer is stationary, the two waves, issuing from a common point on the ring, return to that point with a time difference $\Delta t = 0$. However, if the interferometer is rotating at an angular frequency $\Omega = v/R$, where v is the linear speed of a point on the ring, then the wave propagating in the direction in which the interferometer is circulating will complete a revolution in a longer period than the counter-propagating wave. The difference in periods is then given by

$$\Delta t = 2\pi R \left(\frac{1}{c-v} - \frac{1}{c+v} \right) = \frac{4\pi R v}{c^2 - v^2} \approx \frac{4A\Omega}{c^2} \,, \tag{7.56}$$

in which the ring area is $A = \pi R^2$, and a nonrelativistic speed ($v \ll c$) is assumed in the last relation of (7.56). The phase difference between the two beams is

$$\phi = \omega \Delta t \approx \frac{2\pi c}{\lambda} \frac{4A\Omega}{c^2} = \frac{8\pi A\Omega}{\lambda c} \,. \tag{7.57}$$

Analyses of ring laser performance [215] have shown that a ring of approximately $1\ \mathrm{m}^2$ area, operating at a single mode at 633 nm (the red line of a helium–neon laser) with state-of-the-art dielectric mirror coatings, ought to be able to detect variations in the rotation of the Earth at the level of parts per million, or, in other words, phase shifts

$$\Delta\phi = \frac{8\pi A}{\lambda c} \Delta\Omega \,, \tag{7.58}$$

where $\Delta\Omega \sim 10^{-6}\Omega_{\mathrm{Earth}}$, with $\Omega_{\mathrm{Earth}} = 7.3 \times 10^{-5}$ rad/s. Insertion of the preceding numerical values into (7.58) indicates that a state-of-the-art ring laser interferometer ideally should be capable of detecting phase differences as small as $\sim 10^{-11}$ radians. If the interferometer is stationary but contains a circularly birefringent sample ($\Delta n = |n_{\mathrm{L}} - n_{\mathrm{R}}|$), then the phase shift that results takes the form

$$\Delta\phi = \frac{2\pi d}{\lambda} \Delta n \,, \tag{7.59}$$

in which the optical path length difference $d = 2\pi R$. It then follows from (7.59) and (7.58) that the birefringence equivalent to a given variation in angular rotation rate is

$$\Delta n = \frac{2R}{c} \Delta\Omega \,, \tag{7.60}$$

which leads to a resolvable index difference of $\Delta n \sim 3 \times 10^{-19}$. This value is a factor 10 smaller than the predicted anisotropy realizable by the Earth's rotation for nonresonant radiation with 'detuning' parameter $\omega/\Omega_0 \sim 0.95$.

Ring-laser interferometers of this projected sensitivity and better are under construction or in development at various laboratories in North America, Europe, and New Zealand. For example, as of 2001, the University of Canterbury's UG1 ring laser with area of approximately 367 m^2, constructed in the Cashmere Cavern in Christchurch, New Zealand, became fully operational and has been used to investigate the subtleties in the Earth's rotation.[7] An even larger laser of 800 m^2 is in planning. Size alone, of course, does not lead to enhanced resolutions; there are critical issues of thermal and mechanical stability, and numerous sources of noise to overcome. Nevertheless, it would seem that the prospects of detecting the optical activity engendered in atoms by the Earth's rotation should be feasible.

7.4 'Electron Activity' in a Chiral Medium

If left and right circularly polarized light interact differently with chiral matter, then one may also expect to find a difference in the scattering of 'spin-up' and 'spin-down' electrons (or other massive spin-1/2 particles like neutrons or protons) by chiral matter. Although photons are bosons and electrons are fermions, the state of polarization of both kinds of particles is representable by a mathematical formalism employing two-dimensional spinors and 2×2 density matrices. The reason for this, as has been pointed out previously, is that a massless particle is characterized by two helicity states, i.e., the projection $\boldsymbol{S} \cdot \boldsymbol{p}/|\boldsymbol{p}| = \pm S$ of intrinsic angular momentum $\boldsymbol{S}$ on linear momentum $\boldsymbol{p}$, and not by $2S + 1$ spin substates. We will now examine again the process of forward scattering, but in a more general way so as to include other chiral processes besides that of the rotation of the plane of polarization.

Forward elastic scattering is a coherent process in that there is a well-defined phase relationship between the waves issuing from each scattering center in the target material, as illustrated in Fig. 7.8. As a simple system illustrating the connection between optical parameters and scattering parameters, consider first a beam of scalar particles incident on a thin slab (thickness d) of achiral molecules of number density η. The incident wave function $\psi_{\mathrm{inc}} \sim \mathrm{e}^{\mathrm{i}kz}$ is taken to be a plane wave propagating along the positive z axis with wave vector (i.e., linear momentum in units of $\hbar$) $\boldsymbol{k} = k\boldsymbol{z}$. The total wave at a point a distance z above the thin slab is then the superposition of the incident plane wave and secondary spherical wavelets issuing from each scattering center

$$\psi(z; d) = \mathrm{e}^{\mathrm{i}kz} + \eta d \int\limits_{z}^{\infty} \frac{\mathrm{e}^{\mathrm{i}kr} f(k, \vartheta) 2\pi r \mathrm{d}r}{r} , \qquad (7.61)$$

where the (complex-valued) scattering amplitude $f(k, \vartheta)$ is a function of k and the scattering angle $\vartheta = \cos^{-1}(z/r)$. Upon expanding $f(k, \vartheta)$ in a Taylor

---

[7] The Canterbury Ring Laser [216].

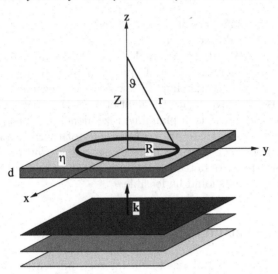

Fig. 7.8. Forward scattering of a beam of particles, described by a plane wave function of wave vector $\mathbf{k}$, incident upon a thin slab of material with η scatterers per unit volume. The amplitude of the wave a distance Z above the plane is a linear superposition of the incident plane wave and secondary spherical wavelets emitted from all points of the slab

series in ϑ, imposing the far-field condition ($\vartheta \to 0$), and discarding the rapidly oscillating contribution of the upper limit leads to the transmitted wave

$$\psi(z; d) = e^{ikz} \left[1 + \frac{2\pi i \eta d}{k} f(k) \right] \equiv (1 + iGd) \psi_{\text{inc}}(z) . \qquad (7.62)$$

(Although it is not mathematically rigorous simply to drop the ill-defined term resulting from the upper limit, the integral can be made convergent by the expedient of adding a positive imaginary term to the wave vector.) From the second expression in (7.62) it is clear that the effect of transmission through a finite slab of thickness D can be derived by regarding the sample as an infinitely large number of differentially thin layers to obtain

$$\psi(z; D) = \lim_{N \to \infty} \left(1 + i\frac{GD}{N} \right)^N \psi_{\text{inc}}(z) = e^{iGD} \psi_{\text{inc}}(z) = e^{ik(\tilde{n}-1)D} \psi_{\text{inc}}(z) . \tag{7.63}$$

The last expression in (7.63) is the general form of an optical wave that has propagated a distance D through a medium of complex refractive index $\tilde{n} = n + i\kappa$.

From (7.62) and the last two relations in (7.63) we see that the scattering amplitude and complex refractive index are related by

$$\tilde{n} = n + i\kappa = 1 + \frac{2\pi\eta}{k^2} f(k) \quad \Longrightarrow \quad \begin{cases} n - 1 = \dfrac{2\pi\eta}{k^2} \text{Re} f(k) , \\[2mm] \kappa = \dfrac{2\pi\eta}{k^2} \text{Im} f(k) . \end{cases} \tag{7.64}$$

The relative intensity of the transmitted wave is then

$$I(k) = \left| \frac{\psi(z;D)}{\psi_{\text{inc}}(z)} \right|^2 = \left| e^{ik(\tilde{n}-1)D} \right|^2 = e^{-2k\kappa D} = e^{-\sigma\eta D} , \qquad (7.65)$$

where the last expression in (7.65) is an outcome of the definition of the total scattering cross-section σ and leads to the relation

$$\kappa = \frac{\eta}{2k}\sigma . \qquad (7.66)$$

Equating the expressions for κ in (7.64) and (7.66) leads directly to the quantum mechanical optical theorem[8]

$$\sigma = \frac{4\pi}{k}\text{Im}f(k) . \qquad (7.67)$$

Let us now return to the problem of particle transmission through a *chiral* medium. The basis states are taken to be spinors

$$|\psi_1\rangle = \begin{pmatrix} 1 \\ 0 \end{pmatrix} , \qquad |\psi_2\rangle = \begin{pmatrix} 0 \\ 1 \end{pmatrix} ,$$

representing respectively 'spin-up' and 'spin-down' states where the axis of quantization (z) and two other mutually orthogonal coordinate axes (x, y) are defined in terms of the wave vectors of the particle k, k' before and after scattering as follows:

$$x = \frac{k' - k}{|k' - k|} , \qquad y = \frac{k \times k'}{|k \times k'|} , \qquad z = \frac{k + k'}{|k + k'|} . \qquad (7.68)$$

It may be shown that the coordinate axes form a right-handed triad with unit vectors satisfying $x \cdot y \times z = 1$ for forward elastic scattering ($|k'| = |k| \equiv k$), which is the case of interest here. Actually, the familiar terminology 'up' and 'down' is not as appropriate as 'parallel' and 'antiparallel' under the present condition where the quantization axis coincides with the direction of propagation. The states $|\psi_1\rangle$ and $|\psi_2\rangle$ are then electron states of helicity $\pm 1/2$ whose optical analogues are left and right circularly polarized light.

The forward scattering amplitude $f(k)$ must now be replaced by an interaction matrix $F(k)$

$$F(k) = \begin{pmatrix} F_{11} & F_{12} \\ F_{21} & F_{22} \end{pmatrix} = F_0 1 + F_1(\sigma \cdot x) + F_2(\sigma \cdot y) + F_3(\sigma \cdot z)$$

$$= F_0 1 + F_1 \sigma_1 + F_2 \sigma_2 + F_3 \sigma_3 , \qquad (7.69)$$

---

[8] For a more rigorous derivation of the optical theorem see, for example, [203] pp. 201–202.

which is expressed above in a basis $(1, \boldsymbol{\sigma})$ of the unit 2×2 matrix and Pauli spin matrices. For the problem under discussion, several symmetries pertain which reduce the general interaction (7.69) to diagonal form in the chiral basis [217]. First, because the origin of chirality lies in the geometric structure of molecular targets which interact with the electron projectiles via the electromagnetic interaction, the scattering matrix F must be time-reversal invariant, i.e., $TFT^{-1} = F$, where T is the time-reversal operator, i.e., an operator reversing the sequence of events in a scattering process and the sense of rotation of angular momenta. T therefore effects the following transformations: $\boldsymbol{k} \rightarrow -\boldsymbol{k}'$, $\boldsymbol{k}' \rightarrow -\boldsymbol{k}$, and $\boldsymbol{\sigma} \rightarrow -\boldsymbol{\sigma}$, whereupon $T(\boldsymbol{\sigma} \cdot \boldsymbol{x})T^{-1} \rightarrow -\boldsymbol{\sigma} \cdot \boldsymbol{x}$, all other terms in (7.69) remaining invariant. Thus the component $F_1 = 0$ for F to be time-reversal invariant. Second, in forward scattering the initial and final wave vectors are parallel, whereupon $\boldsymbol{y} = 0$, and the component F_2 can be eliminated. The final form of the interaction matrix is then

$$F = F_0 1 + F_3 \sigma_3 = \begin{pmatrix} F_0 + F_3 & 0 \\ 0 & F_0 - F_3 \end{pmatrix} . \tag{7.70}$$

Substitution of F in place of f in (7.62) and (7.63), and application of the identity

$$e^{i\boldsymbol{k}\cdot\boldsymbol{\sigma}\theta} = 1 \cos\theta + i(\boldsymbol{k} \cdot \boldsymbol{\sigma}) \sin\theta ,$$

which we have used previously in the book, leads to the state vector

$$|\psi\rangle = e^{ikz} e^{iQ_0} \begin{pmatrix} e^{iQ_3 z} & 0 \\ 0 & e^{-iQ_3 z} \end{pmatrix} |\psi_{\text{inc}}\rangle \tag{7.71}$$

and density matrix

$$\rho = |\psi\rangle\langle\psi| = \begin{pmatrix} e^{-2\text{Im}(Q_0+Q_3)} \rho_{11}^{(0)} & e^{2i\text{Re}Q_3} e^{-2\text{Im}(Q_0-Q_3)} \rho_{12}^{(0)} \\ e^{-2i\text{Re}Q_3} e^{-2\text{Im}(Q_0-Q_3)} \rho_{21}^{(0)} & e^{-2\text{Im}(Q_0-Q_3)} \rho_{22}^{(0)} \end{pmatrix} , \tag{7.72}$$

where the amplitudes appearing in the exponential are defined by

$$Q_0 \equiv \frac{2\pi\eta z}{k} F_0 , \qquad Q_3 \equiv \frac{2\pi\eta z}{k} F_3 , \tag{7.73}$$

and $\rho^{(0)}$ is the density matrix of the incident beam.

From the form of the density matrix (7.72), we can distinguish at least four distinct kinds of chiral signals in forward scattering:

- creation of a longitudinal polarization in an initially unpolarized beam,
- attenuation of the longitudinal polarization in an initially polarized beam,
- difference in the intensity of transmitted beams of initially opposite longitudinal polarizations,
- rotation of the transverse polarization in an initially polarized beam.

We now examine these effects in more detail.

7.4.1 Longitudinal Polarization

Consider an incident unpolarized electron beam characterized by the density matrix

$$\rho_{\text{unp}}^{(0)} = \begin{pmatrix} 1/2 & 0 \\ 0 & 1/2 \end{pmatrix}, \tag{7.74}$$

which contains only relative state populations and no coherence terms. Substitution of the elements $\rho_{ij}^{(0)}$ into (7.72) likewise leads to a density matrix with no coherence terms, but with unequal populations of spin-parallel and spin-antiparallel electrons, giving rise to a longitudinal polarization

$$P_3 \equiv \frac{\rho_3}{\rho_0} = \frac{\rho_{11} - \rho_{22}}{\rho_{11} + \rho_{22}} = - \tanh \left[\frac{2\pi\eta z}{k} \text{Im}(F_3) \right] = - \tanh \left[\frac{1}{2}\eta z(\sigma_p - \sigma_a) \right]. \tag{7.75}$$

The last expression in (7.75) follows from the optical theorem, (7.67), in which σ_p and σ_a are the total elastic scattering cross-sections respectively for helicity $+1/2$ and helicity $-1/2$ electrons.

Two experimental procedures based on (7.75) are to measure:

- the growth of P_3 as a function of path length z through the chiral medium,
- the variation in P_3 as a function of the wave number k (or correspondingly the particle energy).

The latter measurement yields a signal which is the analogue of the optical phenomenon of circular dichroism, i.e., the variation in differential absorption of circularly polarized light with wavelength.

If the incident beam is partially polarized but still incoherent, then the density matrix elements and the initial longitudinal polarization P (with $|P| \leq 1$) are related by

$$\begin{aligned} \text{particle conservation} \quad &\rho_{11}^{(0)} + \rho_{22}^{(0)} = 1, \\ \text{initial polarization} \quad &\rho_{11}^{(0)} - \rho_{22}^{(0)} = P, \end{aligned} \tag{7.76}$$

from which it follows that the initial density matrix can be expressed in the form

$$\rho_{\text{pol}}^{(0)} = \begin{pmatrix} \frac{1}{2}(1 + P) & 0 \\ 0 & \frac{1}{2}(1 - P) \end{pmatrix}. \tag{7.77}$$

Using relations (7.77) and (7.73), one can readily show that the density matrix (7.72) after passage of the beam through a chiral medium of length z and molecular density η, is expressible as

$$\rho = \begin{pmatrix} \frac{1}{2}(1 + P)e^{-\eta z\sigma_p} & 0 \\ 0 & \frac{1}{2}(1 - P)e^{-\eta z\sigma_a} \end{pmatrix}. \tag{7.78}$$

Besides measuring the longitudinal polarization (see Fig. 7.9), which with a little algebraic manipulation can be cast into the form

$$P_3 = \frac{\rho_{11} - \rho_{22}}{\rho_{11} + \rho_{22}} = \frac{P - \tanh\left[\frac{1}{2}\eta z(\sigma_{\mathrm{p}} - \sigma_{\mathrm{a}})\right]}{1 - P\tanh\left[\frac{1}{2}\eta z(\sigma_{\mathrm{p}} - \sigma_{\mathrm{a}})\right]} , \tag{7.79}$$

an equivalently useful experimental procedure in this case is to measure the intensity of the beam

$$I(P) \propto \mathrm{Tr}(\rho) = \frac{1}{2}(1 + P)e^{-\eta z\sigma_{\mathrm{p}}} + \frac{1}{2}(1 - P)e^{-\eta z\sigma_{\mathrm{a}}} \tag{7.80}$$

for initial polarizations $+P$ and $-P$ and take the difference to enhance the chiral contribution to the signal. This yields the differential chiral transmission (DCT)

$$\mathrm{DCT}(P) = \frac{I(P) - I(-P)}{I(P) + I(-P)} = -P\tanh\left[\frac{1}{2}\eta z(\sigma_{\mathrm{p}} - \sigma_{\mathrm{a}})\right] , \tag{7.81}$$

which vanishes for an initially unpolarized beam or for an achiral medium. Under the circumstance (common for naturally optically active organic compounds) that the difference in chiral cross-sections is sufficiently small that

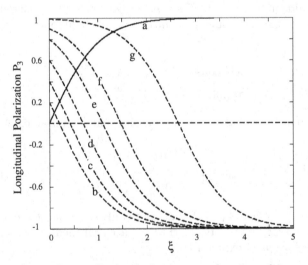

Fig. 7.9. Variation in longitudinal polarization as a function of $\xi = \eta z(\sigma_{\mathrm{p}} - \sigma_{\mathrm{a}})/2$ for (**a**) an initially unpolarized beam (plot of $-P_3 = \tanh\xi$), and polarized beams with initial polarization P equal to (**b**) 0.2, (**c**) 0.4, (**d**) 0.6, (**e**) 0.8, (**f**) 0.9, (**g**) 0.99. [Plot of P_3 given by (7.79)]

one need retain only the first term in a Taylor series expansion of the hyperbolic tangent, the DCT is then linearly proportional to the difference in chiral cross-sections

$$\text{DCT}(P) \approx -\frac{1}{2}P\eta z(\sigma_\text{p} - \sigma_\text{a}) = -P\eta z\overline{\sigma}\frac{\sigma_\text{p} - \sigma_\text{a}}{\sigma_\text{p} + \sigma_\text{a}} \, , \qquad (7.82)$$

where the mean cross-section

$$\overline{\sigma} = \frac{1}{2}(\sigma_\text{p} + \sigma_\text{a})$$

is inferable from the imaginary part of the complex refractive index of an enantiomeric mixture of the two chiral forms.

One sees that (7.75), (7.79), and (7.81) all provide the same information, although the three experimental procedures to which these signals correspond are not necessarily implementable with equal facility or likely to yield the same signal-to-noise ratios.

7.4.2 Transverse Polarization

Consider next an incident beam of electrons whose wave function is a linear superposition of helicity states

$$|\psi_\pm\rangle = \frac{1}{\sqrt{2}}\left(|\psi_1\rangle \pm |\psi_2\rangle\right) \, , \qquad (7.83)$$

resulting in the initial density matrix

$$\rho_\pm^{(0)} = \frac{1}{2}\begin{pmatrix} 1 & \pm 1 \\ \pm 1 & 1 \end{pmatrix} \, . \qquad (7.84)$$

The optical analogue of the states in (7.83) – suggested by the inverse of (7.1) and (7.2) – is a coherent superposition of left and right circular polarizations leading to linearly polarized light oriented along the x and y axes, respectively. Insertion of the elements of (7.84) into the density matrix (7.72) leads to transverse polarization components

$$P_1 \equiv \frac{\rho_1}{\rho_0} = \frac{\rho_{12} + \rho_{21}}{\rho_{11} + \rho_{22}} = \pm\frac{\cos\left[\text{Re}(2Q_3)\right]}{\cosh\left[\text{Im}(2Q_3)\right]} \, , \qquad (7.85)$$

$$P_2 \equiv \frac{\rho_2}{\rho_0} = -\frac{\text{i}(\rho_{21} - \rho_{12})}{\rho_{11} + \rho_{22}} = \mp\frac{\sin\left[\text{Re}(2Q_3)\right]}{\cosh\left[\text{Im}(2Q_3)\right]} \, , \qquad (7.86)$$

and a longitudinal polarization given by (7.75). Equations (7.85) and (7.86) represent a rotation of the transverse polarization by the angle

$$\theta = \text{Re}(2Q_3) = \frac{4\pi\eta z}{k}\text{Re}(F_3) \quad \Longleftrightarrow \quad k(n_\text{p} - n_\text{a})z \, , \qquad (7.87)$$

which, for an observer facing the particle source, is in a clockwise sense for $Re(F_3) > 0$ and counterclockwise for $Re(F_3) < 0$. The last expression in (7.87), based on the relation between scattering amplitude and refractive index in (7.64), indicates the correspondence with the phenomenon of optical rotation in which $n_p - n_a$ is the difference in refractive indices for spin-parallel and spin-antiparallel electrons.

It is of interest to note that the transverse polarization, which is a measurable quantity, rotates through *twice* the angle that the state vector, which is not a measurable quantity, rotates upon passage through the chiral medium. [Compare the angles in (7.87) and in (7.7) which, with replacement of ω/c by k, applies to electrons as well as to light.] There is no inconsistency here. We have encountered this feature before in a different context in the discussion in Chap. 5 of spinor rotations and quantum jumps. Although the electron state vector or wave function is not observable, the phase of the spinorial part of a wave function linearly superposed with a reference wave is an observable quantity. Also to be noted is the contrast in experimental significance between the wave function of an electron and the electric field [e.g., in (7.6)] of a classical light wave. The former is a mathematical function associated with probability, whereas the latter is associated directly with a classical force and is consequently a measurable quantity.

Because the difference in chiral cross-sections for the scattering of electrons from naturally optically active molecules is ordinarily very small, experiments to observe the manifestations of such scattering are difficult. The first reported case [218] of polarized electron transmission through the vapor of optically active camphor was not replicable and has been attributed to instrumental artifacts. A successful observation of electron optical dichroism was eventually achieved [219, 220] by enhancing electron scattering through the addition of an ytterbium atom to camphor-based organic molecules to form the jaw-breaking compound tris[(3-heptafluorpropylhydroxymethylene)camphorato]ytterbium or more simply $Yb(hfc)_3$ – and analogous compounds with other heavy lanthanoid metal atoms. The experimentally observed differential chiral transmission, quantified in (7.81) and (7.82), is shown in Fig. 7.10 as a function of electron energy. Asymmetries in transmission on the order of 10^{-4} were observed, which, for the reported experimental conditions, implied a comparable magnitude for the ratio $(\sigma_p - \sigma_a)/(\sigma_p + \sigma_a)$.

In the experiment cited above, the researchers created an incident beam of transversely polarized electrons with $P \sim 0.4$ by irradiating a photoemissive GaAs cathode with circularly polarized light. The sense of circular polarization of the light, and therefore the sign of P of the electrons, was switched harmonically by means of an electro-optic switch known as a Pockels cell. Before the electron beam entered the cell containing the chiral target molecules, the electron polarization was rotated from the transverse to longitudinal direction by a Wien filter, a device with crossed electric and magnetic fields usually employed as a velocity selector. After passage of the beam through the sample cell, the electron polarization was converted again to a transverse polarization

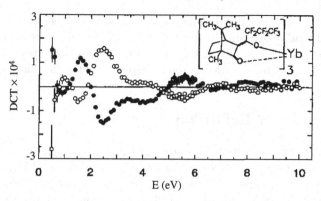

Fig. 7.10. Differential chiral transmission (DCT) of electrons of energy E through a vapor of chiral molecules D-Yb(hfc)$_3$ (*filled circles*) and L-Yb(hfc)$_3$ (*open circles*). *Error bars* indicate the statistical uncertainty whenever it is larger than the symbols denoting the measured asymmetry. (Adapted from Mayer et al. [219])

by a second Wien filter and measured by means of a Mott analyzer. In a Mott detector the electrons are accelerated to energies on the order of 100 eV and made to scatter from a thin metal foil (usually gold) of high atomic number. This Mott scattering – i.e., the Rutherford scattering of charged spin-1/2 particles – is spin-dependent because of a spin–orbit interaction of the projectile electrons with the atoms in the foil.

The foregoing brief description of the first successful detection of electron optical activity gives no hint of the numerous small instrumental effects that had to be eliminated or compensated for. For that the reader must consult the original literature. As one who has himself struggled to detect for the first time another weak chiral signal of comparable difficulty (discussed in the following section), I express my admiration for this elegant experiment. Other experiments have since been done to observe the asymmetric scattering (DCT) of polarized electrons by organic thin films of chiral molecules [221]. As of this writing, I am not aware of experiments that have observed the electron analogue of optical rotation (of the transverse polarization).

Besides its intrinsic interest as a quantum phenomenon or its potential application as a source of polarized particles, the preferential interaction of spin-polarized electrons with chiral molecules has attracted attention as a conceivable mechanism leading to the homochirality of terrestrial organisms. Numerous such mechanisms have been proposed over the years, but as of this writing I know of none that is regarded as definitive. Any viable mechanism must contend with the likelihood that precursors of biologically significant molecules existed in racemic mixtures. However, the production of spin-polarized electrons by beta decay of terrestrial elements could provide one reasonably common process that violates parity conservation. Perhaps it was the case that beta electrons selectively decomposed, or in some way rendered less stable, one of the two enantiomeric forms of certain critical molecules.

Moreover, the efficacy of such a mechanism would be expected to be greater for molecules adsorbed on solid surfaces than in gas phase or solution [222]. Future experiments to ascertain whether polarized electrons engender dissymmetric chemical transformations should help resolve the matter.

7.5 Chiral Light Reflection

Augustin Fresnel (1788–1827) was one of those few scientific geniuses who died very young (39 years), yet accomplished so much as to have his name perpetuated all throughout his field of endeavor (physical optics). A glance through the index of a standard optics textbook would reveal his eponymous diffraction, integral, reflectance and transmittance relations, double-mirror experiment, double-prism device, lens (used for many more purposes than lighthouses), and zone plate, to mention a few. It was Fresnel who, more than anyone else before Maxwell, wrought the change in paradigm from a corpuscular theory to a wave theory of light. Of particular interest here, however, are his researches on light reflection and optical activity.

To my knowledge, Fresnel was the first to:

(a) deduce the relative amplitudes for light reflection and refraction at the interface of two transparent, homogeneous, isotropic, *achiral* media;
(b) conceive of the idea of transverse circularly polarized states of light, demonstrate the existence of these states experimentally by separating an incident beam of unpolarized light into its circularly polarized components (by multiple refraction through a composite prism of alternating left- and right-handed quartz segments), and explain the phenomenon of optical rotation as a manifestation of circular birefringence, i.e., the difference in phase velocities of states of opposite circular polarization.[9]

One might have thought that Fresnel, having achieved (a) and (b) separately, would have turned his attention subsequently to examining their 'intersection', i.e., the reflection of light at the surface of an optically active material, but I have seen no evidence to suggest he ever had.[10] Perhaps the idea never occurred to him; perhaps, in the crush of other thoughts that raced through his fertile mind, he simply placed the project on the back burner and died too soon to follow through; or perhaps, like Isaac Newton, who preferred the income and prestige of becoming a high-ranking civil servant of the Crown, Fresnel may

---

[9] A comprehensive discussion of light polarization and the theory and observation of light reflection from a chiral medium is given in [7].

[10] During the 1980s, when I held the Frederic Joliot Chair of Physics at the Ecole Supérieure de Physique et Chimie in Paris, France, the office in which I worked contained a large old volume of Fresnel's complete works, musty and dusty from lying many years unread. I may have missed it, but I found no mention of a theoretical or experimental exploration of light reflection and refraction at the surface of an optically active medium.

have lost interest in problems of 'natural philosophy' as he concentrated on more practical matters as an employee of the French Lighthouse Commission. In any event, he did not appear to pursue the matter of light reflection from an optically active medium.

What is perhaps more surprising is that by the mid-1980s the experimental observation of light reflection from a naturally optically active material was still unachieved. Furthermore, theoretical descriptions of the phenomenon were not correct and had led to violations of energy conservation. In 1985, shortly after publication of my correct theory of light reflection from a homogeneous, isotropic, transparent chiral medium [223] and its generalization to absorbing chiral media [197], I began work with my students on the experimental procedures to detect the predicted chiral effects [224]. The difficulty of the endeavor may be gauged by the fact that the undertaking, initially thought to require one summer, was successfully concluded with other colleagues seven years later [204]!

Specular reflection of light is a form of coherent light scattering. Indeed, the scattering amplitudes can be calculated quantum mechanically by determining the oscillatory currents created in the receiving medium by the incident electromagnetic field and then coherently superposing the radiated secondary wavelets which, in their totality, comprise the reflected wave and transmitted waves. I am not aware, however, that this calculation has ever been done for light reflection from a chiral medium, since the scattering amplitudes, i.e., the chiral Fresnel relations, can be obtained more simply by solving Maxwell's equations with the appropriate chiral constitutive relations and correct boundary conditions.

For both conceptual and practical reasons, there is considerable interest in specular light reflection from a chiral medium. As the basis for a novel form of spectroscopy of naturally chiral materials, reflection gives rise to a strong light flux that can be detected by standard procedures for measuring voltage and current, rather than by counting photons, and would facilitate investigations of strongly absorbing materials and materials in micro-quantities such as thin films. But there are obstacles. The net optical rotation or circular dichroism of a beam propagating *through* a chiral medium increases with the optical path length, as expressed by (7.7), (7.8) and (7.73). In effect, the chiral parameter [equivalent to the circular birefringence (7.18)] is amplified by the ratio of the optical path length to the optical wavelength, which for 500 nm visible light passing through a 1 cm sample cell is about 20 000. Specular reflection from a chiral medium, on the other hand, is a 'one-shot' process. For most naturally optically active materials of interest, the chiral parameter is of the order of one part in 10^5 or 10^6. The weakness of this signal level is best seen in comparison. On the basis of electrodynamics alone, optical activity is strictly forbidden in centrosymmetric systems such as atoms, as was pointed out in Sect. 7.2. However, parity-nonconserving weak nuclear interactions give rise to a circular dichroism in caesium and thallium atoms on the order of one part in 10^5 or 10^6 and to an optical rotary power in bismuth on the order of one part in 10^7 [225]. Thus the experimental task of observing 'classical' optical activity by reflection is of comparable difficulty to discerning the influence of the nuclear weak interactions on atomic electrons.

Nevertheless, there are ways to achieve stronger chiral effects in reflection by taking advantage of (a) sample symmetry and (b) signal enhancement.

(a) Symmetry

Most chiral materials are anisotropic. Although the intrinsic chirality of these anisotropic substances is in many cases stronger than that of the few naturally occurring isotropic chiral materials (e.g., sodium chlorate or bismuth silicon oxide), the chiral contribution to the index of refraction of the medium is largely dominated by the intrinsic *linear* birefringence, except for light polarization along specific optical axes where the birefringence vanishes. For non-vanishing linear birefringence, the chiral contribution appears as a quadratic term in the index of refraction. Theoretical studies of light reflection from birefringent, optically active media have shown, however, that the differential chiral (or circular) reflection (DCR), i.e., the counterpart to DCT of the previous section, is *linear* in the chiral parameter(s), just as in the case for isotropic media [226]. Thus, one can try to observe the DCR from anisotropic samples with stronger intrinsic chirality.

(b) Enhancement

Theoretical analysis of light reflection at the interface of a transparent achiral dielectric and a chiral medium (either transparent or absorbing) under conditions permitting total reflection have shown that enhancement in the DCR by several orders of magnitude is possible in the vicinity of critical angle [227]. Correspondingly, analysis of total reflection from a chiral *amplifying* medium predicts that not only is the DCR enhanced, but circularly polarized light is amplified [228].

Recognition that chiral effects in transmission are enhanced by increasing the optical path length suggest by analogy that *multiple* reflections from a chiral medium may also lead to an enhancement of the DCR. This, indeed, has been found to be the case [229]. However, the two situations, although related, are not identical, and the conceptual differences have significant implications for experiment.

The net optical rotation of a light beam that has made an even number of passes back and forth through a naturally optically active medium is zero, in contrast to the additive effect of the Faraday rotation of a light beam multiply reflected in a medium with magnetically-induced optical activity. The reason for this was explained previously and is tied to the fact that, in the absence of external fields, the only direction by which to define optical rotation is the wave vector of the light. One sees an example of this in light reflection from a naturally optically active medium at normal incidence. The Fresnel amplitudes derived from Maxwell's equations with the physically correct set of chiral constitutive relations lead to a *null* DCR at normal incidence even though the medium has an intrinsic handedness.

A heuristic explanation of this puzzling feature may be sought in the microscopic model of reflection justified by the Ewald–Oseen extinction theorem.[11] According to the theorem, the incident light does not interact with the reflecting medium at the surface only, but propagates into the medium and is absorbed (i.e., extinguished), thereby inducing molecular dipoles to radiate secondary waves that superpose coherently to form the reflected wave. (This is precisely the quantum picture of coherent light scattering with which I began the section.) At normal incidence this interaction is equivalent to a penetration of the wave by some characteristic thickness followed by reflection from a perfect mirror. The reversal of wave helicity upon reflection leads to two contributions of opposite circular polarization within the penetration depth, and the net chiral effect (like optical rotation) vanishes. However, at larger angles of incidence, the cancellation does not occur, and indeed one finds that the predicted magnitude of the DCR reaches a maximum at a little beyond Brewster's angle, i.e., the angle at which incident unpolarized light becomes 100% linearly polarized upon reflection in the absence of optical activity.

In a similar manner, one might expect that multiple reflection of light from a chiral medium at large angle of incidence could lead to an enhancement of the DCR proportional to the number N of reflections. Theoretical analysis shows, however, that although this enhancement does not occur for ordinary reflection, a significant enhancement does occur in the case of total reflection [228]. For light incident on a transparent chiral medium near critical angle, the enhancement is to a good approximation proportional to N. For absorbing media the functional relationship is more complicated, but the enhancement is nevertheless substantial. Moreover, if the medium is absorbing, even weakly, the range of incident angles over which enhancement occurs is sufficiently wide to be of interest in spectroscopy or ellipsometry. Under these circumstances, the use of a reflection-based method for investigating chiral materials would compare favorably with the classical methods based on light transmission.

Figure 7.11 illustrates the pertinent geometrical features of light reflection at the interface of a transparent achiral medium (within which the light originates) and a chiral medium (transparent or absorbing). The plane of incidence, determined by the incident, reflected, and transmitted wave vectors, is normal to the reflecting surface. Basis states for describing the electric field of the three waves are designated σ or π, respectively, depending on whether they are perpendicular (*senkrecht* in German) or parallel to the plane of incidence. [These states of polarization are also referred to as transverse electric (TE) and transverse magnetic (TM).] The angles of incidence (θ) and trans-

---

[11] This theorem is discussed in detail for achiral media in [210, p. 100]. I am unaware of a comparable mathematical treatment of chiral media.

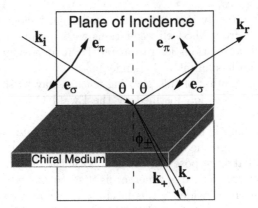

Fig. 7.11. Specular reflection at the interface of a chiral medium and achiral medium. The wave vectors of the incident (k_i), reflected (k_r), and transmitted (k_+, k_-) waves all lie in the plane of incidence. The angles of incidence (θ) and transmission (ϕ_+, ϕ_-) are defined with respect to the normal to the interface. The subscripts σ and π on the electric field basis vectors designate polarizations normal and parallel to the plane of incidence, respectively

mission $(\phi_\pm)$ are defined with respect to the normal to the reflecting surface and are governed by the familiar relations:

$$\text{Law of the mirror: angle of incidence} = \text{angle of reflection}, \qquad (7.88)$$

$$\text{Snell's law of refraction:} \quad n_0 \sin\theta = \tilde{n}_\pm \sin\phi_\pm, \qquad (7.89)$$

where n_0 is the (real-valued) refractive index of the incident medium and

$$\tilde{n}_\pm = \tilde{n} \pm \tilde{g} \qquad (7.90)$$

are the (possibly complex-valued) refractive indices for RCP (upper sign) and LCP (lower sign) states of light in a medium with chiral (also called gyrotropic[12]) parameter $\tilde{g}$. It is assumed that the two media are intrinsically nonmagnetic, whereupon the refractive indices and dielectric functions are related by

$$\varepsilon_0 = n_0^2, \qquad (7.91)$$

$$\varepsilon_\pm \equiv \varepsilon \pm \gamma = \tilde{n}_\pm^2 = \tilde{n}^2 \pm 2\tilde{g}\tilde{n} + \tilde{g}^2. \qquad (7.92)$$

In the familiar case of reflection of σ or π polarized light from a homogeneous, isotropic achiral medium, there is no change in the polarization state of the light. However, upon reflecting from a chiral medium, a σ or π polarized wave acquires a (usually small) component of orthogonal linear polarization. We can describe the reflected waves succinctly in matrix notation

$$\begin{pmatrix} r_\pi \\ r_\sigma \end{pmatrix} = \begin{pmatrix} a_{11} & a_{12} \\ a_{21} & a_{22} \end{pmatrix} \begin{pmatrix} e'_\pi \\ e'_\sigma \end{pmatrix} = \begin{pmatrix} a_1 & -i\delta \\ i\delta & a_2 \end{pmatrix} \begin{pmatrix} e'_\pi \\ e'_\sigma \end{pmatrix}, \qquad (7.93)$$

---

[12] The term 'gyrotropy' derives from Greek roots meaning 'turn in a circle' and refers to chiral optical phenomena like natural optical activity and the Faraday effect.

where the elements a_{ij} are the chiral Fresnel amplitudes, and I have shown in the cited references that the off-diagonal elements created by the chirality of the reflecting medium take the antisymmetric form of the last expression of (7.93). The exact expressions for the chiral Fresnel amplitudes are somewhat cumbersome and will be left to the original literature [197]. In view of the weakness of the chiral parameter for naturally optically active substances, one can reduce these amplitudes to the following simplified expressions

$$a_1 \sim \frac{\cos\theta - q}{\cos\theta + q}, \qquad a_2 \sim \frac{(\varepsilon/\varepsilon_0)\cos\theta - q}{(\varepsilon/\varepsilon_0)\cos\theta + q}, \tag{7.94}$$

$$\delta \sim \frac{(\varepsilon/\varepsilon_0)(z_+ - z_-)\cos\theta}{(\cos\theta + q)\left[(\varepsilon/\varepsilon_0)\cos\theta + q\right]}, \tag{7.95}$$

in which

$$q = \left(\frac{\varepsilon}{\varepsilon_0} - \sin^2\theta\right)^{1/2} \tag{7.96}$$

and

$$z_\pm = \left(1 - \frac{\sin^2\theta}{\varepsilon_\pm/\varepsilon_0}\right)^{1/2}. \tag{7.97}$$

The expressions in (7.94) and (7.96) will be recognized as the standard Fresnel relations for an achiral dielectric in which q is proportional to the normal component (relative to the reflecting surface) of the wave vector of the transmitted wave in the absence of optical activity. In the case of ordinary reflection and transparent media, the dielectric functions and refractive indices are real-valued, and (by use of Snell's law) q corresponds to $(n/n_0)\cos\phi$, where ϕ is the angle of refraction. Under the same circumstances, but for a chiral reflector, $z_\pm$ corresponds to $\cos\phi_\pm$. The chiral influence of the medium, as shown in (7.95), appears principally in the factor $z_+ - z_-$, which can be re-expressed as a difference in chiral dielectric functions

$$z_+ - z_- = \frac{\varepsilon_0(\varepsilon_+ - \varepsilon_-)\sin^2\theta}{\varepsilon_+\varepsilon_-(z_+ + z_-)}. \tag{7.98}$$

Equation (7.98) is an exact relation. For a weakly chiral medium, however, it reduces in the cases of ordinary reflection and total reflection at critical angle[13] to the following approximate relations:

$$\text{Ordinary reflection:} \qquad z_+ - z_- \sim \frac{2gn_0\sin^2\theta}{\varepsilon q}, \tag{7.99}$$

$$\text{Total reflection at critical angle:} \qquad z_+ - z_- \sim 2\sqrt{\frac{g}{n}}. \tag{7.100}$$

---

[13] For a weakly optically active medium, the critical angles for LCP and RCP waves are given approximately by

$$\theta_\pm \sim \theta_c \pm \frac{g}{\sqrt{n_0^2 - n^2}},$$

Since $|g| \ll 1$, one sees from the square-root dependence in (7.100) the advantage of performing chiral reflectance measurements near critical angle.

The Fresnel amplitudes for reflected LCP and RCP waves can be deduced from (7.93) by applying the transformation equivalent to (7.1) and (7.2)

$$\begin{pmatrix} r_{\mathrm{L}} \\ r_{\mathrm{R}} \end{pmatrix} = \frac{1}{\sqrt{2}} \begin{pmatrix} 1 & i \\ 1 & -i \end{pmatrix} \begin{pmatrix} r_{\pi} \\ r_{\sigma} \end{pmatrix} . \tag{7.101}$$

It is not necessary, however, to obtain these amplitudes explicitly, because the differential chiral reflection (DCR),

$$\mathrm{DCR} \equiv \frac{I_{\mathrm{L}} - I_{\mathrm{R}}}{I_{\mathrm{L}} + I_{\mathrm{R}}} = \frac{|r_{\mathrm{L}}|^2 - |r_{\mathrm{R}}|^2}{|r_{\mathrm{L}}|^2 + |r_{\mathrm{R}}|^2} , \tag{7.102}$$

which is the experimental quantity of interest here, is deducible directly from the scalar invariant $r_{\pi}^* \cdot r_{\sigma}$ as follows [197]

$$\mathrm{DCR} \equiv \frac{2\mathrm{Im}(r_{\sigma} \cdot r_{\pi}^*)}{|r_{\sigma}|^2 + |r_{\pi}|^2} = \frac{2\mathrm{Re}(a_1^* \delta + a_2 \delta^*)}{|a_1|^2 + |a_2|^2 + 2|\delta|^2} . \tag{7.103}$$

For weakly gyrotropic media one can neglect the chiral contribution $|\delta|^2$ in the denominator. Examination of the configuration corresponding to a chiral Fabry–Perot interferometer, whereby waves introduced into an achiral layer undergo an arbitrary number N of reflections between upper and lower bounding chiral media, have led to an exact closed-form expression for the DCR(N) as a function of N. For ordinary reflection, the maximum value of the DCR(N) (in the vicinity of Brewster's angle) is about the same as single-pass reflection. However, at critical angle, DCR(N) $\sim 2N|\delta|$.

The first experiment to have observed successfully the DCR from a chiral substance, a solution of the chiral molecule camphorquinone (Fig. 7.12) in methanol, with chiral indices,

$$\tilde{n}_{\mathrm{L,R}} = n_{\mathrm{L,R}} + i\kappa_{\mathrm{L,R}} = (1.3327 + i1.46 \times 10^{-4})\left[1 \mp (0.94 + i6.02) \times 10^{-7}\right] ,$$

where

$$\sin \theta_c = \frac{\sqrt{n^2 - \kappa^2}}{n_0} ,$$

with n the mean refractive index and κ the mean absorption coefficient of the chiral medium. The two critical angles are sufficiently close together that for most practical purposes one can approximate them by the mean critical angle θ_c. However, (7.100) was obtained by evaluating $z_+ - z_-$ at θ_-, whereupon

$$z_+(\theta_-) = \sqrt{1 - \frac{\varepsilon_-/\varepsilon_0}{\varepsilon_+/\varepsilon_0}} \quad \text{and} \quad z_-(\theta_-) = 0 .$$

To have evaluated the expression at θ_c would result in $z_-(\theta_c)$ being complex-valued. For more details, see [226].

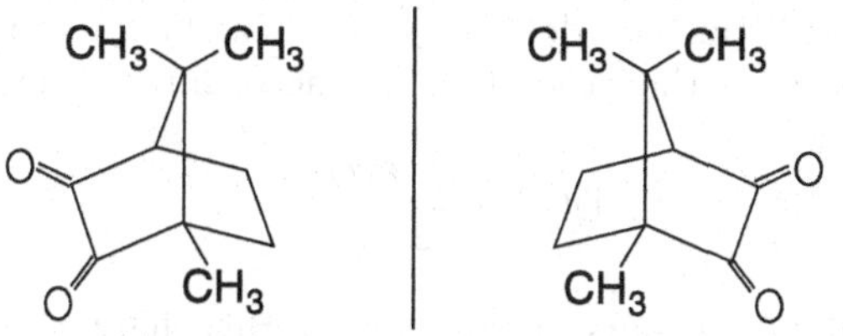

CH₃   CH₃

CH₃

l-2-3-bornanedione

CH₃   CH₃

CH₃

d-2-3-bornanedione

Fig. 7.12. Mirror-image forms (enantiomers) of the molecule camphorquinone (2-3-bornanedione), which is frequently used as a standard in the measurement of optical activity. The chiral base, camphor, lacks the second (*rear*) oxygen atom, which is replaced by two hydrogen atoms as at the other unsubstituted vertices

tested the theoretical predictions of enhancement by total reflection and multiple reflection. The reflection cell, shown in Fig. 7.13, was constructed of fused silica ($n_0 = 1.4637$) with two sample compartments which, depending on whether they were filled with chiral solution or achiral solvent, permitted 0, 1, or 2 light reflections before the incident beam exited the cell. Because the differential reflection, even with enhancement, was very small in magnitude, one could not measure separately the reflection of left and right circularly polarized light beams and take the difference. Instead, this difference was obtained in a single measurement by modulating the phase of the incident beam, derived from a xenon-arc lamp, at a fixed frequency ($f = 50$ kHz) with a photoelastic modulator and using phase-sensitive detection to extract from the output current I of the photodetector (a photomultiplier tube) the compo-

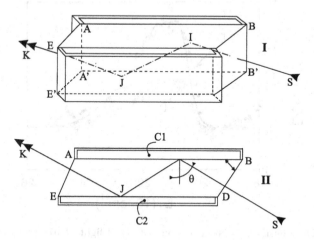

Fig. 7.13. *Horizontal* and *vertical* views of the two-compartment (C1, C2) fused-silica reflection cell. The angle ABD of the rhombus is 68°, which is close to the theoretical critical angle 65.58°. The DCR was measured under conditions where none, one, and both of the compartments contained a chiral solution. (Adapted from Silverman et al. [204])

nents oscillating at 0 Hz (i.e., dc) and f. The sought-for experimental signal was then obtained from the ratio $I(f)/I(0)$ according to the formula

$$\text{DCR} = \frac{I(f)/I(0)}{2J_1(\phi_m)} \qquad (7.104)$$

(whose origin [197, 222] is outside the scope of this chapter), where $J_1(\phi_m)$ is a Bessel function of order 1, and $\phi_m \sim 2.4$ (the first zero of the Bessel function J_0) is the modulation amplitude.

The results are shown in Fig. 7.14 which records the variation in DCR at $\theta = 67°$ for one and two reflections as a function of light energy expressed in the common spectroscopic unit of wavenumbers: $10^4 \text{ cm}^{-1} \sim 1.24$ eV. At 21 000 cm^{-1} (476 nm), the DCR(1) was measured to be 17×10^{-5} with a rms noise of 10^{-6} when θ was adjusted as closely as experimentally possible to $\theta_c = 65.58°$. Thus the measured signal represented an enhancement of nearly 300 times the signal level expected for a single-pass ordinary reflection. The excellent match of theoretical prediction (7.102) and (7.103) and experimental outcome (7.104), as shown in the figure, confirms the identity of the signal.

No DCR signal above noise was observed for reflection from compartments filled with pure solvent. Furthermore, to ensure that the observed chiral signal did not arise from artifacts related to the absorption band of the camphorquinone, measurements were also made for methanol and a concentrated solution of the achiral dye Rhodamine 6G, which has nearly the same absorption spectrum as camphorquinone. No DCR was observed in this case either.

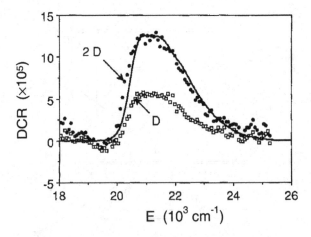

Fig. 7.14. Differential circular reflection (DCR) of light at 67° to the interface of a chiral solution of camphorquinone dissolved in methanol and achiral fused silica. D denotes a single chiral reflection; 2D denotes two chiral reflections. The *solid curve* is the theoretical DCR calculated from the optical constants of the solution. The energy of the light is expressed in wavenumbers ($10^4 \text{ cm}^{-1} = 1.24$ eV). (Adapted from Silverman et al. [204])

Successful observation of the difference with which a chiral medium reflects left and right circularly polarized light completed, in a certain sense, the 'circle' of Fresnel's studies on light reflection and optical activity and has opened the door to new spectroscopic and ellipsometric possibilities at a time when the chirality of molecules is of especial concern for both practical reasons (e.g., in production of homochiral pharmaceutical products) and theoretical reasons (e.g., the origin of homochirality as a hallmark of life on Earth). My colleagues and I, however, did not attempt to separate spatially the two reflected circularly polarized components of an incident linearly polarized beam, as this would have required a compound prism, like Fresnel's, with a large number of compartments alternately filled with chiral solutions of opposite handedness. Fresnel had constructed his compound prism out of alternating enantiomorphic forms of quartz. The term enantiomorphic – rather than enantiomeric – is employed because the optical activity of quartz derives from the helical arrangement of intrinsically achiral silicon dioxide molecules. When melted, a quartz crystal loses its optical activity. That was why the sample cell in Fig. 7.13 could be constructed of fused quartz. Interestingly, as this chapter was being written, a modern replication of Fresnel's experiment employing intrinsically chiral molecules of opposite handedness, was published [230]. Intensity plots recorded with a CCD camera showed separation of LCP and RCP components from an incident linearly polarized light beam after passage through 20 interfaces between enantiomerically opposite solutions.

7.6 Chirality in a Medium with Broken Symmetry

Quantum mechanics, as we have seen in Sect. 7.2, applied to the interaction of light with a system that can undergo simultaneous electric and magnetic dipole transitions – e.g., an enantiomeric form of a chiral molecule – leads to left and right circularly polarized eigenstates of light propagating with different phase velocities. The circular birefringence of this chiral medium is expressed macroscopically by indices of refraction of the form shown in (7.19). An intrinsically achiral medium subject to a chiral condition like rotation or an external magnetic field, as discussed in Sect. 7.3, likewise gives rise to circular birefringence, characterized for example by the refractive indices in (7.49). The symmetries underlying the origin of circular birefringence in these two cases are different as can be revealed by the different net optical rotation of a linearly polarized light beam made to pass back and forth through the medium.

We conclude this chapter by considering interesting features exhibited by light passing through a medium that displays both kinds of optical activity- i.e., an intrinsically isotropic, homogeneous chiral medium whose symmetry is broken by the presence of an external field [231]. The most expedient way to handle this case is to examine the constitutive relations of the medium. Consider first the simpler case already treated of a field-free naturally optically

active medium. As a consequence of the quantum interaction with light leading
to the perturbation, (7.30), it can be shown that the electric displacement field
D of the medium depends not only on the electric field E of the light, but
also on the variation in time of the magnetic field H [232]. Correspondingly,
the magnetic induction B of the medium depends not only on the magnetic
field H, but also on the variation in time of the electric field E. With use of
Maxwell's equations to replace the time-derivatives of fields, the two relations
can be expressed in the compact form

$$D = \varepsilon(E - \mathrm{i}f\hat{k}\times E)\,, \qquad B = \mu(H - \mathrm{i}f\hat{k}\times H)\,, \qquad (7.105)$$

in which the dimensionless quantitities ε and μ are respectively the mean
permittivity (dielectric function) and permeability of the medium, f is an
empirical chiral parameter ($f \ll 1$) arising from the interaction (7.30), and
the unit vector $\hat{k}$ gives the direction of light propagation. The relations in
(7.105) are valid to first-order in f. We limit this discussion to an intrinsically nonmagnetic medium, for which $\mu = 1$, but will retain the symbol μ
throughout to manifest a certain symmetry in the resulting expressions.

Substitution of D and B, as given in (7.105), into the Maxwell equations
corresponding to Faraday's law and the Ampere–Maxwell law

$$\nabla \times E = -\frac{1}{c}\frac{\partial B}{\partial t}\,, \qquad \nabla \times H = \frac{1}{c}\frac{\partial D}{\partial t}\,, \qquad (7.106)$$

with assumption of transverse plane-wave fields leads to the eigenvalue equation

$$(1 - m^2)E - \mathrm{i}f(1 + m^2)\hat{k}\times E = 0 \qquad (7.107)$$

for the refractive index n defined by the dimensionless quantity $m = n/\sqrt{\varepsilon\mu}$.
The two independent solutions of the equation are

$$m_{\pm}^2 = \frac{1 \pm f}{1 \mp f} \quad \Longrightarrow \quad n_{\pm} \approx \sqrt{\varepsilon\mu}(1 \pm f)\,, \qquad (7.108)$$

corresponding to right (upper sign) and left (lower sign) circular polarizations
(i.e., helicities ∓ 1). We see, then, that the quantum and phenomenological
descriptions of light in a chiral medium are consistent.

A significant point worth noting explicitly is that, although the medium
is assumed to be intrinsically nonmagnetic, the constitutive relations (7.105)
show a light-induced magnetization. The origin of this induced magnetization,
of course, is obvious from the quantum mechanical derivation. Nevertheless,
since magnetic interactions of light with matter are ordinarily much weaker
than electric interactions, it is not uncommon to find in optics or electrodynamics books [233] the modified relations

$$D = \varepsilon(E - \mathrm{i}f\hat{k}\times E)\,, \qquad B = \mu H\,, \qquad (7.109)$$

which omit the light-induced magnetization. Repeating the preceding analysis with the constitutive relations (7.109) results in chiral refractive indices $n'_\pm \approx \sqrt{\varepsilon\mu}(1 \pm f/2)$ that differ from those of (7.108) merely by a scale factor. Since f is a phenomenological parameter anyway, there is no harm in describing light *propagation* through a chiral medium by means of the relations (7.109). However, (7.109) is not in accord with the quantum picture of optical activity and, upon implementation of Maxwellian boundary conditions, does not lead to correct chiral Fresnel amplitudes for coherent light *reflection*.

Because this section is concerned primarily with effects of chirality on light propagation, I will adopt (7.109) as the simpler starting point for examining the consequences of symmetry breaking through the addition of an external field $g = g\hat{g}$. The total gyrotropic effect on light can then be described to a good approximation by the relations

$$D = \varepsilon(E - if\hat{k}\times E - ig\hat{g}\times E)\,, \qquad B = \mu H\,, \qquad (7.110)$$

where the real-valued parameters f and g characterize, respectively, the strengths of natural and induced optical activity. The phenomena of interest here derive exclusively from the electric relation which can be cast in the form $D = \tilde{\varepsilon}E$ with dielectric tensor

$$\tilde{\varepsilon} = \varepsilon \begin{pmatrix} 1 & i(g + f\gamma) & -if\beta \\ -i(g + f\gamma) & 1 & 0 \\ if\beta & 0 & 1 \end{pmatrix}, \qquad (7.111)$$

defined with respect to a right-handed Cartesian coordinate system in which g is oriented along the z axis and x is normal to the plane defined by g and the wave vector k. The direction cosines of k with respect to the y and z axes are respectively β and γ, i.e., the unit propagation vector is $\hat{k} = (0, \beta, \gamma)$.

Assuming plane waves of angular frequency ω with propagation vector $k = n(\omega/c)\hat{k}$ for the light fields E and H in the medium and substituting these waveforms into Maxwell's equations, one arrives at a set of three linearly coupled equations for the components of E, which can be written as

$$\begin{pmatrix} m^2 - 1 & -i(g + f\gamma) & if\beta \\ i(g + f\gamma) & m^2\gamma^2 - 1 & -m^2\gamma\beta \\ -if\beta & -m^2\gamma\beta & m^2\beta^2 - 1 \end{pmatrix} \begin{pmatrix} E_x \\ E_y \\ E_z \end{pmatrix} = 0\,, \qquad (7.112)$$

where again $m = n/\sqrt{\varepsilon\mu}$. A nontrivial solution exists only if the determinant of the coefficient matrix vanishes, which yields a secular equation quadratic in m^2 from which the refractive indices $n^2(\hat{k})$ are obtained

$$n^2_\pm = \varepsilon\mu \left[1 - \frac{1}{2}g^2\beta^2 \pm \sqrt{(f + g\gamma)^2 + \left(\frac{1}{2}g^2\beta^2\right)^2} \right]. \qquad (7.113)$$

From the preceding general result we note the following special cases:

- inactive dielectric ($f = g = 0$):

$$n_\pm^2 = \varepsilon\mu \,, \tag{7.114}$$

- natural optical activity ($f = 0$, $g \neq 0$):

$$n_\pm^2 = \varepsilon\mu(1 \pm f) \,, \tag{7.115}$$

- pure induced gyrotropy ($f = 0$, $g \neq 0$):

$$n_\pm^2 = \varepsilon\mu \left(1 - \frac{1}{2}g^2\beta^2 \pm g\sqrt{\gamma^2 + \frac{1}{4}g^2\beta^4} \right) \,, \tag{7.116}$$

- weak simultaneous natural and induced gyrotropy ($f \sim g \ll 1$):

$$n_\pm^2 = \varepsilon\mu\left[1 \pm (f + g\gamma)\right] \,. \tag{7.117}$$

We consider next the implications of the above relations for a plane wave propagating through the medium and reflecting at a flat boundary, as shown in Fig. 7.15. Our concern is not with the reflection amplitudes for which a more appropriate set of constitutive relations would be required, but only with the angles of reflection – in essence, Snell's law for a chiral medium with broken symmetry. Note that the direction cosines γ, γ' of the incident and reflected light have opposite signs, whereas the corresponding direction cosines β, β' have the same sign. For example, consider waves propagating normal to the surface either parallel or antiparallel to $\boldsymbol{g}$; then $\gamma = -\gamma' = 1$. From (7.113) it is seen that the indices of refraction for the incident and reflected light must

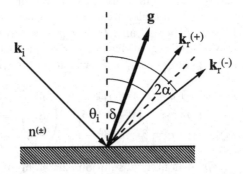

Fig. 7.15. Geometric configuration of specular reflection within a naturally optically active medium with external field $\boldsymbol{g}$ at angle δ to the normal to the reflecting surface. The medium is circularly birefringent with refractive indices $n^{(\pm)}$ that depend on the relative orientation of $\boldsymbol{g}$ and the wave vectors $\boldsymbol{k}_i$ and $\boldsymbol{k}_r^{(\pm)}$, and therefore differ for incident and reflected light. The angular interval (greatly exaggerated for visibility) between $\boldsymbol{k}_r^{(+)}$ and $\boldsymbol{k}_r^{(-)}$ is $\theta_r^{(-)} - \theta_r^{(+)} = 2\alpha$

in general differ in a medium with both natural and induced gyrotropy, in contrast to the widely assumed (but not universally applicable) 'law of the mirror' according to which angles of incidence and reflection must be equal in specular reflection.

Quantitatively, for $\hat{g}$ oriented at an angle δ to the surface normal, the direction cosines are related to the angles of incidence θ_i and reflection θ_r by

$$
\begin{aligned}
\beta &= \sin(\theta_i + \delta) , & \beta' &= \sin(\theta_r - \delta) , \\
\gamma &= -\cos(\theta_i + \delta) , & \gamma' &= \cos(\theta_r - \delta) .
\end{aligned}
\tag{7.118}
$$

Continuity of the tangential component of E across the boundary requires that the phases of the incident and reflected waves at the boundary be equal, which leads to the generalization of Snell's law

$$
n_i^{(\pm)} \sin \theta_i = n_r^{(\pm)} \sin \theta_r^{(\pm)} .
\tag{7.119}
$$

To estimate the size of the difference between θ_i and $\theta_r^{(\pm)}$, we can employ (7.117) for a weakly gyrotropic medium, express the reflection angle in the form $\theta_r^{(\pm)} = \theta_i \pm \alpha$ which is substituted into (7.119). Expansion of the resulting expression in a Taylor series to first order in small quantities (f, g, α) leads to

$$
\alpha = \frac{\left[n_i^{(\pm)} - n_r^{(\pm)} \right] \sin \theta_i}{n_r^{(\pm)} \cos \theta_i} \approx \mp g \sin \theta_i \cos \alpha .
\tag{7.120}
$$

The inequality of the angles of incidence and reflection illustrates a superficially puzzling aspect of light interactiion with a gyrotropic medium. It appears that the system violates the reciprocity theorem,[14] or, equivalently, time-reversal invariance. If light propagates through the medium with $\hat{k}$ parallel to $\hat{g}$, the natural and induced gyrotropic interactions may, for example, act in concert to rotate the plane of polarization; for $\hat{k}$ antiparallel to $\hat{g}$, the two interactions would then act in opposition to rotate the polarization by a different amount. The state of a light wave therefore depends on whether it has moved 'forward' or 'backward' through the medium. There is no paradox here, however. Examination of Maxwell's equations shows that they are time-reversal invariant when the velocities of the field sources are reversed along with the transformation $t \rightarrow -t$. This implies that time-reversal invariance of the system now under consideration should be restored if, in addition to reversal of the light propagation directions, the symmetry-breaking field g were also reversed, since the current flow giving rise to g is reversed. Equations (7.112) and (7.113)–(7.117) show that the indices of refraction and electric fields are indeed invariant under the transformation $k \rightarrow -k$ and $g \rightarrow -g$.

---

[14] See, for example, [210, p. 381], which gives the reciprocity theorem of Helmholtz: A point source at P_0 will produce at P the same effect as a point source of equal intensity at P will produce at P_0.

We will conclude this chapter by examining the properties of the light fields that propagate in selected directions relative to the symmetry-breaking field:

- Case A: Propagation along the gyroelectric axis $\hat{k} \parallel (\hat{g} = z)$. With $\beta = 0$, $\gamma = 1$, the refractive indices reduce to

$$n_\pm = \sqrt{\varepsilon\mu}\left[1 \pm (f + g)\right]^{1/2} , \tag{7.121}$$

which is of the same form as (7.115). Substitution into (7.112) leads to circularly polarized waves with right- and left-handed sense, respectively[15]

$$\begin{aligned}
E^{(+)}(x, t) &= E_0(x - iy)e^{i\omega[n_+(z/c)-t]} , \\
E^{(-)}(x, t) &= E_0(x + iy)e^{i\omega[n_-(z/c)-t]} .
\end{aligned} \tag{7.122}$$

The waves are transverse (normal to the propagation direction) and homogeneous (planes of constant phase coincide with planes of constant amplitude).

- Case B: Propagation normal to the gyroelectric axis $\hat{k} \perp (\hat{g} = x)$. With $\beta = 1$, $\gamma = 0$, the refractive indices reduce to

$$n_\pm = \sqrt{\varepsilon\mu}\left(1 - \frac{1}{2}g^2 \pm \sqrt{f^2 + \frac{1}{4}g^4}\right)^{1/2} . \tag{7.123}$$

If the medium were intrinsically inactive ($f = 0$), there would result a wave linearly polarized along the gyrotropic axis

$$E^{(+)}(x, t) = E_0 z e^{i\omega[n_+(y/c)-t]} , \tag{7.124}$$

with index $n_+ \approx \sqrt{\varepsilon\mu}$ analogous to the 'ordinary' wave in a calcite crystal, and an elliptically polarized wave in the plane determined by g and k

$$E^{(-)}(x, t) = E_0(x + igy)e^{i\omega[n_-(y/c)-t]} , \tag{7.125}$$

with index $n_- = \sqrt{\varepsilon\mu}(1 - g^2)^{1/2}$ analogous to the 'extraordinary' wave in a calcite crystal. The two rotating components are 90° out of phase and trace out in time the ellipse $x\cos\omega t + gy\sin\omega t$ with semimajor axis $a = 1$ and semiminor axis $b = g$ (for $g < 1$). The elliptical polarization is characterized by the ellipticity $e \equiv b/a = g$.

---

[15] The handedness of the fields here are opposite those in my original paper because of the difference in convention employed. In the latter, the handedness was determined by pointing the thumb of the designated hand in the direction of the propagation vector; the fingers then curled in the direction of the field rotation. Thus the fingers of the right hand designate the rotational sense of a left-circularly polarized wave. In this book, I employ the convention widely used by optical physicists in which the electric field of a right (left) circularly polarized wave rotates towards the right (left) of an observer facing the source.

If the medium has a weak intrinsic chirality $f < g$, then the fields in both cases are elliptically polarized, nontransverse, with waveforms and refractive indices that can be worked out but will not be given here.

The time-averaged energy flux in a gyrotropic medium is calculable from the Poynting vector $\langle S \rangle = (c/8\pi)\mathrm{Re}(E \times H^*)$ exactly as in the case of a non-grytropic medium. The hermiticity of the dielectric tensor (7.111) leads to an energy conservation expression of the same form (Poynting's theorem) as that to which the symmetric dielectric tensor of a nongyrotropic medium gives rise. Using the fields derived from this tensor, one finds that the energy flow is along the direction of propagation $\hat{k}$ only if f is null or if k and g are parallel. The interpretation of this is that $\hat{k}$ is normal to the wave fronts (planes of constant phase), but that the wave fronts advance in the direction of $\langle S \rangle$, the ray direction. That the directions of $\langle S \rangle$ and k need not coincide is a familiar circumstance in the optics of anisotropic, optically inactive crystals. That it may also occur for intrinsically isotropic optically active liquids subject to an external field is not well known.

8

Condensates in the Cosmos: Quantum Stabilization of Degenerate Stars

8.1 Stellar End States

How stars form may still be a matter of some controversy [234], but the essential features of how stars end are considered fairly well established. Although many details remain to be understood better about the terminal evolution of stars that have fully exhausted their nuclear fuel, there is nevertheless a broad consensus regarding the following basic conclusions. A star whose collapse can be halted by electron degeneracy pressure ends its life as a white dwarf – in effect, a giant atom with approximately the mass of the Sun compacted to the size of a planet ($\sim 5\,000$ km). If, on the contrary, electron degeneracy pressure cannot achieve hydrostatic equilibrium, the collapse continues until inverse beta decay ($\mathrm{p} + \mathrm{e} \longrightarrow \mathrm{n} + \nu$) converts the constituent electrons and protons into neutrons whose degeneracy pressure halts the collapse, creating in effect a giant nucleus with approximately the mass of the Sun compacted to the size of a city (~ 20 km) [235]. (The resulting neutrinos escape the star.) But if neutron degeneracy pressure cannot halt the collapse, then it is believed nothing can spare the star from its subsequent fate: collapse to a singularity in the space-time continuum detectable only through its gravitational field, i.e., a stellar black hole.[1]

The collapse of a massive relativistic degenerate star to a stellar black hole poses what is perhaps the most serious challenge to the laws of physics as they are currently understood. Prevailing theory, starting with the earliest comprehensive analyses of the problem by Chandrasekhar [236], who investigated stellar collapse under Newtonian gravity, and Oppenheimer and his group [237], who employed general relativity, holds that all matter and energy, once having passed through the event horizon – a spherical surface defined by the Schwarzschild radius $R_\mathrm{S} = 2GM/c^2$ with M the total mass–energy, c the speed of light, and G the universal gravitational constant – fall irreversibly into a central singularity. Nevertheless, to the extent that a black hole is

---

[1] I am referring here to an isolated black hole, i.e., a star without an accretion disk of infalling matter which can radiate.

a real physical object and not merely a mathematical solution to the differential equations of general relativity, it is difficult to believe that 10^{30} or more kilograms of matter can actually collapse to a geometric point – or to a region of the size of the so-called Planck length $l_P = \sqrt{G\hbar/c^3} \sim 1.6 \times 10^{-35}$ m, when quantum principles and Planck's constant $\hbar \equiv h/2\pi$ are taken into account. Clearly something important has been omitted from the relevant physics.

Although the physics of the terminal stages of a relativistic degenerate star beyond the Chandrasekhar or Oppenheimer–Volkoff (OV) mass limits is highly complex, the basic conclusion that total collapse is unavoidable derives essentially from two fundamental results, one classical and the other quantum. The first result, a kinematic consequence of general relativity, is that, within the region bounded by the Schwarzschild surface, the roles of spatial and temporal coordinates are interchanged. Thus, for an object at radial coordinate $r < R_S$ to reverse direction and pass outward through the horizon would be tantamount to moving backward in time, which is not permissible. The authors Misner, Thorne, and Wheeler, in their weighty tome on gravity, have described this restriction dramatically, if not grimly, as follows [238]:

> That unseen power of the world which drags everyone forward willy-nilly from age twenty to forty and from forty to eighty also drags the rocket in from time coordinate $r = 2M$ to the later value ... $r = 0$. No human act of will, no engine, no rocket, no force ... can make time stand still. As surely as cells die, as surely as the traveler's watch ticks away 'the unforgiving minutes,' with equal certainty, and with never one halt along the way, r drops from $2M$ to 0.

The second result, a dynamical consequence of quantum statistics, follows from the dependence of the relativistic fermion degeneracy pressure on the 4/3 power of the mass density, in contrast to the 5/3 power dependence of the nonrelativistic degeneracy pressure. Thus, for masses beyond a certain limit, the relativistic degeneracy pressure within a collapsing neutron star is insufficient to match the overpowering weight of infalling matter, whereupon it is presumed that nothing further can halt the collapse.

A simple model, which will be developed in greater detail shortly, illustrates this critical modification in the density dependence of the degeneracy pressure. Consider a system of N noninteracting fermions of mass m within a spherical volume $V \sim R^3$. Since, by the Pauli principle, no two fermions can occupy the same quantum state (i.e., cell of phase space), the momentum of a fermion is then approximately $p \sim \hbar/(R/N^{1/3})$, where $R/N^{1/3}$ is the mean distance between fermions. The total energy of the system, $U \sim N\sqrt{p^2c^2 + m^2c^4}$, can be expressed in dimensionless quantities as

$$\frac{U}{mc^2} = N\left(1 + N^{2/3}\frac{\lambdabar^2}{R^2}\right)^{1/2} , \tag{8.1}$$

with $\lambdabar = \hbar/mc$ the reduced fermion Compton wavelength ($\lambdabar \equiv \lambda/2\pi$). The preceding expression leads to the following nonrelativistic and relativistic approximations for energy and corresponding degeneracy pressures $P = -(\partial U/\partial V)_N$:

$$\frac{U_{\mathrm{nr}}}{mc^2} = N + \frac{1}{2}N^{5/3}\frac{\lambda^2}{R^2} \,, \qquad \frac{P_{\mathrm{nr}}}{P_0} = \frac{1}{3}n^{5/3}\lambda^5 \,, \tag{8.2}$$

and

$$\frac{U_{\mathrm{r}}}{mc^2} = N^{4/3}\frac{\lambda}{R} \,, \qquad \frac{P_{\mathrm{nr}}}{P_0} = \frac{1}{3}n^{4/3}\lambda^4 \,, \tag{8.3}$$

in which $n = N/V$ is the fermion number density and

$$P_0 = \frac{mc^2}{\lambda^3} = \frac{m^4c^5}{\hbar^3} \tag{8.4}$$

is the pressure (or mass–energy density) in a system with one particle per cubic Compton wavelength. For a system of compressed neutrons (mass $m_{\mathrm{n}} = 1.67 \times 10^{-27}$ kg), this pressure is $P_0 = 1.63 \times 10^{37}$ Pa. (For comparison, recall that 1 atm $\sim 10^5$ Pa.)

For a uniformly dense self-gravitating system of fermions interacting via Newtonian gravity, the addition to (8.1) of the gravitational binding energy (derived in Appendix 8A) leads to the expression

$$\frac{U}{mc^2} = N\left(1 + N^{2/3}\frac{\lambda^2}{R^2}\right)^{1/2} - \frac{3}{5}\left(\frac{M}{m_{\mathrm{P}}}\right)^2\frac{\lambda}{R} \,, \tag{8.5}$$

where $m_{\mathrm{P}} = \sqrt{\hbar c/G} \sim 2.2 \times 10^{-8}$ kg is the Planck mass. (The Planck length is the reduced Compton wavelength of a particle with Planck mass.[2]) One sees immediately that the condition for hydrostatic equilibrium, $dU = 0$, in the relativistic case, (8.3), can be satisfied only for a single value of the stellar mass, $M_{\max} \sim N^{2/3}m_{\mathrm{P}}$, because the fermion energy and the gravitational energy both depend on the same power of the stellar radius R, and the total energy consequently has no minimum. Since the nonrelativistic expression for total energy *has* a minimum, the mass $M_{\max}$ must be an upper limit above which no equilibrium state exists. Approximating the mass of a neutron star by $M = Nm_{\mathrm{n}}$ and equating M to $M_{\max}$ leads to the limiting number of nucleons $N_{\max} \sim (m_{\mathrm{P}}/m_{\mathrm{n}})^3 = 2.3 \times 10^{57}$ and a limiting mass $M_{\max}$ of approximately 1.9 solar masses ($M_{\mathrm{S}} = 2 \times 10^{30}$ kg), in basic accord with previous results of more refined calculations.

It is good, however, to bear in mind Alfred North Whitehead's wise observation that [239]:

---

[2] The Compton wavelength λ is the de Broglie wavelength of a particle whose relativistic kinetic energy ($pc = hc/\lambda$) is equal to its rest-mass energy (mc^2), or $\lambda = h/mc$. The reduced Compton wavelength is then $\lambda = \hbar/mc$. The Planck mass m_{P} is defined here by the equality $m_{\mathrm{P}}c^2 = Gm_{\mathrm{P}}^2/\lambda$, which leads to $m_{\mathrm{P}} = \sqrt{\hbar c/G} = 2.18 \times 10^{-8}$ kg. The Planck length is the Compton wavelength corresponding to the Planck mass: $l_{\mathrm{P}} = \hbar/m_{\mathrm{P}}c = \sqrt{\hbar G/c^3} = 1.62 \times 10^{-35}$ m.

> There is no more common error than to assume that, because pro-
> longed and accurate mathematical calculations have been made, the
> application of the result to some fact of nature is absolutely certain.

Neither of the preceding two objections need be definitive when one recognizes
the intrinsically quantum nature of a stellar black hole. (If a neutron star is
like a giant nucleus, then a stellar black hole must assuredly represent an even
more fundamentally quantum mechanical kind of system.) Thus, with respect
to classical kinematics, the coordinatization of the interior of a black hole
and subsequent general relativistic analysis of collapse trajectories can be no
more valid than would be a semi-classical (Bohr-like) description of electron
orbits within a hydrogen atom. And with respect to the equation of state,
hitherto neglected quantum processes must take place within a collapsing relativistic degenerate star to alter the pressure–density dependence and restore
hydrostatic equilibrium. Such processes are elucidated in this chapter.

Of the various processes that may intervene to stabilize the collapse of
a degenerate star beyond the OV mass limit I have investigated the following:

1. phase transitions that transform neutrons into other forms of matter with
   a different equation of state [240];
2. long-range magnetic interactions among nucleons [241], which, among
   other things, affect the radial dependence of the internal energy;
3. existence of a vacuum quantum field process referred to as particle resorption [242], which is complementary to the process responsible for Hawking
   radiation [243] and affects the fermion number density.

These are processes which are accountable within the framework of currently
known principles of physics. There have also been attempts to suppress the
formation of an event horizon by appeal to as yet incomplete and untested
theories of quantum gravity [244].

The possibility of stabilizing a collapsing star (and preserving the laws of
physics) by means of an appropriate phase change is of particular interest in
light of very recent advances in the investigation of degenerate atomic Fermi
gases at ultra-low temperature [245]. This possibility concerns the pairwise
(or other even-numbered) condensation of fermions to produce an equilibrium
two-phase system of fermions and composite bosons. It may seem surprising
at first that such processes occur, as one is long used to thinking of fermions
and bosons as entirely distinct quantum entities obeying radically different
statistical laws (except in the high-temperature regime where both forms of
quantum statistics resemble classical Maxwell–Boltzmann statistics).

Currently under intensive theoretical and experimental investigation, ultra-
cold gases of fermionic atoms, such as ^{40}K and ^{6}Li, with magnetically tunable
interactions provide a means of exploring the predicted transition between two
extreme types of fermionic associations: Bardeen–Cooper–Schrieffer (BCS)
superfluidity and Bose–Einstein condensation (BEC) into composite bosons.
Both kinds of behavior have been observed with ultra-cold fermionic atoms.
Theorists have predicted, and experiments are confirming, that behavior of

the fermionic atoms becomes universal in the strong-interaction regime; i.e., that the interactions become independent of the details of the atomic potential [246]. Thus, degenerate atomic fermi gases provide experimentally accessible model systems for exploring processes that may also occur in dense nuclear matter such as in relativistic degenerate stars and in a quark–gluon plasma [247].

The occurrence of BCS-type superfluids in neutron stars (primarily neutron superfluidity and proton superconductivity) is believed to account for sudden period changes ('glitches' or 'starquakes') and their subsequent slow relaxation [248], although some questions remain concerning the exact nature and distribution of these hadronic superfluids [249]. The universality of fermionic condensation in the strong-interaction regime, based on the theory and laboratory investigations of fermionic atoms, would suggest, therefore, the possibility that BEC-type behavior could occur as well under conditions of even greater compression. Qualitatively, if a sufficiently massive gas of degenerate fermions should undergo an equilibrium transformation to a condensate of composite bosons, one would expect a subsequent (and likely rapid) phase separation as the fermion component continued to collapse inward, whilst the condensate phase, by virtue of the quantum laws to which it is subject, cannot collapse to a domain much smaller than the volume defined by its coherence length.[3] The collapsing fermion sphere, therefore, need no longer terminate in a singularity because, having fewer particles and correspondingly lower Fermi energy than initially as a consequence of the phase transition, it can then achieve hydrostatic equilibrium at a stellar radius of finite (and, as we shall see, macroscopic) size.

Although the formation of boson condensates comprising pions or kaons at the center of a neutron star, as well as promotion of neutrons into more massive baryons like Λ-hyperons, have been proposed in the past [250], the physical circumstances of these models were such as to lead to a weaker degeneracy pressure that facilitated, rather than prevented, stellar collapse. These earlier proposals are quite different from the process outlined above and examined in the following sections.[4]

8.2 Quantum Properties of a Self-Gravitating Condensate

A distinctive feature of the fermion condensation process to stabilize the collapse of a relativistic degenerate star otherwise destined to terminate in a singularity is the formation of a condensate of composite bosons. As we shall see, the condensate is at a temperature so far below the theoretical transition temperature that for all practical purposes it can be considered at absolute zero,

---

[3] I discuss this property of boson condensates in [82].

[4] The idea of a fermion condensation to bosons within a collapsing neutron star were first presented in the brief reports [251].

i.e., all bosons have condensed into the ground state. As is well known, an ideal condensate of noninteracting bosons at 0 K exerts no pressure because all particles are in a state of zero relative momentum [252]. (The pressure of a nonrelativistic Bose gas approaches zero with temperature as $T^{5/2}$ independent of volume.) An essential feature of a self-gravitating condensate, however, recognized in the author's previous investigations of the nature of galactic dark matter as a condensate of very low mass bosons [82, 253], is that a quantum condensate cannot be localized to a region much smaller than its coherence length; i.e., it cannot collapse to a singularity. Thus, the condensate can contribute comparatively little to the weight that must be supported by fermion degeneracy pressure if the bulk of the condensate at equilibrium lies outside the core of collapsed fermionic matter. The maintenance of a finite condensate size on the order of the condensate coherence length may be understood heuristically to follow from the Heisenberg uncertainty principle. Equivalently, were the condensate to collapse radially inwards, the particles would acquire relative momentum and therefore kinetic energy, but such an excitation cannot occur at 0 K. Moreover, direct observation of condensate phase separation in an ultra-cold, strongly interacting Fermi gas of ^{6}Li atoms gives reason to believe that a phase separation would also occur in a system of dense nuclear matter even in the absence of gravity [254].

To estimate the size of a spherical self-gravitating condensate, consider a system of N_b bosons of mass m_b (reduced Compton wavelength λ_b) for which the quantum equation of motion is the nonlinear wave equation known as the Gross–Pitaevskii (GP) equation

$$\left[-\frac{\hbar^2}{2m_b}\nabla^2 + V(r) + g|\psi|^2 \right]\psi = \mu_b\psi . \tag{8.6}$$

$V(r)$ is the confining potential energy function, μ_b is the chemical potential, and

$$g = \frac{4\pi\hbar^2 N_b a}{m_b} = 4\pi N_b m_b c^2 \lambda_b^2 a \tag{8.7}$$

is the coupling parameter containing the scattering length a [255]. The solution to the GP equation in the Thomas–Fermi (TF) approximation, where kinetic energy is negligible compared with potential energy, takes the simple form [256]

$$|\psi(r)|^2 = A^2\left(1 - \frac{r^2}{R^2}\right) = \frac{\mu_b}{g}\left[1 - \frac{V(r)}{\mu_b}\right] , \tag{8.8}$$

in which the first equality results from normalization of the wave function (approximated by a gaussian of effective size R with normalization constant $A^2 = 15/(8\pi R^3)$, and the second equality follows directly from the GP equation with neglect of the kinetic energy (i.e., Laplacian) term. Comparison of (8.7) and (8.8) with substitution of the normalization constant leads to the relation

$$\mu_b = \frac{R^2 V(r)}{r^2} = \frac{15}{8\pi}\frac{g}{R^3} . \tag{8.9}$$

In the table-top experiments on atomic condensates, external magnetic fields created a harmonic oscillator trapping potential energy $V(r) = m\omega^2 r^2/2$ with angular frequency ω. In a formally similar way (although the physics is quite different), a boson at the surface ($r = R$) of a self-gravitating condensate mass M of uniform density ρ_b, has a gravitational potential energy of magnitude $GMm_b/R = m_b\omega^2 R^2/2$ with angular frequency parameter $\omega = \sqrt{8\pi G\rho_b/3}$. Substitution of ω into $V(r)$ in (8.9) yields an expression for the condensate size, which can be recast in the form

$$R = \sqrt{\frac{15}{2}a\lambda_b}\,\frac{m_P}{m_b}\,. \tag{8.10}$$

An alternative, but basically equivalent, procedure is to minimize the internal energy

$$U_b(r) = \rho_b\left(\frac{4}{3}\pi R^3\right)c^2 - \frac{3}{5}\frac{GM^2}{R} \tag{8.11}$$

with respect to R, in which it will be shown later that the condensate mass density (excluding gravitational binding energy) is related to the condensate number density n_b by

$$\rho_b = m_b n_b(1 + 4\pi\lambda_b^2 an_b)\,, \tag{8.12}$$

and the total condensate mass is here approximated by $M = N_b m_b$. This approach, employing uniform density $n_b = 3N_b/4\pi R^3$, yields a condensate radius

$$R = \sqrt{15a\lambda_b}\,\frac{m_P}{m_b}\,, \tag{8.13}$$

which differs from (8.10) only by a factor of $\sqrt{2}$.

If, for example, we consider a more or less minimal magnitude of a to be comparable to λ_b for a composite boson comprising a pair of neutrons, then, upon inserting $m = 2m_n$ and $a = \lambda_b = \hbar/(2m_n)c$ in (8.13), we obtain $R \sim 2.64$ km, which is numerically close to the Schwarzschild radius

$$R_S = \frac{2GM_{Sun}}{c^2} \sim 2\lambda_n\left(\frac{m_n}{m_P}\right)^2 N_{Sun}\,, \tag{8.14}$$

of a solar-mass neutron star ($N_{Sun} \sim 1.2 \times 10^{57}$, $R_S \sim 2.97$ km).

It is worth noting explicitly that the exact numerical value of R estimated above is *not* the point of emphasis. In a later section we will determine the size, mass, and other stellar properties of hybrid fermion-condensate stars more rigorously and precisely. What is significant now is that the radius of a self-gravitating condensate sphere is expected to be comparable to the macroscopic-sized Schwarzschild radius. If the collapse of the neutron component is arrested at some substantial fraction of the condensate radius, then the terminal equilibrium state of the entire star will be of macroscopic size and finite density, and there will be no singularity. This will be demonstrated in detail shortly. It is also to be noted that the quantum relation for condensate number density, $n = N|\psi|^2$, derived from the TF approximation to the

GP equation is not uniform, but takes the shape of an inverted parabola with central density $n_0 = N\mu/g \sim N/R^3$. This refinement does not invalidate the preceding conclusion, and the assumption of uniform number density will be retained for analytical convenience wherever possible.

Consider next the transition temperature T_C at which a self-gravitating boson condensate can form. Standard statistical theory for an ideal noninteracting Bose gas [257] relates the condensate number density n to the thermal de Broglie wavelength $\lambda_T = h/\sqrt{2\pi m k_B T_C}$ by $n = \zeta(3/2)/\lambda_T^3$, where $\zeta(3.2) = 2.612$ is a zeta function. From this follows the relation

$$k_B T_C \sim \frac{3.313\hbar^2 n^{2/3}}{m} , \tag{8.15}$$

where k_B is Boltzmann's constant. For a condensate of $N_b = 10^{57}$ composite bosons of mass $m = 2m_n$ in a volume of radius $R = 2.64$ km, one estimates $k_B T_C \sim 377$ MeV or $T_C \sim 4.4 \times 10^{12}$ K. This theoretical deduction, however, is not strictly applicable to a condensate subject to a harmonic trapping potential. In that case the theoretical transition temperature has been shown to be [258]

$$kT_C \sim 0.94\hbar\omega N^{1/3} , \tag{8.16}$$

which, for the same parameters assumed above, leads to the even higher critical temperature $T_C \sim 1.1 \times 10^{13}$ K. From numerical studies of cold Fermi atom gases [259], a critical temperature $T_C \sim 0.23T_F$ was deduced, in which T_F is the Fermi temperature. The Fermi temperature of a degenerate neutron star with baryon number comparable to that of the Sun but collapsed to a radius approximately that of the Schwarzschild radius is of the order of 10^{12}–10^{13} K, as will be demonstrated in the following section. Since the temperature within the core of a neutron star ordinarily falls within the range 10^6–10^8 K, well below T_C, and since the number of bosons in the ground state relative to the total number of bosons in a condensate varies with temperature as $1 - (T/T_C)^{3/2}$ for $T < T_C$, it is safe to conclude that both fermionic and bosonic phases within the collapsed star can be treated as if at absolute zero, and that the condensate is entirely in its ground state for the purposes of this analysis.

8.3 Quantum Properties of a Self-Gravitating System of Degenerate Fermions

A relativistic degenerate star at its thermonuclear end point prior to fermion condensation can be modeled as a self-gravitating system of N fermions (assumed here to be neutrons) of uniform number density n_n in volume V at absolute zero and stable to beta decay. The internal energy function of this system, according to standard quantum statistics and general relativity (Schwarzschild geometry) [260, 261] can be expressed in the form

$$U_n = U_f + U_g = \frac{m_n c^2 g(y_F)V}{8\pi^2 \lambda_n^2} + \left[1 - f(x)\right]Mc^2 = Mc^2 , \tag{8.17}$$

in which M is the total mass–energy of the star. We shall examine the two terms, U_f and U_g, in sequence.

The fermionic kinetic energy (including mass–energy) term U_f is a function of the dimensionless Fermi parameter y_F

$$y_F \equiv \frac{p_F}{m_n c} = \left(\frac{3\pi^2 N_n}{V}\right)^{1/3} \lambda_n = (3\pi^2 n_n)^{1/3} \lambda_n , \qquad (8.18)$$

where p_F is the Fermi momentum. It is to be recalled that in a fully degenerate system of fermions all quantum states from the lowest ($m_n c^2$) to the Fermi energy (ε_F) are filled with one particle of specified spin orientation per state. Thus, for spin-1/2 particles, there are two particles per state, and the highest energy state reached by the system of N particles is

$$\varepsilon_F = m_n c^2 \sqrt{1 + y_F^2} . \qquad (8.19)$$

The quantum statistical function $g(y)$

$$g(y) \equiv y(2y^2 + 1)\sqrt{y^2 + 1} - \sinh^{-1}(y) \qquad (8.20)$$

is obtained by summing the energy (i.e., integrating the momentum) of all the particles in the system. The expansions of $g(y)$ for both nonrelativistic ($y < 1$) and relativistic ($y > 1$) cases are

$$g(y) \sim \begin{cases} \dfrac{8}{3}y^3 \left(1 + \dfrac{3}{10}y^2\right) & (y < 1) , \\ 2y^4 & (y > 1) , \end{cases} \qquad (8.21)$$

leading to the respective fermion energy expressions

$$\frac{U_f}{m_n c^2} \approx \begin{cases} N\left(1 + \dfrac{3}{10}y_F^2\right) \sim N + \text{const.}\dfrac{N^{5/3}}{R^2} & (y_F < 1) , \\ \dfrac{3}{4}Ny \sim \text{const.}\dfrac{N^{4/3}}{R} & (y_F > 1) , \end{cases} \qquad (8.22)$$

which justify the previously used simplified relation of (8.1).

The fermion pressure (i.e., the neutron degeneracy pressure) P_n and chemical potential μ_n follow from the energy function U_f by taking appropriate derivatives. The exact expression for pressure

$$P_f = -\left(\frac{\partial U_f}{\partial V}\right)_N = \frac{m_n c^2}{24\pi^2 \lambda_n^3}\left[y_F(2y_F^2 - 3)\sqrt{y_F^2 + 1} + 3\sinh^{-1} y_F\right] \qquad (8.23)$$

can be expanded in the nonrelativistic and relativistic cases to yield

$$P_n \approx \begin{cases} \dfrac{(3\pi^2)^{2/3}}{5}\dfrac{\hbar^2}{m_n}n_n^{5/3} & (y_F < 1) , \\ \dfrac{(3\pi^2)^{1/3}}{4}\hbar c n_n^{4/3} & (y_F > 1) , \end{cases} \qquad (8.24)$$

producing again the expected dependences on particle number density n_n. The neutron degeneracy pressure factor in (8.23) [to be compared to that of (8.4)] is

$$P_0 = \frac{m_n c^2}{24\pi^2 \lambda_n^3} = 6.9 \times 10^{34} \text{ Pa}. \tag{8.25}$$

Also, as expected, the chemical potential at $T = 0$

$$\mu_n = \left(\frac{\partial U_f}{\partial N_n}\right)_V = m_n c^2 \sqrt{1 + y_F^2} \tag{8.26}$$

is identical to the Fermi energy.

From (8.24), it follows that the threshold value of the Fermi parameter y_F at which the nonrelativistic and relativistic expressions for degeneracy pressure are equal is determined by the particle density $n_n = (5/4)^3/3\pi^2\lambda^3 \sim 7.2 \times 10^{45}$ m^{-3} and is evaluated to be $y_F^0 \sim 1.25$. This is equivalent to a Fermi temperature $T_F = \mu_n(y_F^0)/k_B \sim 1.7 \times 10^{13}$ K, as indicated in the previous section. For a solar mass neutron star, comprising very nearly 10^{57} neutrons, of radius 3 km (which is marginally larger than the Schwarzschild radius of the Sun), the number density is 8.8×10^{45} m^{-3} and the Fermi parameter is $y_F = 1.34$. We shall see shortly that the equilibrium Fermi parameter *after* fermion condensation lies unmistakably in the nonrelativistic domain.

Consider next the gravitational term U_g in (8.17). In the Schwarzschild solution to the equations of general relativity, the total mass M of the system, i.e., the manifestation of the energy of the star to a distant observer, is defined by the relation

$$M \equiv \int_0^R 4\pi r^2 \rho dr = \frac{4\pi}{3} R^3 \rho, \tag{8.27}$$

which is evaluated above for uniform density ρ. There is a subtlety to the foregoing expression, however, because it is *not* an integral over a spherical volume in the Schwarzschild geometry for which the actual volume element would be

$$dV = \frac{4\pi r^2 dr}{\sqrt{1 - \dfrac{2GM(r)}{rc^2}}} = \frac{4\pi r^2 dr}{\sqrt{1 - \dfrac{8\pi G\rho}{3c^2} r^2}}.$$

Integration over this volume element leads to the volume

$$V = \left(\frac{4\pi}{3} R^3\right) f(x_S) \tag{8.28}$$

of a sphere of coordinate radius R, in which the function $f(x)$ is

$$f(x) \equiv \frac{3}{2x^2}\left(\frac{\sin^{-1} x}{x} - \sqrt{1 - x^2}\right), \tag{8.29}$$

and the dimensionless Schwarzschild parameter x_S, defined by

$$x_S \equiv \sqrt{\frac{2GM}{Rc^2}} = \sqrt{\frac{8\pi G\rho}{3c^2}}R \,, \tag{8.30}$$

must lie in the range $0 \leq x_S \leq 1$; $f(x)$ approaches 1 for vanishing x and equals $3\pi/4$ for $x = 1$. Equation (8.27) for the mass M is, in effect, an integral over volume of the quantity

$$\rho\sqrt{1 - \frac{8\pi G\rho}{3c^2}r^2} \,,$$

which is less than the density by virtue of the (negative) gravitational binding energy. The general relativistic expression for gravitational potential energy is then the fraction $1 - f(x)$ of the total mass energy Mc^2. Justification of this interpretation is obtained by expanding $f(x) \sim 1 + 3x^2/10$ to obtain the Newtonian gravitational binding energy of a uniformly dense sphere (Appendix 8A)

$$U_g^{(\text{Newton})} = -\frac{3}{10}x^2 Mc^2 = -\frac{3}{5}\frac{GM^2}{R} \,. \tag{8.31}$$

The precise relation between mass density ρ and the product mn, which were treated as equal peviously, can be deduced from (8.17) in the tractable case of uniform density by adding the gravitational and total mass energy terms and dividing all terms of the equation by the volume V to obtain

$$\rho = \frac{m_n g(y_F)}{8\pi^2\lambda^3} \approx \begin{cases} m_n n_n \left(1 + \frac{3}{10}y_F^2\right) = m_n n_n + \frac{3}{2}\frac{P_n}{c^2} & (y_F < 1) \,, \\ \frac{3}{4}m_n n_n y_F = \frac{3P_n}{c^2} & (y_F > 1) \,, \end{cases} \tag{8.32}$$

where the nonrelativistic and relativistic expansions follow from (8.21). The second equalities in the half-bracketed expression above follow readily from general thermodynamic relations linking density, chemical potential, and pressure:[5]

$$\mu \equiv \frac{d\rho c^2}{dn} \,, \tag{8.33}$$

$$P \equiv -\frac{d(\rho c^2/n)}{d(1/n)} = n\mu - \rho c^2 \,. \tag{8.34}$$

It follows immediately from the preceding set of equations that the pressure must vanish in a system for which $\rho = mn$.

---

[5] At $T = 0$, the first law of thermodynamics $dU = -PdV + \mu dN$ leads to $\mu = (\partial U/\partial N)_V$. Since the volume is held constant, one can divide both numerator and denominator by V to obtain a derivative involving energy and particle densities $\mu = (d\varepsilon/dn)_V$, which, upon setting $\varepsilon = \rho c^2$, is equivalent to (8.33). From the property that energy (and entropy) are first-order homogeneous functions of the extensive parameters of a system, one can write that $U = -PV + \mu N$ at $T = 0$. Dividing both sides by V leads immediately to $\varepsilon = -P + \mu n$, which is equivalent to (8.34). See, for example, [262].

Within the framework of general relativity, the condition of hydrostatic equilibrium is expressed by the Tolman–Oppenheimer–Volkoff (TOV) equation, which, in the case of uniform density, takes the form

$$-\frac{dp}{dr} = \frac{4\pi G}{3} \frac{(\rho + P/c^2)(\rho + 3P/c^2)r}{1 - \dfrac{8\pi G\rho}{3c^2}r^2} , \tag{8.35}$$

or

$$-\frac{dz}{dx} = \frac{1}{2} \frac{(1+z)(1+3z)x}{1-x^2}$$

when written in terms of the dimensionless quantities $z \equiv P/\rho c^2$ and $x = (8\pi G\rho/3c^2)^{1/2}r$. Evaluated at the stellar surface ($r = R$), x becomes the Schwarzschild parameter of (8.30). Equation (8.35) can be integrated to yield the pressure-density relation at the center of the star ($x = 0$)

$$z_0 = \frac{1 - \sqrt{1 - x_S^2}}{3\sqrt{1 - x_S^2} - 1} . \tag{8.36}$$

When the ratio of pressure to mass–energy density is much less than unity, and all points of the stellar surface lie outside the Schwarzschild radius, (8.35) and (8.36) reduce to the Newtonian relations

$$-\frac{dp}{dr} = \frac{4\pi}{3}G\rho^2 r \qquad \text{or} \qquad -\frac{dz}{dx} = \frac{1}{2}x , \tag{8.37}$$

$$z_0 = \frac{1}{4}x_S^2 . \tag{8.38}$$

Figure 8.1 shows the variation of z_0 with x for the TOV and Newtonian relations. As deducible from (8.36), the TOV central pressure becomes comparable to mass–energy density at $x_S^2 = 3/4$ and approaches infinity as x_S^2 approaches $8/9 \sim 0.889$. One might expect the Newtonian relation to prove at least qualitatively reliable, if not actually quite good, for x_S^2 below about 0.6, or, equivalently, $R/R_S > 1.7$.

In the following section, we will employ the Newtonian condition for hydrostatic equilibrium for several reasons. First, as a matter of justification in hindsight, it will be seen that implementation of Newtonian hydrostatic equilibrium, together with other applicable equilibrium conditions, leads to stellar densities and radii for which the Newtonian reduction of the TOV relation is largely applicable. Second, it is in fact problematical whether the TOV equation actually would apply in its current form to the processes discussed here, since the conditions of its derivation, as summarized particularly clearly by Oppenheimer and Volkoff, require fermion conservation and positive pressure, whereas the number of fermions markedly changes in the formation of the boson condensate and, as we shall see shortly, circumstances can arise for which the condensate pressure is negative. Third, as was noted previously, general relativity is a classical theory, and it is not at all clear whether there

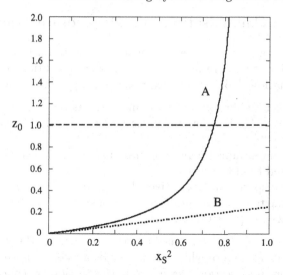

Fig. 8.1. Variation of the central pressure-to-mass density ratio with the square of the Schwarzschild parameter for (**A**) the general relativistic TOV equation and (**B**) the Newtonian equation of hydrostatic equilibrium

can continue to be a well-defined event horizon in a fully quantum theory of ultra-dense self-gravitating nuclear matter. And fourth, even if the TOV equation can be made to apply to models with variable fermion number and negative pressure, there may be little point in utilizing a general relativistic equation of hydrostatic equilibrium without taking account comprehensively of all circumstances in which a highly curved space-time may influence the physical components of the model. For example, in evaluating the partition function for a self-gravitating system of fermions, one calculates the Fermi energy, Fermi parameter, and other thermo-statistical quantities by integrating separately over regions of coordinate space and momentum space. Such a separation would not be possible, however, and the resulting integrals would be incomparably more difficult, if fermion momentum depended on location within a space-time region of high curvature.

Such issues, as raised above, are being investigated, but little progress would be made to begin with them. It may be argued that if the TOV equation, or, more broadly, general relativity itself, does not apply, then almost certainly neither would Newtonian theory. Nevertheless, as a matter of practicality, one has to begin somewhere, and one may hope to make more progress with a simpler theory whose limitations are understood, than with a mathematically much more difficult one which, in the present case, raises deep conceptual issues. In this regard it is interesting to note that Schrödinger initially failed in describing the hydrogen atom by a relativistic quantum theory, but succeeded brilliantly only when forced to revert to what at the time may have seemed to him a less inclusive nonrelativistic approach.

8.4 Fermion Condensation in a Degenerate Star

In an evolving star, such as the Sun, which supports itself against collapse by nuclear fusion, there are four basic equilibrium conditions in the form of differential equations [263], which, expressed simply in words, are:

1. Hydrostatic Equilibrium: (a) Mass increases as one moves out from the center [(8.27)]. (b) Pressure increases as one moves in towards the center [(8.35) or (8.37)].
2. Thermal Equilibrium: Energy produced in the core must balance the energy radiated from the surface.
3. Energy Transport: (a) Radiative: The more opaque the interior, the steeper is the temperature gradient. (b) Convective: The temperature gradient follows the pressure gradient.

These conditions must be maintained at each point within the star, the mathematical implementation of which requires substantial computing power.

The equilibrium conditions developed in this section are considerably simplified as a consequence of two assumptions. First, because the initial state of the matter in the star is at its thermonuclear end point and is assumed to be purely neutronic and beta stable at a temperature of 0 K, there is no energy generation, transport, or radiation to deal with. Second, the two resulting phases (fermion and condensate) are assumed to be quantum gases of uniform density, and pressure is proportional to a power of the density in both the nonrelativistic and relativistic domains. Thus the conditions of chemical and hydrostatic equilibrium to be developed take the form of algebraic, rather than differential, equations, an approximation which, while not strictly rigorous, is often adopted in astrophysics to obtain a mathematically tractable understanding of the physical behavior of a complex system.

Consider, first, the condition of chemical equilibrium in the reversible transformation $sn \leftrightarrow b$, whereby an even number s of fermions (assumed to be neutrons) associate to produce a composite boson with mass parameter $Q = m_b/sm_n$. Studies of hadron superfluidity in neutron stars [264] have generally assumed $s = 2$, but other combinations are conceivable given the enormous densities occurring in stars destined to form black holes. Assumption of neutron conservation requires that

$$N_n + sN_b = N_n^0 \,, \tag{8.39}$$

where N_n^0 is the initial number of neutrons before onset of condensate formation, and N_n and N_b are respectively the equilibrium numbers of neutrons and composite bosons. The condition of chemical equilibrium is expressed by the equality of chemical potentials

$$s\mu_n = \mu_b \,, \tag{8.40}$$

in which (8.26) gives the chemical potential of the fermion component. Investigations of ultra-cold atomic condensates described earlier indicate that the chemical potential of a condensate of interacting particles is proportional to the number density n_b, which, with addition of the mass–energy term (not pertinent to tabletop atomic experiments, but of critical importance in the present astrophysical context), lead to the form

$$\mu_b = m_b c^2 (1 + \varepsilon n_b) , \qquad (8.41)$$

with parameter [258, p. 106]

$$\varepsilon = 4\pi \lambda_b^2 a \qquad (8.42)$$

defined in terms of the reduced Compton wavelength $\lambda_b = \hbar/m_b c$ and interaction parameter (scattering length) a of the bosons. Equation (8.41) is equivalent to (8.6) and (8.7) in the TF approximation in absence of gravity, which is accounted for separately in the relations of hydrostatic equilibrium.[6] Substitution of (8.26) and (8.41) into (8.40) leads to the condition of chemical equilibrium:

$$\sqrt{1 + y_F^2} = Q(1 + \varepsilon n_b) . \qquad (8.43)$$

It is to be noted that a, and therefore ε, can be either positive or negative, signifying either a repulsive or attractive interaction between condensate particles. The physical implications of this difference will made apparent shortly.

Consider next the conditions of hydrostatic equilibrium at the boundary of radius R between the neutron core and condensate shell and at the outer condensate boundary of radius R_0 defining the size of the star, as shown in Fig. 8.2. Insertion of chemical potential (8.41) into relations (8.33) and (8.34) leads to the following expressions for the condensate mass density and pressure

$$\rho_b = m_b n_b (1 + \varepsilon n_b) , \qquad (8.44)$$

$$P_b = \frac{1}{2} \varepsilon m_b c^2 n_b^2 . \qquad (8.45)$$

Pursuant to the approximations outlined at the beginning of this section, implementation of hydrostatic equilibrium at the inner boundary leads to the relation

$$P_n(0) = P_b(R) + \frac{2\pi}{3} G \rho_n^2 R^2 , \qquad (8.46)$$

and, assuming for the present an interaction $a > 0$, the relation

$$P_b(R) = \frac{4\pi}{3} G \rho_b^2 R_0^2 (1 - \zeta) \left[\left(\frac{\rho_n}{\rho_b} - 1 \right) \zeta^2 + \frac{1}{2}(1 + \zeta) \right] , \qquad (8.47)$$

---

[6] Gravitational potential energy is not ordinarily included in the chemical potential, because, in accordance with the equivalence principle, one can always consider chemical equilibrium in a local Lorentz frame where gravity vanishes. However, comparison of chemical potentials at different locations in a star does require taking account of the gravitational potential, and this can influence the condition of chemical equilibrium. See [265].

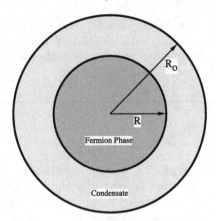

Fig. 8.2. Schematic cross-section through a degenerate star comprising two separated phases (fermionic and bosonic) in chemical and hydrostatic equilibrium after fermion condensation

with

$$\zeta \equiv \frac{R}{R_0} \tag{8.48}$$

at the outer boundary, where the boundary condition $P_b(R_0) = 0$ has been imposed.

To implement the model, one specifies at the outset the initial neutron number N_0 and condensate parameters Q, s, a, and solves the equations of equilibrium to obtain the equilibrium neutron fraction $X \equiv N_n/N_0$, the Fermi parameter y_F, and radial ratio ζ, from which follow all the pertinent stellar properties formulated in Table 8.1. Re-expressed dimensionlessly in terms of the three convenient independent variables, the equations of equilibrium take the form:

- Chemical equilibrium:

$$\frac{\sqrt{1 + y_F^2} - Q}{Q y_F^3} = \frac{1}{3\pi^2 s} \frac{\varepsilon}{\lambda_n^3} \frac{1 - X}{X} \frac{\zeta^3}{1 - \zeta^3} . \tag{8.49}$$

- Hydrostatic equilibrium at R:

$$y_F^2 \left[y_F(2y_F^2 - 3)\sqrt{y_F^2 + 1} + 3\sinh^{-1} y_F \right]$$

$$= \frac{1}{4\pi} \left(\frac{m_n}{m_P}\right)^2 \left(\frac{9\pi}{4} N_0\right)^{2/3} \left[y_F(2y_F^2 + 1)\sqrt{y_F^2 + 1} - \sinh^{-1} y_F \right] X^{2/3}$$

$$+ 12\pi^2 \frac{\lambda_n^4}{\lambda_b \varepsilon} y_F^2 \left(\frac{\sqrt{y_F^2 + 1} - Q}{Q}\right)^2 . \tag{8.50}$$

Table 8.1. Evaluation of equilibrium stellar properties from X, y_F, ζ

Fermion number	$N_n = X N_0$
Boson number	$N_b = \dfrac{1}{s}(N_0 - N_n)$
Inner radius	$R = \left(\dfrac{9\pi}{4} N_0\right)^{1/3} \dfrac{X^{1/3} \lambda_n}{y_F}$
Outer radius	$R_0 = R/\zeta$
Fermion number density	$n_n = \dfrac{N_n}{4\pi R^3/3} = \dfrac{1}{3\pi^2}\left(\dfrac{y_F}{\lambda_n}\right)^3$
Boson number density	$n_b = \dfrac{N_b}{4\pi(R_0^3 - R^3)/3} = \dfrac{\sqrt{1+y_F^2} - Q}{\varepsilon Q}$
Fermion mass density	$\rho_n = \dfrac{m_n g(y_F)}{8\pi^2 \lambda^3}$
Boson mass density	$\rho_b = m_b n_b(1 + \varepsilon n_b)$
Gravitational binding energy	$U_g = -\dfrac{3}{5}\left(\dfrac{4\pi}{3}\right)^2 G \rho_b^2 R_0^5 \left\{ 1 + \dfrac{5}{2}\left(\dfrac{\rho_n}{\rho_b} - 1\right)\zeta^3(1 - \zeta^2) \right.$ $\left. + \left[\left(\dfrac{\rho_n}{\rho_b}\right)^2 - 1\right]\zeta^5 \right\}$
Total fermion mass	$M_n = \dfrac{4\pi}{3} R^3 \rho_n$
Total boson mass	$M_b = \dfrac{4\pi}{3}(R_0^3 - R^3)\rho_b$
Total mass	$M = M_n + M_b + U_g/c^2$
Schwarzschild radius	$R_S = 2GM/c^2$

- Hydrostatic equilibrium at R_0:

$$\frac{1-\zeta}{\zeta^2}\left\{ \left[\frac{\varepsilon Q^2 \lambda_b}{8\pi^2 \lambda_n^4} \frac{y_F(2y_F^2+1)\sqrt{y_F^2+1} - \sinh^{-1} y_F}{1 + y_F^2 - Q\sqrt{1+y_F^2}} - 1 \right]\zeta^2 + \frac{1}{2}(1+\zeta) \right\}$$
$$= \frac{\dfrac{3\varepsilon}{8\pi \lambda_b \lambda_n^2}\left(\dfrac{m_p}{m_b}\right)^2 Q^2 y_F^2}{\left(\dfrac{9\pi}{4} N_0 X\right)^{2/3}(1 + y_F^2)}. \tag{8.51}$$

The three coupled equations are highly nonlinear in the sought-for variables and must be solved numerically. For this purpose an iterative Levenberg–Marquardt method was employed [266] to obtain solutions to a tolerance of 10^{-10} for nearly all the model conditions presented in Tables 8.2 through 8.7. The iterative procedure requires an estimation of the unknowns as input, but in most of the cases summarized in the tables a 'ballpark' guess was sufficient for rapid convergence. In a few noted instances, however, convergence was very slow, and results were obtained to a lower tolerance of 10^{-4}.

The fixed parameters Q, s, a of the model define the physics of the fermion condensation process and are not specified by the model. In principle, they would be deducible from a (currently unknown) comprehensive theory of nuclear or subnuclear matter at extreme pressures and densities. Nevertheless, it is worth noting that the model does, in fact, impose informatively restrictive limits on the values of the parameters, since initial choices of parameter values well outside these limits did not lead to viable solutions or, in most instances, to any solutions at all. Because an iterative method of solution can require very closely estimated initial values, it was not always possible to say with certainty that failure to find a solution to the equations of equilibrium for a particular set of condensate parameters meant that no solution existed. Graphical analysis, i.e., plotting the three equilibrium relations on one graph as a function of y_F for selected values of X and ζ and searching for a three-line intersection, could be executed comparatively rapidly and was helpful in locating and confirming solutions when the iterative routine failed to find a solution at a tolerance of 10^{-10}.

An important question is how one would recognize that a solution represents a state of *stable* equilibrium. In this regard it is helpful to consider first a self-gravitating system of fermions only, i.e., a fermion sphere of radius R and no condensate shell. The condition (8.50) of hydrostatic equilibrium at R (the only boundary) can then be reduced to an equality between a universal function $F(y_F)$ of the Fermi parameter and a constant term determined by the number of fermions, as follows:

$$
\begin{aligned}
F(y_F) &= \frac{y_F^2 \left[y_F(2y_F^2 - 3)\sqrt{y_F^2 + 1} + 3\sinh^{-1}(y_F) \right]}{\left[y_F(2y_F^2 + 1)\sqrt{y_F^2 + 1} - \sinh^{-1}(y_F) \right]^2} \\
&= \frac{1}{4\pi} \left(\frac{9\pi}{4} \right)^{2/3} \left(\frac{m_n}{m_P} \right)^2 N_n^{2/3} .
\end{aligned} \tag{8.52}
$$

Figure 8.3 shows a plot of the above function, which reaches a maximum value of 0.11 at $y_F = 0.905$. A horizontal straight line [the constant on the right hand side of (8.52)] can intersect this plot at two points, or at one point (the peak), or else lie above and not intersect at all. The first case leads to two equilibrium solutions of which the intersection at the lower value of y_F, i.e., at a point of positive slope, is the stable equilibrium. Since the plot represents the ratio of degeneracy pressure to hydrostatic pressure, a small departure from the first intersection to larger y_F results in an increase of degeneracy pressure over gravity, and therefore a resistance to collapse. At the second intersection (i.e., higher value of y_F), gravity prevails, and the system will experience irreversible collapse. Intersection at the point of zero slope (peak) represents the threshold beyond which there are no equilibrium solutions. From the maximum value of $F(y_F)$ one can calculate the maximum number of particles that can form a stable system in the absence of any mechanism of fermion reduction, which, for the case of neutrons treated as a degenerate fermi gas, is $N_0 = 5.07 \times 10^{56}$.

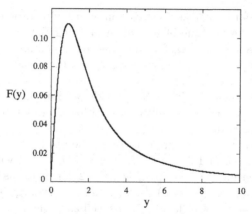

Fig. 8.3. Universal curve $F(y)$ for the ratio of fermion degeneracy pressure to hydrostatic pressure as a function of the Fermi parameter y. The maximum value of the function is 0.11 at the coordinate $y = 0.905$

An example of a comparable plot when fermion condensation occurs and the full set of equations (8.49)–(8.51) are required is shown in Fig. 8.4 for the case of $N_0 = 4 \times 10^{57}$ neutrons and the condensate parameters specified in the figure caption. The chemical equilibrium condition (8.49) has reduced the number of fermions although the shape of the plot of the ratio of degeneracy pressure to the *sum* of condensate and hydrostatic pressures is substantially the same as the plot in Fig. 8.3. A dashed horizontal line in Fig. 8.4 at unit

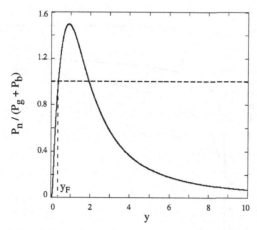

Fig. 8.4. Ratio of the neutron degeneracy pressure at the stellar center to the sum of the condensate pressure and hydrostatic pressure at R for the equilibrium state characterized by parameters $N_0 = 4 \times 10^{57}$, $Q = 0.9$, $s = 2$, and $a = 1 \times 10^{-12}$ m. The *dashed line* indicates a ratio of unity, and therefore the value y_F at which hydrostatic equilibrium occurs. The rising section of the plot (positive slope) is the region for stable equilibrium

ordinate intersects the plot at the two points of hydrostatic equilibrium, the first intersection (lower value of y_F) again marking the state of stable equilibrium. In Fig. 8.5 is plotted, for the same initial number of neutrons and condensate parameters as in the preceding figure, the variation with y_F of the separate pressures, i.e., the logarithm of the ratio of these pressures, which is more convenient for display. Figure 8.6 provides a companion plot for the same parameters, showing the logarithm of the ratio of fermion-to-boson number density and mass density.

We consider now the principal features of the resulting stellar equilibrium states, as summarized in Tables 8.2 through 8.7, beginning with a condensate formed by exothermic ($Q < 1$) pairing ($s = 2$) of neutrons with a positive interaction parameter ($a = 1 \times 10^{-12}$ m) which is very large in comparison to the (reduced) Compton wavelength λ_b, as well as to the mean separation between bosons $n_b^{-1/3}$ (a circumstance establishable only after the equations are solved). The choice of large a approximates what is called the unitarity limit, signifying a short-range, strong two-body interaction between constituents. The unitary limit is thought to be realized approximately in the inner crust of neutron stars, where the neutron–neutron scattering length is at least a factor of ten larger than the mean interparticle separation [267]. As pointed out at the beginning of this chapter, the behavior of the fermion condensates in the unitarity limit is expected to be applicable to a wide range of systems from cold atomic gases to dense nuclear matter.

An examination of Table 8.2, which records equilibrium values of the stellar properties summarized in Table 8.1 for initial neutron numbers N_0 increasing sequentially from approximately 4 to 10 times that of the Sun shows, first

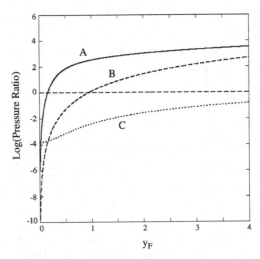

Fig. 8.5. Variation with y of log of the following pressure ratios: (**A**) P_n/P_b, (**B**) P_n/P_0, (**C**) P_b/P_0. The *dashed horizontal line* indicates a unit ratio. The fermion number and condensate parameters are the same as for Fig. 8.4

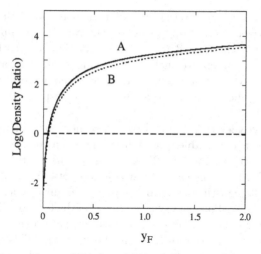

Fig. 8.6. Variation with y_F of the logarithm of the neutron-to-condensate density ratios: (**A**) mass density ρ_n/ρ_b, (**B**) number density n_n/n_b. The fermion number and condensate parameters are the same as for Fig. 8.4

Table 8.2. Equilibrium states for condensate parameters $a = 10^{-12}$, $Q = 0.9$, $s = 2$

$N_0/10^{57}$	4	5	6	7	8	9	10
y_F	0.383	0.338	0.297	0.261	0.228	0.200	0.176
N_n/N_0	0.068	0.046	0.031	0.021	0.013	6.33(−03)	1.66(−03)
N_b/N_0	0.932	0.954	0.969	0.979	0.987	9.94(−01)	9.98(−01)
ζ	0.093	0.088	0.085	0.083	0.078	0.070	0.050
R [km]	6.81	7.30	7.74	8.10	8.21	7.75	5.83
R_0 [km]	73.64	82.53	90.59	98.00	104.80	111.00	116.56
n_n [m^{-3}]	2.07(44)	1.41(44)	9.62(43)	6.49(43)	4.35(43)	2.92(43)	2.00(43)
n_b [m^{-3}]	1.12(42)	1.01(42)	9.34(41)	8.70(41)	8.20(41)	7.81(41)	7.53(41)
ρ_n [kg/m^3]	3.60(17)	2.44(17)	1.65(17)	1.11(17)	7.37(16)	4.93(16)	3.38(16)
ρ_b [kg/m^3]	3.67(15)	3.31(15)	3.03(15)	2.81(15)	2.64(15)	2.50(15)	2.41(15)
M/M_{Sun}	3.16	3.90	4.64	5.36	6.08	6.80	7.51
R_S [km]	9.37	11.57	13.75	15.90	18.03	20.15	22.26
R_0/R_S	7.86	7.13	6.59	6.17	5.81	5.51	5.24

of all, that an equilibrium state is actually reached in each case with a stellar radius comparable to the radius of a neutron star and larger than the associated Schwarzschild radius. Yet in all cases the resulting stellar masses exceed the Chandrasekhar and Oppenheimer–Volkoff limits for white dwarf and neutron stars, respectively. In other words, the predicted final state is not a black hole and has no central singularity, although, in the absence of the mechanism herein proposed, a black hole would assuredly have been the expected outcome. Basic trends predicted by the theory, as one progresses from lowest to highest values of N_0, are (a) the Fermi parameter falls, as initially expected, from a relativistic value[7] to a nonrelativistic value; (b) the equilibrium neutron fraction decreases; (c) the radius of the neutron core increases up to approximately 8 times N_{Sun} and then begins to fall; (d) the outer radius of the condensate shell increases monotonically; and (e) the neutron number density and mass density, both of which decrease with increasing initial neutron number, remain larger than the corresponding densities of the condensate.

The fermion condensation model presented here also allows for an exotic, but by no means inconceivable, possibility whereby a condensate can form by endothermic ($Q > 1$) pairing ($s = 2$) with large negative interaction parameter ($a = -1 \times 10^{-12}$ m). In this case the interaction between composite bosons is attractive, but the constituent fermions require input of energy to overcome repulsive interactions. This energy is supplied by the fermion kinetic energy as accounted for quantitatively by the equilibrium relation (8.49). Nevertheless, one might think such a composite would be unstable. This is not the case, however, for systems in a structured environment in the absence of dissipation, and, in fact, repulsively bound atom pairs have been reported [268] (while this chapter was being written) for a boson condensate of ^{87}Rb. Such systems are expected to occur ubiquitously in cold-atom physics, a circumstance that encourages the belief, therefore, that comparable processes may also take place in ultra-dense nuclear matter.

Equally significant as the value of Q, are the implications for a negative-valued interaction parameter a, which, according to (8.42) and (8.45), results in a negative condensate pressure or a kind of suction at the outer boundary. A heuristic way to envision this would be to imagine a droplet of water immersed in a hydrophobic environment. Since stable hydrostatic equilibrium requires that the condensate pressure increase from the outer boundary inwards, and since the composite bosons interact only very weakly with the neutrons, we must modify the boundary conditions by setting $P(R) = 0$ and $P(R_0) = -P_{\text{b}}$. The resulting equations to be solved then take the form:

- Chemical equilibrium:

$$\frac{Q - \sqrt{1 + y_{\text{F}}^2}}{Q y_{\text{F}}^3} = \frac{1}{3\pi^2 s} \frac{|\varepsilon|}{\lambda_{\text{n}}^3} \frac{1 - X}{X} \frac{\zeta^3}{1 - \zeta^3}. \tag{8.53}$$

- Hydrostatic equilibrium at R:

$$\frac{y_{\text{F}}^2 \left[y_{\text{F}} (2y_{\text{F}}^2 - 3)\sqrt{y_{\text{F}}^2 + 1} + 3 \sinh^{-1} y_{\text{F}} \right]}{\left[y_{\text{F}} (2y_{\text{F}}^2 + 1)\sqrt{y_{\text{F}}^2 + 1} - \sinh^{-1} y_{\text{F}} \right]} = \frac{1}{4\pi} \left(\frac{m_{\text{n}}}{m_{\text{P}}} \right)^2 \left(\frac{9\pi}{4} N_0 \right)^{2/3} X^{2/3}. \tag{8.54}$$

---

[7] If the hypothesized fermion condensation process did not occur, and a neutron star collapsed to the density in which the mean interparticle separation is the reduced Compton wavelength λ_{n}, then the Fermi parameter would be $y_{\text{F}} = (3\pi^2)^{1/3} \sim 3.094$, which is highly relativistic.

- Hydrostatic equilibrium at R_0:

$$\frac{1-\zeta}{\zeta^2}\left\{\left[\frac{|\varepsilon|Q^2\lambda_b}{8\pi^2\lambda_n^4}\frac{y_F(2y_F^2+1)\sqrt{y_F^2+1}-\sinh^{-1}y_F}{Q\sqrt{1+y_F^2}-(1+y_F^2)}-1\right]\zeta^2+\frac{1}{2}(1+\zeta)\right\}$$

$$=\frac{\frac{3|\varepsilon|}{8\pi\lambda_b\lambda_n^2}\left(\frac{m_p}{m_b}\right)^2Q^2y_F^2}{\left(\frac{9\pi}{4}N_0X\right)^{2/3}(1+y_F^2)}. \qquad (8.55)$$

Table 8.3 records the properties of the stellar equilibrium states resulting from solution of the foregoing equations for condensate parameters $Q = 1.1$, $s = 2$, $a = -1 \times 10^{-12}$ m and the same sequence of initial fermion numbers as recorded in Table 8.2. Basic trends predicted by the theory for increasing values of N_0 are that (a) the Fermi parameter decreases, as before, to nonrelativistic values; (b) the equilibrium neutron fraction decreases; (c) the radius of the neutron core increases monotonically; (d) the outer radius of the condensate shell increases monotonically, remaining larger than the corresponding Schwarzschild radius; (e) the neutron number density and mass density are larger than the corresponding densities of the condensate, but by smaller factors until the trend is reversed at approximately 10 times N_{Sun}. In all cases the resulting stellar masses exceed the Chandrasekhar and OV limits, but, again, no stellar black hole is formed.

The preceding calculations were also carried out for magnitudes of the interaction parameter a comparable to the boson Compton wavelength, and therefore smaller (by about $1/10$ to $1/5$) than the mean separation between

Table 8.3. Equilibrium states for condensate parameters $a = -10^{-12}$, $Q = 1.1$, $s = 2$

$N_0/10^{57}$	4	5	6	7	8	9	10
y_F	0.196	0.175	0.152	0.128	0.101	0.071	0.046
N_n/N_0	0.031	0.021	0.014	9.46(−03)	5.85(−03)	3.08(−03)	1.73(−03)
N_b/N_0	0.969	0.979	0.986	0.991	0.994	0.997	0.998
ζ	0.113	0.113	0.116	0.122	0.133	0.154	0.198
R [km]	10.16	10.81	11.63	12.72	14.37	17.19	22.73
R_0 [km]	89.58	95.30	100.09	104.23	107.94	111.42	115.04
n_n [m^{-3}]	2.78(43)	1.96(43)	1.29(43)	7.68(42)	3.77(42)	1.31(42)	3.53(41)
n_b [m^{-3}]	6.45(41)	6.76(41)	7.05(41)	7.32(41)	7.57(41)	7.77(41)	7.89(41)
ρ_n [kg/m^3]	4.69(16)	3.30(16)	2.18(16)	1.29(16)	6.32(15)	2.18(15)	5.89(14)
ρ_b [kg/m^3]	2.28(15)	2.39(15)	2.49(15)	2.58(15)	2.66(15)	2.73(15)	2.77(15)
M/M_{Sun}	3.41	4.23	5.04	5.83	6.62	7.40	8.18
R_S [km]	10.10	12.53	14.93	17.29	19.63	21.94	24.24
R_0/R_S	8.87	7.60	6.70	6.03	5.50	5.08	4.75

particles. Table 8.4 shows the results for increasing N_0 for condensate parameters $Q = 1.1$, $s = 2$, and $a = -1 \times 10^{-16}$ m. Most noticeable here is that a smaller interaction parameter has led to lower equilibrium stellar masses and much lower equilibrium stellar radii than before for corresponding values of N_0. The ratios of stellar radius (i.e., the outer radius of the condensate shell) to Schwarzschild radius are marginally larger than unity. Also, because the outer radius of the condensate is not much larger than the inner radius ($\zeta \sim 0.9$), the condensate number and mass densities are significantly larger than the neutron densities. Because this is a less familiar occurrence, one might question whether a distribution of self-gravitating matter with a lower density core and higher density shell can be stable, since it is well known that in a stratified system of immiscible fluids the least dense fluid floats on top. The two situations, however, are not at all parallel, for, within the framework of the model calculation, the condensate exerts a *negative* pressure which vanishes at the surface of the neutron sphere. In Appendix 8B it is shown that an object of density ρ immersed in a spherical self-gravitating environment of density ρ_0 with negative pressure always sinks to the center irrespective of the relative density ρ/ρ_0.

The consequences of a negative interaction parameter are followed systematically in Table 8.5, which records the equilibrium properties of a degenerate star with $N_0 = 4 \times 10^{57}$ neutrons initially and magnitudes $|a|/\lambda_b$ increasing from approximately 1 to 10^4. Properties evolve monotonically with increasing $|a|/\lambda_b$, with decreasing densities and condensate-to-neutron density ratios greater than unity, until a value of $|a|/\lambda_b$ somewhere between 10^3–10^4 is reached at which point there is a reversal of trend, the neutron density surpassing that of the condensate, and the stellar radius markedly increasing. What interaction parameter actually may pertain to a condensed neutron sys-

Table 8.4. Equilibrium states for condensate parameters $a = -10^{-16}$, $Q = 1.1$, $s = 2$

$N_0/10^{57}$	4	5	6	7	8	9	10
y_F	0.320	0.296	0.276	0.257	0.241	0.226	0.212
N_n/N_0	0.058	0.042	0.032	0.025	0.020	0.016	0.014
N_b/N_0	0.942	0.958	0.968	0.975	0.980	0.984	0.986
ζ	0.930	0.931	0.933	0.935	0.936	0.938	0.940
R [km]	7.73	8.09	8.43	8.77	9.10	9.43	9.75
R_0 [km]	8.32	8.68	9.04	9.28	9.72	10.05	10.37
n_n [m^{-3}]	1.20(44)	9.56(43)	7.69(43)	6.24(43)	5.10(43)	4.21(43)	3.50(43)
n_b [m^{-3}]	3.99(45)	4.54(45)	5.00(45)	5.38(45)	5.70(45)	5.97(45)	6.20(45)
ρ_n [kg/m^3]	2.07(17)	1.64(17)	1.31(17)	1.06(17)	8.67(16)	7.14(16)	5.92(16)
ρ_b [kg/m^3]	1.43(19)	1.63(19)	1.78(19)	1.92(19)	2.03(19)	2.12(19)	2.20(19)
M/M_{Sun}	2.41	2.72	2.95	3.10	3.17	3.19	3.14
R_S [km]	7.14	8.07	8.74	9.17	9.40	9.44	9.31
R_0/R_S	1.16	1.08	1.03	1.01	1.03	1.06	1.11

Table 8.5. Equilibrium states for parameters $N_0 = 4 \times 10^{57}$, $Q = 1.1$, $s = 2$

a [m]	$-1.00(-16)$	$-1.00(-15)$	$-1.00(-14)$	$-1.00(-13)$	$-1.00(-12)$
y_F	0.320	0.204	0.112	0.098	0.196
N_n/N_0	0.058	0.032	0.014	0.011	0.031
N_b/N_0	0.942	0.968	0.986	0.989	0.969
ζ	0.93	0.832	0.659	0.362	0.113
R [km]	7.732	9.965	13.621	14.595	10.163
R_0 [km]	8.315	11.979	20.668	40.27	89.576
n_n [m^{-3}]	1.20(44)	3.10(43)	5.16(42)	3.44(42)	2.78(43)
n_b [m^{-3}]	3.99(45)	6.34(44)	7.47(43)	7.59(42)	6.45(41)
ρ_n [kg/m^3]	2.07(17)	5.24(16)	8.64(15)	5.75(15)	4.69(16)
ρ_b [kg/m^3]	1.43(19)	2.24(18)	2.63(17)	2.67(16)	2.28(15)
M/M_{Sun}	2.41	2.72	3.02	3.25	3.41
R_S [km]	7.14	8.06	8.96	9.62	10.10
R_0/R_S	1.16	1.49	2.31	4.19	8.87

tem, and how that parameter depends on density, are questions that await a comprehensive theory of ultra-dense nuclear matter.

Similar calculations were attempted to investigate the equilibrium properties resulting from assumption of a positive interaction parameter comparable in magnitude to the Compton wavelength. In these instances, however, the Levenberg–Marquardt procedure failed to converge for tolerances stricter than 10^{-2}, although graphical analysis showed a 'crossing' of the three equations of equilibrium at the point $y_F = 0.0802$, $X = 0.00104$, $\zeta = 0.962$ to within approximately $\pm 1 \times 10^{-6}$. The equilibrium properties represented by this point characterize a terminal state similar to that resulting from a negative interaction parameter of the same magnitude, namely: a small star ($R = 8.07$ km, $R_0 = 8.39$ km), slightly larger than its Schwarzschild radius ($R_0/R_S = 1.24$), with condensate densities (in MKS units) ($n_b = 7.37 \times 10^{45}$, $\rho_b = 2.35 \times 10^{19}$) greatly larger than neutron densities ($n_n = 1.89 \times 10^{42}$, $\rho_n = 3.17 \times 10^{15}$), and a total mass $M/M_{Sun} = 2.28$, although $N_0/N_{Sun} = 3.33$. As in the previous cases, the terminal state is not a black hole, although the proximity in size to the Schwarzschild radius and the huge condensate density suggests that an appropriate general relativistic analysis would be justified. In any event, because of the difficulty of obtaining solutions in the case of a positive interaction parameter comparable in magnitude to the Compton wavelength, and the question of whether such results obtained by graphical analysis actually represent true solutions, the search for additional solutions was not pursued.

In all the foregoing cases, the condensation of neutrons was assumed to occur in pairs. A preliminary investigation was also made of fermion condensation in clusters ranging from $s = 4$ to $s = 10$, which is summarized in Table 8.6 for a range of N_0 and condensate parameters $s = 4$, $Q = 1.1$, $a = -1 \times 10^{-12}$ m, and in Table 8.7 for fixed $N_0 = 4 \times 10^{57}$ over a range of s. As N_0 increases sequentially (for $s = 4$) over the same range as before, one finds that (a) the

Table 8.6. Equilibrium states for condensate parameters $a = -10^{-12}$, $Q = 1.1$, $s = 4$

$N_0/10^{57}$	4	5	6	7	8	9	10
y_F	0.104	0.068	0.048	0.034	0.026	0.021	0.018
N_n/N_0	0.012	0.005	0.003	1.30(−03)	7.78(−04)	4.93(−04)	3.79(−04)
N_b/N_0	0.988	0.995	0.997	0.999	0.999	1.000	1.000
ζ	0.327	0.378	0.431	0.474	0.510	0.539	0.564
R [km]	14.13	17.59	21.44	25.03	28.49	31.55	34.50
R_0 [km]	43.27	46.53	48.75	52.81	55.80	58.50	61.13
n_n [m^{-3}]	4.16(42)	1.14(42)	4.00(41)	1.39(41)	6.48(40)	3.37(40)	2.21(40)
n_b [m^{-3}]	3.02(42)	3.12(42)	3.15(42)	3.17(42)	3.18(42)	3.18(42)	3.19(42)
ρ_n [kg/m^3]	6.96(15)	1.90(15)	6.69(14)	2.31(14)	1.08(14)	5.63(13)	3.68(13)
ρ_b [kg/m^3]	2.12(16)	2.19(16)	2.21(16)	2.22(16)	2.23(16)	2.23(16)	2.23(16)
M/M_{Sun}	3.26	4.03	4.78	5.53	6.26	6.99	7.71
R_S [km]	9.67	11.93	14.17	16.38	18.57	20.72	22.86
R_0/R_S	4.47	3.90	3.44	3.22	3.01	2.82	2.67

Table 8.7. Equilibrium states for parameters $a = -1 \times 10^{-12}$, $N_0 = 4 \times 10^{57}$, $Q = 1.1$

s	2	4	6	8	10
y_F	0.196	0.104	0.018	0.017	0.012
N_n/N_0	0.031	0.012	6.48(−05)	4.40(−05)	1.39(−05)
N_b/N_0	0.969	0.988	1.000	1.000	1.000
ζ	0.113	0.327	0.485	0.592	0.657
R [km]	10.16	14.13	14.19	13.47	12.38
R_0 [km]	89.58	43.27	29.27	22.78	18.84
n_n [m^{-3}]	2.78(43)	4.16(42)	2.17(40)	1.72(40)	6.99(39)
n_b [m^{-3}]	6.45(41)	3.02(42)	7.16(42)	1.27(43)	1.99(43)
ρ_n [kg/m^3]	4.69(16)	6.96(15)	3.62(13)	2.87(13)	1.17(13)
ρ_b [kg/m^3]	2.28(15)	2.12(16)	7.54(16)	1.79(17)	3.49(17)
M/M_{Sun}	3.41	3.26	3.15	3.06	2.97
R_S [km]	10.10	9.67	9.33	9.06	8.80
R_0/R_S	8.87	4.47	3.14	2.51	2.14

Fermi parameter decreases monotonically to nonrelativistic values; (b) the equilibrium neutron fraction decreases; (c) the radius of the neutron core increases monotonically; (d) the outer radius of the condensate shell increases monotonically, remaining larger than the corresponding Schwarzschild radius; and (e) the neutron number density and mass density become significantly lower than the corresponding condensate densities. The stellar masses again exceed the Chandrasekhar and OV limits, but no black hole forms.

For a star of initial neutron number $N_0 = 4 \times 10^{57}$, an increase in cluster size from $s = 2$ to $s = 10$ leads to (a) a decrease in Fermi parameter by an order of magnitude; (b) a decrease in the equilibrium neutron fraction by two orders of magnitude; (c) a relatively small change in the radius of the neutron core; (d) a decrease in the outer radius of the condensate shell by approximately a factor of five; (e) a decrease in the neutron number density and mass density and an associated increase in the condensate densities by about two orders of magnitude. In general, for a large negative interaction parameter, condensation into larger neutron clusters results in a denser star, comprising nearly 100% condensate with an outer radius approaching (but not falling below) the Schwarzschild radius.

Use of the Levenberg–Marquardt algorithm failed to yield solutions for the condensation into composite neutron quartets with large positive interaction parameter ($a = 1 \times 10^{-12}$ m). Indeed, cursory attempts to solve the equations of equilibrium for positive a and fermion combinations other than $s = 2$ generally failed, even for tolerances as low as 10^{-4}.

8.5 Fermicon Stars vs Black Holes

The objective of the previous sections has been to examine the implications of a relatively recently observed quantum phenomenon – the condensation of fermions to form systems of composite bosons – as a possible mechanism preventing the unremitting collapse to a singularity of a relativistic degenerate star that exceeds the Chandrasekhar and OV mass limits. It is to be noted that the designation 'black hole' need not imply a central singularity, for it is conceivable that a process can be found whereby a star collapses to a volume within its event horizon, yet nevertheless reaches a state of nonvanishing radius and finite density. Such a situation is not permitted by general relativity, but general relativity and quantum mechanics are not mutually compatible. In the following section I discuss such a possibility arising from the ultra-strong magnetic fields associated with neutron stars. In any event, the solutions to the equations of equilibrium for neutronic matter undergoing fermion condensation to composite bosons describe end states of finite size and finite mass density for degenerate stars that would otherwise have collapsed to a black hole.

The terminal equilibrium states predicted by the model, however, are neither black holes, since the predicted stellar surfaces lie outside the associated Schwarzschild horizons, nor ordinary neutron stars, since the stars comprise separated neutron and condensate phases, with the latter more often than not (for the calculations discussed here) the major component. I have proposed the term 'fermicon hybrid stars', or simply fermicon stars, to designate terminal equilibrium states of thermonuclear end-point stars produced by partial fermion condensation (either neutron or quark) into a condensate of composite bosons.

Bosonic pairing of neutrons with either large positive or large negative interaction parameters lead in both cases to fermicon stars with a neutron core denser than the condensate shell, although a fermicon star with negative condensate pressure may also have, for sufficiently large initial neutron number, a neutron core less dense than the condensate shell. In all cases so far studied with interaction parameters much larger in magnitude than the interparticle separation, the resulting stellar radii are approximately 4 to 8 times the associated Schwarzschild radii, yielding Schwarzschild parameters x_S^2 sufficiently small to justify use of Newtonian gravity. In the cases involving interaction parameters comparable to the boson Compton wavelength, however, the resulting stars are marginally larger than their Schwarzschild radii with thin ($\zeta \sim 0.9$) highly dense shells, and a general relativistic analysis would be warranted here although a more appropriate condition of hydrostatic equilibrium than the classical TOV equation is probably needed.

Looking beyond neutron pairs to even-numbered bosonic clusters of 4 or more neutrons, one finds that, all other parameters held fixed, the larger the cluster, the denser and thicker is the condensate shell, the smaller is the Schwarzschild radius, and the closer it is approached from above by the stellar radius. For clusters of 6 or more, a general relativistic analysis would again be appropriate, although what is needed most is advancement of nuclear theory to ascertain whether, and under what circumstances, these clusters can form.

In considering the fermion condensation process in this chapter, I have focused attention on neutrons, in contrast to quarks, because neutron stars, after all, are known to exist, and quark stars are, as yet, largely hypothetical, albeit conceivable, stellar end states.[8] The dissolution of nucleons into quarks is expected to occur at densities corresponding approximately to 3 to 10 times the saturation density of nuclear matter [270] ($n_N \sim 1.5 \times 10^{44}$ m^{-3}), which is comparable to the densities predicted by the fermicon model for certain values of the condensate interaction parameter, both positive and negative. At the time this chapter is being written there is some evidence, based on inferences regarding the rate of cooling and the ratio of mass to radius, that at least two stars containing quark cores may have been observed,[9] although the arguments for this conclusion are not definitive and other interpretations of the stars' properties have been given. Moreover, recent observation of the gravitational redshift of absorption lines in the X-ray burst spectra of a neutron star with relatively low magnetic field (10^7–10^9 gauss) seems to indicate that the star is made up of normal nuclear matter, rather than exotic matter such as quarks [272]. In time, more redshift measurements of more neutron stars should help clarify the question of their composition. In Appendix 8C I discuss a simple model of the neutron–quark phase transition which predicts the pressure and density at which such a transition should occur. Such results are highly model dependent and are by no means to be regarded as conclusive. Moreover, the predicted occurrence of a neutron–quark phase transition does not necessarily mean that a stable stellar end state results. Nevertheless, the fact the various models employing different equations of state seem to support the 'melting' of nucleons to form an equilibrium free quark phase encourages

---

[8] There is a substantial literature on the possibility of a quark phase in neutron stars. See, for example, [269].

[9] Several quark stars may possibly have been found by the Chandra and Hubble space telescopes [271].

further investigation into the consideration of quark pairs to form composite bosons and possible equilibrium stellar end states. These investigations are in progress.

The question also arises of how one might determine observationally whether a degenerate star of mass greater than the OV limit is a black hole with an event horizon and interior singularity or a type of fermicon hybrid star predicted here. The usual distinguishing criteria of mass (estimated from orbital data for stars in binary associations or from gravitational microlensing events) and size (estimated from fluctuations in X-ray emissions by accreting matter) may be inconclusive. What is necessary are astronomical observations that can reveal the characteristics of the stellar surface, in particular to determine the presence of a a physical crust in contrast to a mere mathematical horizon at the Schwarzschild radius. Evidence for a degenerate star more massive than about 3 solar masses with a physical surface would be definitive, as such a system cannot exist as a neutron star according to classical general relativity.

Hawking radiation aside (which has not been confirmed observationally or experimentally), neither an isolated black hole nor an isolated fermicon star can illuminate itself, and therefore cannot be observed directly. However, a compact degenerate star in binary association with a much larger (although not necessarily more massive) normal star with gaseous atmosphere is likely to be surrounded by dense swirls of infalling matter drawn from the normal companion. The accumulation of this energetic matter, arriving at the compact star at close to the speed of light, gives rise at irregular intervals to diffuse thermonuclear explosions that can illuminate not only the surface but part of the interior of the compact star as well. Events of this kind have been observed [272] for the neutron star EXO 0748-676 by the European Space Agency's XMM Newton X-ray satellite and provide important details about the equation of state the matter comprising the star [273]. Such a process cannot occur for black holes because they have no physical surface upon which infalling matter could accumulate. However, self-illuminating thermonuclear surface explosions could occur for fermicon stars, if they exist. Besides appropriate instrumentation, one needs the good fortune to be looking at the right star at the right moment.

Although the model discussed in this section disregards the attribute of stellar rotation, it is likely that a precursor to a black hole or fermicon star has angular momentum. Thus, if by continued good fortune, one could observe a star undergoing the predicted phase transformation, then a change in stellar moment of inertia may also reveal the nature of the terminal state.

8.6 Can Ultra-Strong Magnetic Fields Prevent Collapse?

Neutron stars and their variants give evidence of the strongest magnetic fields in the cosmos [274]. For example, mean field strengths typical of radio pulsars are of the order of 10^{12}–10^{13} gauss; soft-gamma-ray repeaters and 'anomalous' X-ray pulsars are thought to be neutron stars with superstrong magnetic fields exceeding 10^{14} gauss. How such fields came into existence is not known with certainty. Neutron stars derive from the cores of main-sequence stars with masses greater than about 8 solar masses that terminate in a supernova explosion upon the exhaustion of their nuclear fuel supply. One possibility, therefore, is that the considerably weaker magnetic fields present in the progenitor star become greatly amplified by magnetic flux conservation during core collapse. As a rough order-of-magnitude estimate, the magnetic field B_0 of a progenitor with radius 10 times the solar radius ($R_{Sun} = 7 \times 10^8$ m), collapsing to a neutron star with radius $10^{-5} R_{Sun}$, would be amplified by the factor 10^{12}. Thus an initial field B_0 in the range of 1–10^3 gauss would lead to the range of enormous fields manifested by neutron stars.

An alternative and possibly more likely scenario is that the magnetic field is generated by a sort of convective dynamo in the first minute or so after the neutron star has formed, during which time the differential rate of flow of charged particles (protons, electrons, ionized atoms) in the predominantly neutral matter is high. In such a dynamo, the kinetic energy of stellar material of mass density ρ undergoing convective motion at speed v_{con} is transformed into magnetic energy according to $B^2/8\pi \sim \rho v_{con}^2/2$, leading to an estimate $B_{max} \sim \sqrt{4\pi\rho} v_{con}$ for the upper limit of the dipolar magnetic field strength of the neutron star [275]. Although a wide range of circumstances can characterize the formation of neutron stars, it is perhaps not far off the mark to say that a newly born neutron star may rotate with a period t_{rot} approximately ten times longer than the convective turn-over time t_{con}. If at formation a solar-mass neutron star of radius 10 km rotates at approximately one-tenth the angular frequency at which it would break up,[10] then its rotational period would be about 10 ms, and therefore $t_{con} \sim 1$ ms. Supposing that the size of a convective cell is approximately one-tenth the radius of the star, then one obtains a convective speed $v_{con} \sim (1 \text{ km/ms}) = 10^8$ cm/s. Under the preceding assumptions, a convecting fluid at the saturation density of nuclear matter, $\rho_N \sim m_n n_{sat} \sim 2.5 \times 10^{14}$ g/cm^3, could generate maximum field strengths of about $B_{max} \sim 3 \times 10^{15}$ gauss. (Mechanical quantities are expressed in cgs units above because the magnetic energy density is expressed in the esu–emu system of units.)

However the magnetic field of neutron stars is formed, the essential point here is that they can be extremely large, and one could reasonably expect the

---

[10] A lump of matter of unit mass at the equator of a rotating sphere of mass M and radius R can no longer rotate as part of the sphere when the gravitational force GM/R^2 attracting it towards the center is insufficient to provide the necessary centripetal acceleration $\omega^2 R$ at angular frequency ω. This leads to a maximum angular frequency $\omega_{max} = \sqrt{GM/R^3}$ or minimum period $T_{min} = 2\pi\sqrt{R^3/GM}$. Symbolically, the relation is the same as that of Kepler's third law relating the square of the period to the cube of the orbital radius (or, more accurately, the semi-major axis of the orbit).

field to become even stronger and have a significant effect on the dynamics
of a compact star that cannot achieve hydrostatic equilibrium by means of
fermion degeneracy pressure alone. A heuristic argument [241] shows that this
may be so. We develop the argument in the context of Newtonian gravity for
the reasons previously given in Sect. 8.3, but also because the claim is frequently made that, even within the framework of Newtonian gravity, a cold
star of noninteracting fermions of total mass greater than a certain limit cannot reach stable equilibrium [276]. The explanation given, as discussed at the
beginning of this chapter, refers to the weaker dependence on density of the
degeneracy pressure of relativistic fermions than of nonrelativistic fermions.
This conclusion, however, does not take account of quantum interactions between fermions which conceivably may increase pressure and lead to stable
equilibria. It evokes a mental image of the collapse process – unbound, uncorrelated nuclear particles hurtling at breakneck speeds from all directions
towards a bottomless abyss at the center of a collapsing sphere – which may
be unfounded under the circumstances to be examined.

We consider in this section the affect of an ultra-strong magnetic field on
the internal energy of a compact star modeled, as in Sect. 8.1, as an ideal gas
of N fermi particles of mass m, each now endowed with a magnetic moment
of magnitude $\mu_N = g_N(e\hbar/2mc)$. For neutrons, the nuclear g-factor is -1.91
(compared to $g_N = 1$ for a classical particle and $g_N = 2$ for a spin-1/2 Dirac
particle). Although a compact star at the densities to be considered would
hardly constitute an ideal fermion gas, we will use that equation of state
nevertheless because it is the model previously employed by others to reach
the conclusion we are now contesting. Moreover, that model, augmented by
the magnetic interaction, will be seen to lead to stable equilibrium states of
macroscopic size. Thus a more accurate equation of state for which pressure
depends more strongly on density should also lead to stable equilibrium states.

Expressed in dimensionless quantities as in (8.5), the total energy of the
star (excluding rest mass) as a function of its radius R takes the form

$$\frac{U}{mc^2} = N\left[\left(1 + \frac{N^{2/3}\lambda^2}{R^2}\right)^{1/2} - 1\right] + \delta N^2 \frac{3g_N^2\alpha_f}{16\pi}\frac{\lambda^3}{R^3} - \frac{3}{5}\left(\frac{M}{m_P}\right)^2\frac{\lambda}{R}, \quad (8.56)$$

in which the three terms on the right represent respectively the relativistic
kinetic energy U_k, the fermion magnetic dipole coupling energy U_m, and the
Newtonian gravitational potential energy U_g of a star of mass M. The dimensionless fine-structure constant $\alpha_f = e^2/\hbar c \sim 1/137$ is a measure of the
electromagnetic coupling strength. The expression for the magnetic energy
term derives from the magnetic dipole interaction $U_m = N\mu^2/r^3$, where (a)
$r = R/N^{1/3}$ approximates the mean distance between particles or, equivalently, the coherence length of the fermion wave function, (b) only nearest neighbor interactions have been taken into account, and (c) contiguous
moments are either aligned ($\delta = +1$) for minimum entropy or antiparallel
($\delta = -1$) for minimum magnetic energy. It will be seen shortly that in a de-

generate star, for which the fermions are effectively at 0 K, stable equilibrium results from the parallel alignment of magnetic moments.

The assumption of maximally aligned magnetic moments is motivated by the following considerations. If a neutronic star – let us say of solar mass – were to collapse to a sphere of radius equal to its Schwarzschild radius ($\sim$ 3 km), the density of the star, assuming it to be uniform and applying (8.27), would be approximately $\rho_S = 1.8 \times 10^{19}$ kg/m^3. To estimate the corresponding density of particles (which we will assume to be all neutrons), one needs to employ a particular equation of state. The simplest and least accurate estimate is to divide the foregoing density by the mass of a neutron, in effect to apply an equation of state describing a system with null pressure, to arrive at $n_n = 11.1$ fm^{-3}. A better estimate is to employ the ideal fermi gas equation of state and solve the relation [see (8.20)]

$$g(y) = 8\pi^2 \frac{\rho_S}{m_n/\lambda_n^3} ,$$

to obtain $y_S \equiv (3\pi^2 n_S)^{1/3}\lambda_n = 1.1$, from which follows $n_n = 4.97$ fm^{-3}. A still more reliable estimate would be to use the nuclear equation of state employed in Appendix 8C (to study the disruption of neutrons into quarks), which leads to $n_n = 3.86$ fm^{-3}. There is no need to go further; the basic outcome is that at this stage of collapse there are about 4 or 5 neutrons within the volume (a sphere of radius $\sim$ 1 fm) ordinarily occupied by approximately 0.15 neutrons in normal nuclear matter.

Under such circumstances of tight compaction, it is not unreasonable to assume that the wave functions of the neutrons overlap significantly to form a highly-correlated magnetic fluid rather than a turbulent gas of uncorrelated dipoles, presuming for this discussion that baryon disruption into quarks does not occur. Previous studies of intense magnetic fields in gravitationally collapsed bodies have predicted that a degenerate fermion gas can give rise to a state of ferromagnetism which is the sum of all microscopic magnetic moments associated with particles in their respective Landau levels, while the Landau levels of the system are in turn maintained by the macroscopic magnetization [277]. Although the analysis was performed for degenerate fermionic systems terminating as white dwarfs or neutron stars, it is plausible to assume that an equilibrium state reached by further collapse would also manifest a strong ferromagnetism arising from dipole alignment. This would seem inescapable as the strong poloidal magnetic field characteristic of neutron stars must become even stronger (by flux conservation) during a more or less isotropic contraction. The alternative possibility of an anti-ferromagnetic state is highly unlikely, for such a state would require the reversal of some 10^{57} nuclear moments (for a solar-mass star). In that regard, it is worth noting explicitly that the phenomenon of stellar ferromagnetism, if it exists, is a macroscopic quantum effect analogous to the sustained rotation of a superfluid or the sustained circulation of current in a superconductor, the cessation of either process requiring a nearly simultaneous change in quantum state of a macroscopic number of elementary constituents.

In the absence of the magnetic interaction ($\delta = 0$), the nonrelativistic and relativistic reductions of U_k respectively lead to $N^{5/3}\lambda^3/2R^2$ and $N^{4/3}\lambda/R$ as pointed out previously. Since the radial dependence of the relativistic Fermi energy is the same as that of the gravitational term U_g, the minimization of total energy U for constant M and N leads to an upper limit of the mass of a relativistic degenerate star, as is well known.

However, with inclusion of the magnetic interaction, the energy U in the relativistic limit takes the form

$$U = \frac{a}{R} + \frac{b}{R^3},$$

leading to a real-valued stable equilibrium radius $R_{eq} = \sqrt{-3b/a}$ under the conditions $a < 0$ and $b > 0$, which requires that $\delta = +1$ in (8.56). Expressed in terms of the stellar parameters, the equilibrium radius is

$$R_{eq} = N\lambda \left[\frac{\beta}{(3/5)(M/m_p)^2 - N^{4/3}} \right]^{1/2}, \tag{8.57}$$

where $\beta = 9g^2\alpha_f/16\pi$. Expressing the mass of the star in terms of the solar mass ($M = \sigma M_S$) and the number of fermions in terms of the number of neutrons in a solar-mass neutron star $[N = \tau(M/m_n)]$ reduces (8.57) to the form

$$\frac{R_{eq}}{\lambda} = \frac{(1.2 \times 10^{18})\tau}{\sqrt{1 - 2.52(\tau^2/\sigma)^{2/3}}}. \tag{8.58}$$

According to (8.57), the model of a self-gravitating ferromagnetic Fermi gas leads to stable compact stars with masses greater than a minimum threshold: $M > \sqrt{5/3}N^{2/3}m_p$. Note, too, that the fermion mass appears only in the Compton wavelength and therefore does not affect the stability criteria.

Applied to the case of a fully spin-aligned star comprised primarily of neutrons ($m_n = 1.67 \times 10^{-27}$ kg, $\lambda = 2.1 \times 10^{-16}$ m, $\tau = 1$), (8.58) leads to stellar radii

$$R_{eq}(\text{km}) = \frac{0.245}{\sqrt{1 - 2.52\sigma^{-2/3}}}, \tag{8.59}$$

and a threshold mass parameter $\sigma > 4$, i.e., a minimum of 4 solar masses. For $\sigma = 5$ for example, the predicted stellar radius would be ~ 637 m (compared with the associated Schwarzschild radius of 14.9 km) and the mass density $\sim 9.2 \times 10^{18}$ g/cm^3. Such end products of stellar evolution, if they exist, would be cold, highly magnetized, degenerate systems like neutron stars, but with masses greater than the currently accepted Chandrasekhar or Oppenheimer–Volkoff mass limits ($\sigma < 2$) and radii that can fall below the corresponding classical Schwarzschild radii. In contrast to fermicon stars, they would be like black holes in the sense that light could not escape, but there would be no central singularity. However, unlike black holes whose mass and size (i.e., Schwarzschild radius) are linearly proportional, the stars described by (8.57)–(8.59) are smaller in size the greater the mass. In this regard, they would resemble nonrelativistic neutron stars for which the mass varies inversely with the cube of the radius.[11]

It must be borne in mind, however, that the model presented here is a heuristic one meant only to illustrate that extreme magnetism can modify the stability conditions for a compact star so that it does not necessarily follow that a relativistic Fermi gas of sufficient mass will ineluctably collapse to a singularity in space. Although the use of Newtonian gravity has led to stable structures smaller than the Schwarzschild radius, a fully general relativistic quantum calculation may conceivably support stable equilibrium states of radius larger than the Schwarzschild radius. Also, in some successful future merger of quantum theory and gravity, the very concept of the Schwarzschild horizon may need to be modified, if not eliminated.

8.7 Gravitationally-Induced Particle Resorption into the Vacuum

Everything comes from something. This broad principle of evolution applies not only to the development of life on Earth but to the cosmos as a whole. Looking into space is literally like looking back in time, and the views of deep-field objects through the Hubble Space Telescope and other land- and satellite-based telescopes have revealed that the size and content of the Universe today is far different than it was 5 billion years ago or 10 billion years ago. We do not live in a steady-state universe. Stars evolve; galaxies evolve; matter and energy undergo transformations.

The Standard Model of particle physics together with Einstein's theory of general relativity and the hypothesis of inflation provides a cosmological 'concordance model' that describes the evolution of the universe from its earliest moments ($< 10^{-30}$ s?) with astonishing success, as gauged by the capacity of the model to account for such details as the frequency distribution and power spectrum of the cosmic microwave background radiation, and the relative abundance of light elements (primarily isotopes of H, He, and Li).[12] The

---

[11] The relationship between mass and radius for a neutron star may be deduced readily as follows. From the Newtonian equation for hydrostatic equilibrium one obtains the relation $P \propto \rho^2 R^2$ for pressure, density, and radius. But the equation of state for a nonrelativistic degenerate Fermi gas leads to $P \propto \rho^{5/3}$ where by definition $\rho \propto M/R^3$. The three relations yield $\rho \propto R^{-6}$ or $M \propto R^{-3}$. Applying the same reasoning to the relativistic degenerate Fermi gas leads to M independent of R, as we have seen.

[12] The temperature of the expanding (and therefore cooling) universe remained high enough for formation of isotopes primarily of H, He, and Li. Heavier elements through Fe and Ni, which have the highest binding energy per nucleon, were forged much later by sequential fusion reactions within stars. The elements beyond the

currently known laws of physics even permit an answer – albeit still speculative – to the ultimate question of origins: Whence arose all the matter and energy of the universe in the first place?

From the perspective of quantum physics, the ultimate source of matter and energy is the quantum vacuum. Far from the image of nothingness evoked by vernacular use of the term, the vacuum is a pervasive roiling background from which virtual particles continually emerge and back to which they subsequently decay. The 'big-bang' origin of the universe could be thought of as one of those improbable, yet not impossible, occurrences when a quantum fluctuation of the vacuum occurred on a sustainable scale and continued expanding. However, if the total energy of the universe, which is initially zero, is to remain a conserved quantity, then there must be a corresponding source of negative energy to compensate for the positive mass–energy created by the big bang. Negative energy is provided by the universal gravitational attraction of all forms of matter and radiation. If this supposition is correct, then the total gravitational potential energy within the universe should balance the total energy of matter and radiation. An estimate (see Appendix 8D) of the energy balance within the accessible universe appears to support this hypothesis.

The creation of matter from the vacuum, even with conservation of energy, violates other conservation laws that are believed to hold rigorously under less extreme circumstances. Most prominent among these violations would be the conservation of baryon number and lepton number, since both are assumed to be zero for the vacuum prior to the big bang. Had particles of matter and antimatter been produced in equal numbers, then baryon number and lepton number would be conserved quantities. However, the apparent predominance of matter over antimatter, as inferred, for example, from the fact that in no interstellar region of the spectroscopically visible universe is found a significant 511 keV gamma-ray flux from electron–positron annihilations, calls for an explanation. Either the creation process was asymmetric or else the asymmetry arose dynamically afterwards. Although the precise mechanism is not understood, it is believed that nonequilibrium processes that violated charge-conjugation symmetry (C) and the combined symmetry (CP) of charge conjugation and space inversion (parity) resulted in a universe with nonzero baryon and lepton numbers very shortly after formation [278]. It seems reasonable to assume, therefore, that baryon-number violating processes can also occur under other circumstances involving extreme gravity, density, and pressure, such as prevail in the collapse of a degenerate compact star to a black hole.

One of the consequential predictions concerning the quantum properties of black holes is that these stars should give rise to gravitationally-induced thermal radiation (Hawking radiation) at the Schwarzschild horizon [243]. Hawking radiation refers to the predicted flux of particles from the immediate vicinity of a black hole, a surprising revelation when it was first made

---

iron family were created in supernova explosions by which sufficiently massive stars ended their lives.

because black holes were long considered to be 'black', i.e., to have so great a gravitational attraction that nothing could escape. A heuristic explanation of the origin of this radiation is that the gravitational field of a black hole perturbs the quantum vacuum giving rise to virtual particle–antiparticle pairs. One of the particles escapes with positive energy and appears as a real particle, while the other with negative energy falls into the black hole, thereby reducing the black hole mass. Hawking radiation, if it occurs, would eventually result in black hole evaporation, although for black holes of stellar mass the time scale of complete evaporation would be orders of magnitude longer than the age of the universe.

If extreme gravity can lead to the emission of particles from the vacuum, then it is to be expected on thermodynamic grounds (principle of detailed balance), as supported by the Hermitian structure of the transition amplitudes of standard quantum field theory, that there must exist a corresponding process by which, under appropriate circumstances, an intense gravitational field induces the vacuum to absorb particles. I have termed such a process particle *resorption* since, from a field-theoretical perspective, the particles are returned to the vacuum whence they arose initially. Resorption would be expected to occur significantly when the density of gravitational energy released by in-falling particles of mass m becomes comparable to mc^2/λ^3. For neutrons, this threshold energy density would be about 103 GeV/fm^3, corresponding to a particle density of approximately 110 fm^{-3}. In a collapsing neutronic star, neutron resorption would raise the (increasingly negative) gravitational potential and lower the (increasingly positive) particle Fermi energy, thereby conceivably leading to an equilibrium state of macroscopic extent. Interestingly, the process of gravitationally-induced resorption would be an example of Le Chatelier's principle operating, not just in chemistry, but under the most extreme physical circumstances imaginable, apart from the big bang.

At the densities cited above, the disruption of hadrons into quarks would almost certainly have taken place. Nevertheless, as a heuristic exploration of particle resorption into the vacuum in the absence of all interactions except for gravity in the tractable case of a flat space-time, let us examine again the variation of the total energy function U of a spherically symmetric system of fermions, assumed to be neutrons, with arbitrary particle number N and radius R. As in the previous section, but without the magnetic interaction, we write the total energy (now including mass–energy) in the dimensionless form

$$\frac{U}{m_n c^2} = N \left(\frac{N^{2/3}\lambda^2}{R^2} + 1 \right)^{1/2} - \frac{3}{5} \left(\frac{m_n}{m_P} \right)^2 N^2 \frac{\lambda}{R}, \tag{8.60}$$

in which $m_P = \sqrt{\hbar c/G}$ is the Planck mass, and we have approximated the stellar mass in the gravitational binding energy by $M = m_n N$. In the relativistic limit $y \equiv p_F/m_n c > 1$, where p_F is the Fermi momentum, (8.60) reduces to

$$\frac{U}{m_n c^2} = \frac{N^{4/3} - aN^2}{z}, \tag{8.61}$$

with constant

$$a = \frac{3}{5}\left(\frac{m}{m_P}\right)^2$$

and dimensionless radius $z = R/\lambda$. For a star with fixed baryon number, the minimum of the internal energy $U = U_k + U_g$ locates the equilibrium state. However, as pointed out several times already, (8.61) does not have a minimum because both the kinetic and gravitational energy terms vary with the same power of the stellar radius. Instead, there is an upper limit to the baryon number of the relativistic degenerate star, $N_{max} = a^{-3/2} \sim (m_P/m)^3 \sim 1.85 N_{Sun}$. [Worked out more exactly, this limit is the Chandrasekhar limit for white dwarf stars and Oppenheimer–Volkoff (OV) limit for neutron stars.] The foregoing conclusion does not follow, however, when N is a function of z.

In the model examined here, we assume that contraction of the star by an amount dR, leading to a release of gravitational energy dU_g results in the resorption of $dN = dU_g/mc^2$ particles. The hypothesized connection between collapse and resorption yields differential equations for the radial variation of the particle number

$$(z^2 + 2aNz)\frac{dN}{dz} - aN^2 = 0 \tag{8.62}$$

and energy

$$\frac{1}{mc^2}\frac{dU}{dz} = \left(\frac{aN^2}{z^2} - \frac{N^{5/3}}{z^3\sqrt{1 + N^{2/3}z^{-2}}}\right) - \left(\frac{2aN}{z} - \frac{1 + \frac{4}{3}N^{2/3}z^{-2}}{\sqrt{1 + N^{2/3}z^{-2}}}\right)\frac{dN}{dz}, \tag{8.63}$$

which reduces in the relativistic limit ($y = N^{1/3}z^{-1} \gg 1$) to

$$\frac{1}{mc^2}\frac{dU}{dz} = \frac{aN^2 - N^{4/3}}{z^2} - \frac{2aN - 4N^{1/3}/3}{z}\frac{dN}{dz}. \tag{8.64}$$

At equilibrium, one has $dU = 0$ and $d^2U < 0$, from which follows from (8.62) and (8.63) an implicit relation for the equilibrium stellar radius ($R_{eq} = \lambda z_{eq}$)

$$az_{eq}^2 y_{eq}\left(\sqrt{y_{eq}^2 + 1} + 1 - \frac{2}{3}y_{eq}^2\right) = 1 \qquad (y_{eq} = N_{eq}^{1/3}z_{eq}^{-1}). \tag{8.65}$$

In the relativistic limit, (8.65) reduces to the explicit solution

$$z_{eq} = \frac{2aN_{eq}/3}{aN_{eq}^{2/3} - 1}, \tag{8.66}$$

in terms of the equilibrium number of particles N_{eq}. The latter, however, must be obtained by solving the nonlinear equation (8.62). This was done numerically by use of a fourth-order Runge–Kutta method. Figure 8.7 shows the variation with radius of the particle number N and internal energy U determined

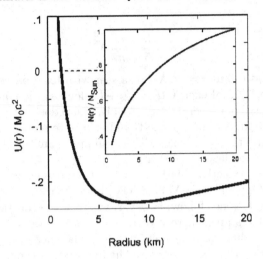

Fig. 8.7. Plot of particle number N (in units of the solar baryon number) and resulting internal energy U (in units of the initial stellar mass energy) versus radius (in km). As particle resorption into the vacuum occurs, neutrons are removed from the system, the Fermi energy decreases, and the potential energy of the star develops a minimum

from the relativistic approximation (8.64) for a degenerate fermion star of initially 10 solar masses. As the star collapses within the Schwarzschild horizon, the resorption of neutrons into the vacuum raises the gravitational potential, leading to an equilibrium state of 8.2 km (compared with the Schwarzschild radius of 23.1 km) and fermion ratio $N_{eq}/N_0 = 0.78$, where N_0 is the initial number of neutrons. Figure 8.8 shows a plot comparing the equilibrium radius (8.66) and Schwarzschild radius of the star as a function of initial mass.

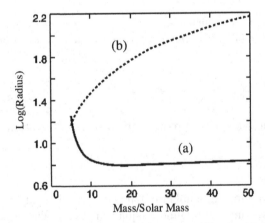

Fig. 8.8. Logarithmic plot of (a) equilibrium radius (R_{eq}) and (b) Schwarzschild radius (R_{Sch}) (in km) as a function of stellar mass (in units of solar mass)

One sees that for masses beyond about 10 solar masses, where the relativistic reduction leading to (8.66) should hold reasonably well, the equilibrium radii vary relatively slowly.

Because dN/dz in (8.62) is always positive, any sufficiently massive degenerate fermion star will collapse to an equilibrium state[13] rather than to a singular point, although it is to be noted that the masses of black hole progenitor stars are limited to about 150 solar masses [279].

I have focused attention on a *relativistic* degenerate system in this section because this is the case that leads to singular collapse when baryon number is conserved and the stellar mass exceeds the OV limit. Since a *non*-relativistic degenerate fermion star always settles into an equilibrium state, it is to be expected – and calculations confirm – that the exact energy function based on (8.63) has a minimum. The equilibrium radius corresponding to this minimum is reproduced by the relativistic approximation with greater accuracy, the more massive the star is. For example, for $M/M_{Sun} = 10, 20, 40$, the equilibrium radius determined graphically from a plot of the exact energy function (8.63) and the radius given by the relativistic approximation (8.66) are respectively (in km): (5.2, 9.5), (5.6, 6.5) and (6.4, 6.2). It is to be emphasized, however, that the significant outcome of the analysis presented here is *not* the precise numerical value of the radius, but the fact that, for a relativistic degenerate fermion star initially over the OV mass limit, there *is* a macroscale equilibrium radius.

Other resorption mechanisms leading to a particle variation law different from (8.65) and (8.66) have been examined, but it is beyond the scope of this chapter to discuss them in detail. Let it suffice to say that all such variations which I have investigated lead to stable equilibrium states of macroscopic extent. These calculations, however, were based on the use of Newtonian gravity. There remains the task of demonstrating rigorously whether a cold fermion star, supported by particle resorption against complete collapse in a Newtonian gravitational potential, will also be supported in the Schwarzschild space-time of general relativity. A preliminary examination of this question is presented in Appendix 8E.

In concluding this description of proposed quantum stabilization processes, it is perhaps pertinent to recall the desperate comment made by Sir Arthur Eddington concerning Chandrasekhar's startling discovery that degenerate stars of mass greater than about 1.4 solar masses cannot end as a white dwarf, but inevitably must collapse to a singular point. Eddington remarked in dismay that [280]:

> Various accidents may intervene to save the star, but I want more protection than that. I think there should be a law of nature to prevent a star from behaving in this absurd way.

---

[13] The relativistic approximation (8.66) requires $N_{eq} > (5/3)^{3/2}(m_P/m)^3 \sim 4N_{Sun}$ for positive radius $z_{eq} > 0$. This restriction does not apply to the exact relation (8.65).

If subsequent research can demonstrate convincingly that self-gravitating ultra-dense fermionic nuclear matter, either as baryons or quarks, must indeed undergo the bosonic condensation or ferromagnetic transition or particle resorption herein proposed, then astrophysics will at long last have the 'law of nature' that Eddington wished for.

Appendix 8A   Gravitational Binding Energy of a Uniform Sphere of Matter

The gravitational binding energy of an object is the work required to assemble it against the force of gravity or, equivalently, the gravitational potential energy of the assembly. To assemble matter into a uniformly dense sphere of radius R, one can add concentric spherical shells of mass one shell at a time. The gravitational potential at the surface of a uniform sphere of mass density ρ and radius r is

$$\phi(r) = -\frac{GM(r)}{r} , \tag{8A.1}$$

in which

$$M(r) = \frac{4}{3}\pi r^3 \rho \tag{8A.2}$$

is the total mass contained within the surface. The differential change in potential energy of the sphere resulting from addition of a concentric spherical shell of thickness dr and mass $dm = 4\pi r^2 \rho dr$ is

$$dU = \phi(r)dm = -\frac{GM(r)dm}{r} = -3G\left(\frac{4}{3}\pi\rho\right)^2 r^4 dr . \tag{8A.3}$$

Integration of (8A.3) over the radial coordinate from the center of the sphere to the final radius R leads to the binding energy

$$U = -\frac{3}{5}G\left(\frac{4}{3}\pi\rho\right)^2 R^5 = -\frac{3}{5}\frac{GM^2}{R} , \tag{8A.4}$$

where M is the total mass of the sphere.

Upon expressing G in terms of the Planck mass $m_\mathrm{P} = \sqrt{\hbar c/G}$ and dividing both sides of (8A.4) by the rest-mass energy mc^2 of the constituent baryons, one obtains the dimensionless expression for gravitational binding energy

$$\frac{U}{mc^2} = -\frac{3}{5}\left(\frac{M}{m_\mathrm{P}}\right)^2 \frac{\lambda}{R} , \tag{8A.5}$$

in which $\lambda = \hbar/mc$ is the reduced Compton wavelength. A useful approximation (which neglects the gravitational binding energy in second order) is to regard the mass M as the sum of the masses m of the N constituent baryons, whereupon the preceding equation becomes a function of baryon number

$$\frac{U}{mc^2} = -\frac{3}{5}\left(\frac{m}{m_\mathrm{P}}\right)^2 \frac{\lambda}{R}N^2 . \tag{8A.6}$$

Appendix 8B   Stability in a Self-Gravitating System with Negative Pressure

Consider a self-gravitating spherically symmetric ball of matter of uniform density ρ_0 and radius R and a small cylindrical object of uniform density ρ aligned along a radial line and initially at rest near the surface, as shown in Fig. 8.9. For comparative purposes, we will examine first the familiar case of positive pressure.

Application of Newton's second law to the immersed object leads to the relation

$$\frac{1}{\rho}\frac{dP}{dr} + g(r) = -a ,\tag{8B.7}$$

in which

$$g(r) = \frac{4\pi}{3}G\rho_0 r \tag{8B.8}$$

is the gravitational acceleration at radius r, a is the acceleration of the object, and the pressure at any point r is given by

$$P(r) = \frac{2\pi}{3}G\rho_0^2(R^2 - r^2) .\tag{8B.9}$$

(8B.9) is obtained by integrating the equation of hydrostatic equilibrium with boundary condition $P(R) = 0$. Substitution of (8B.8) and (8B.9) into (8B.7)

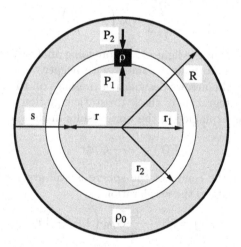

Fig. 8.9. Schematic diagram of a cylindrical object of mass density ρ immersed in a self-gravitating sphere of radius R and density ρ_0. The *heavy arrows* indicate the pressures P_1 and P_2 at the bottom and top of the object, respectively. In a positive-pressure environment (as illustrated), P_2 acts towards the center, parallel to the gravitational acceleration. In a negative-pressure environment, the direction of each pressure arrow is reversed. The coordinate s measures radial displacement from the surface

and evaluation of the integral yields the acceleration

$$a(r) = \ddot{r} = -\frac{4\pi}{3}G\rho_0\left(1 - \frac{\rho_0}{\rho}\right)r \ . \tag{8B.10}$$

Inspection of (8B.10) shows immediately that an object of greater density than the surrounding matter ($\rho_0/\rho < 1$) will accelerate towards the center; i.e., $a < 0$. By contrast, if $\rho_0/\rho > 1$, then $a > 0$, and the object will accelerate away from the center.

Re-expression of (8B.10) in terms of the displacement s from the surface, and integration with boundary conditions $s(0) = 0$ and $\dot{s}(0) = 0$ lead to the displacement and velocity functions (positive directive is radially inwards)

$$s(t) = R\left[1 - \cos(\alpha t)\right] \qquad (R \geq s \geq 0) \ , \tag{8B.11}$$

$$\dot{s}(t) = \alpha R \sin(\alpha t) \ , \tag{8B.12}$$

with real-valued rate

$$\alpha = \sqrt{\frac{4\pi}{3}G\rho_0\left(1 - \frac{\rho_0}{\rho}\right)} \ , \tag{8B.13}$$

for $\rho_0/\rho < 1$. (For $\rho_0/\rho > 1$, the trigonometric functions become hyperbolic trigonometric functions.) The time τ required for the object to sink to the center is determined from $\alpha\tau = \pi/2$, or[14]

$$\tau = \frac{\pi/2}{\sqrt{\frac{4\pi}{3}G\rho_0\left(1 - \frac{\rho_0}{\rho}\right)}} \ . \tag{8B.14}$$

We consider next, in the light of the foregoing analysis, the case of the object immersed in a uniform environment with negative pressure and negative pressure gradient. The arrows designating pressure on the object in Fig. 8.9 are reversed, and integration of the hydrostatic equation with boundary condition $P(0) = 0$, as explained in the text, results in the radial dependence

$$P(r) = -\frac{2\pi}{3}G\rho_0^2 r^2 \ , \tag{8B.15}$$

so that pressure at the center of the sphere is still greater than pressure at the surface. This leads to the acceleration

$$a(r) = \ddot{r} = -\frac{4\pi}{3}G\rho_0\left(1 + \frac{\rho_0}{\rho}\right)r \ . \tag{8B.16}$$

---

[14] It is interesting to note that the kinematics of uniform accelaration ($s = \frac{1}{2}at^2$) applied to a free fall displacement of magnitude R lead to the close result

$$\tau_{approx} = \frac{\sqrt{2}}{\sqrt{\frac{4\pi}{3}G\rho_0\left(1 - \frac{\rho_0}{\rho}\right)}} \sim \frac{1.414}{\sqrt{\frac{4\pi}{3}G\rho_0\left(1 - \frac{\rho_0}{\rho}\right)}} \ .$$

It is clear from (8B.16) that the object sinks towards the center ($a < 0$) irrespective of the relative density. One may have expected this on the grounds that at equilibrium the tension on the lower surface of the object is greater than the tension on the upper surface, but now directed radially inwards. The time of free fall to the center, determined in the same way as that leading to (8B.14), is

$$\tau = \frac{\pi/2}{\sqrt{\dfrac{4\pi}{3}G\rho_0\left(1 + \dfrac{\rho_0}{\rho}\right)}} . \tag{8B.17}$$

Appendix 8C   Quark Deconfinement in a Neutron Star

A relativistic degenerate star ending its life as a neutron star does not necessarily consist exclusively of neutrons, although other constituents have been neglected in the development of the fermicon model described in the text. For sufficiently high densities ($n_n > n_N$, where the saturation density of nuclear matter is $n_N \sim 0.15$ fm^{-3} with 1 fm $= 10^{-15}$ m), the neutron again becomes a beta-unstable particle and can undergo weak decay

$$n \longrightarrow p + e + \overline{\nu}$$

to a proton, electron, and electron antineutrino. Naive application of Le Chatelier's principle might suggest that the reaction is thermodynamically unfavorable because it leads to an increased number of particles. However, the reaction proceeds under the circumstances because it reduces the neutron Fermi energy and increases the total entropy of the system. Assuming that all particles are describable as constituents of ideal relativistic Fermi gases and that the neutrinos escape the system, one can determine the relative number of protons and neutrons from the relations for:

(A)   chemical equilibrium $\mu_n = \mu_p + \mu_e$,
(B)   charge conservation $n_p = n_e$,

where, according to the relativistic reduction of (8.18) and (8.19), the chemical potential and particle density of each particle type are related by

$$\mu = \hbar c(3\pi^2 n)^{1/3} . \tag{8C.18}$$

It is then a matter of simple algebra to deduce that $n_p/n_n = 1/8$.

If the system is opaque to (anti)neutrinos, then the antineutrino chemical potential must be added to the right side of condition (A), and one has the additional constraint of:

(C)   lepton number conservation $n_{\overline{\nu}} = n_e$.

Solution of the three equilibrium relations leads to an even greater preponderance of neutrons $n_p/n_n = 1/27$. The neglect of neutron beta decay in the fermicon model is justified.

In contrast to the electron, the neutron (as well as the proton) is not an elementary particle, but is regarded as a bound state of three quarks: one 'up' quark (u) of charge $+2e/3$ and two 'down' quarks (d) of charge $-e/3$, each 'flavor' of quark being characterized by a spin angular momentum $S_q = \hbar/2$ and one of three possible strong-interaction charges referred to as 'color'. According to quantum chromodynamics (QCD), the prevailing theory of the strong interactions, quarks combine to form hadrons of zero color charge. (It is this analogy to the formation of white light, i.e., no color, from the combination of the three primary colors that led to the whimsical nomenclature for the strong-interaction charge.) Moreover, QCD predicts that under ordinary circumstances free quarks do not exist; quarks will be bound to form hadrons, i.e., the strongly interacting baryons and mesons that comprise normal nuclear matter. The question addressed in this appendix is whether extraordinary thermodynamic conditions within a collapsed relativistic degenerate star can lead to the deconfinement of quarks, in effect the rupture of the neutron via the reaction

$$n \longrightarrow 2d + u ,$$

and the formation of a separate quark core in equilibrium with a surrounding neutron shell. The hypothesized system is analogous to the fermicon model of a compact star comprising a neutron core and boson condensate shell. The neutron–quark phase transformation requires that three equilibrium conditions be satisfied:

(A)   chemical equilibrium $\mu_n = 2\mu_d + \mu_u$,

(B)   charge conservation $\dfrac{2}{3}n_u - \dfrac{1}{3}n_d = 0 \implies n_u = \dfrac{1}{2}n_d$,

(C)   hydrostatic equilibrium $P_n = P_d + P_u$.

At the density and pressure at which a neutron–quark phase transition might be expected to occur, the system of neutrons is unlikely to behave like an ideal degenerate Fermi gas. For illustrative purposes, therefore, we adopt an equation of state [281] taking account of neutron interactions and leading to a stronger dependence of pressure on density. The neutron energy density, chemical potential, and pressure can be expressed in the respective forms

$$\varepsilon_n = m_n c^2 n_n \left[1 + \left(\frac{n_n}{n_0} \right)^{\nu-1} \right] , \tag{8C.19}$$

$$\mu_n = \frac{d\varepsilon_n}{dn_n} = m_n c^2 \left[1 + \nu \left(\frac{n_n}{n_0} \right)^{\nu-1} \right] , \tag{8C.20}$$

$$P_n = n_n \mu_n - \varepsilon_n = m_n c^2 n_0 (\nu - 1) \left(\frac{n_n}{n_0} \right)^{\nu} , \tag{8C.21}$$

where the parameters of the model have the approximate values $\nu = 2.5$ and $n_0 = 15 n_{\mathrm{N}}$ (with saturation density $n_{\mathrm{N}} \sim 0.15 \ \mathrm{fm}^{-3}$ of normal nuclear matter). Thus the neutron pressure increases with particle density as approximately $n^{7.5/3}$, in comparison to $n^{5/3}$ or $n^{4/3}$ for the nonrelativistic or relativistic limits of the ideal degenerate Fermi gas.

One of the striking features of QCD, which simplifies the analysis of the quark phase, is the phenomenon of asymptotic freedom. Although QCD is an exceedingly complex, nonlinear theory of interacting quarks and gluons, the effective quark–gluon coupling parameter α_{c} (which plays the same role in QCD as does the fine structure constant $\alpha_{\mathrm{f}} \sim 1/137$ in QED) is not constant but vanishes in the limit of an infinitely high interaction energy.[15] A simple, but useful, model accounting for this behavior is the so-called MIT 'bag model', in which quarks in a hadron (such as a neutron) are confined by a bag pressure B, but otherwise behave like an ideal gas of relativistic fermions. However, whereas the ordinarily observable spin-1/2 fermions (e, p, n) are doubly degenerate, a spin-1/2 quark is sixfold degenerate because of the three additional color degrees of freedom. The quantum statistical expressions for an ideal fermion gas of arbitrary degeneracy g are

$$\varepsilon = \frac{gmc^2}{16\pi^2 \lambdabar^3} \left[y(2y^2 + 1)\sqrt{1 + y^2} - \sinh^{-1} y \right] , \qquad (8C.22)$$

$$\mu = mc^2 \sqrt{1 + y^2} , \qquad (8C.23)$$

$$P = \frac{gmc^2}{48\pi^2 \lambdabar^3} \left[y(2y^2 - 3)\sqrt{1 + y^2} + 3\sinh^{-1} y \right] , \qquad (8C.24)$$

with

$$y \equiv \frac{p_{\mathrm{F}}}{mc} = \left(\frac{6\pi^2 n}{g} \right)^{1/3} \lambdabar , \qquad (8C.25)$$

and reduce in the case of relativistic quarks ($g = 6$) to

$$\varepsilon = \frac{3mc^2}{4\pi^2 \lambdabar^3} y^4 = \frac{3}{4} \pi^{2/3} \hbar c n^{4/3} , \qquad (8C.26)$$

$$\mu = mc^2 y = \hbar c \pi^{2/3} n^{1/3} , \qquad (8C.27)$$

$$P = \frac{mc^2}{4\pi^2 \lambdabar^3} y^4 = \frac{1}{4} \pi^{2/3} \hbar c n^{4/3} , \qquad (8C.28)$$

---

[15] It is to be noted that the electron–photon coupling parameter α_{f} in QED is not constant either, but *increases* with increasing energy of interaction or, equivalently, with decreasing distance at which a charged particle is probed.

with

$$y \equiv \frac{p_F}{mc} = \pi^{2/3} n^{1/3} \lambda \,. \tag{8C.29}$$

In implementing the conditions of equilibrium, it is useful to express the particle density, energy density, and pressure of each relevant species as a dimensionless function of its respective chemical potential. For the neutron component, the relations are

$$n_n = n_0 \left[\frac{1}{\nu} (\zeta - 1) \right]^{1/(\nu-1)} \,, \tag{8C.30}$$

$$\frac{\varepsilon_n}{m_n c^2 / \lambda_n^3} = n_0 \lambda_n^3 \left[\frac{1}{\nu} (\zeta - 1) \right]^{1/(\nu-1)} \left[1 + \frac{1}{\nu} (\zeta - 1) \right] \,, \tag{8C.31}$$

$$\frac{P_n}{m_n c^2 / \lambda_n^3} = n_0 \lambda_n^3 (\nu - 1) \left[\frac{1}{\nu} (\zeta - 1) \right]^{\nu/(\nu-1)} \,, \tag{8C.32}$$

where we have defined the dimensionless variable

$$\zeta = \frac{\mu_n}{m_n c^2} \,. \tag{8C.33}$$

Corresponding expressions for the total quark component comprising d and u quarks whose ground-state energy density relative to the vacuum is B are

$$n_q = n_d + n_u = \frac{1}{\pi^2} \frac{\mu_d^3 + \mu_u^3}{(\hbar c)^3} = \frac{3(1 + 2^{-4/3})^{-3}}{16\pi^2} \left(\frac{\zeta}{\lambda_n} \right)^3 \,, \tag{8C.34}$$

$$\frac{\varepsilon_q}{m_n c^2 / \lambda_n^3} = \frac{3}{64\pi^2} (1 + 2^{-4/3})^{-3} \zeta^4 + \frac{B}{m_n c^2 / \lambda_n^3} \,, \tag{8C.35}$$

$$\frac{P_q}{m_n c^2 / \lambda_n^3} = \frac{1}{64\pi^2} (1 + 2^{-4/3})^{-3} \zeta^4 - \frac{B}{m_n c^2 / \lambda_n^3} \,. \tag{8C.36}$$

In deriving the preceding set of equations, we have used the conditions of equilibrium (A) and (B) which connect the quark and neutron chemical potentials by

$$\mu_u = 2^{-1/3} \mu_d \tag{8C.37}$$

and

$$\mu_d = \frac{\mu_n}{2(1 + 2^{-4/3})} \,. \tag{8C.38}$$

The pressure and density at which the neutron–quark transition occurs is deduced by equating pressures (8C.32) and (8C.36) and solving for ζ either numerically by an appropriate algorithm or graphically as shown in Fig. 8.10 for a bag parameter $B^{1/4}/(\hbar c)^3 = 200$ MeV (or $B = 3.36 \times 10^{34}$ Pa) characteristic of the upper limit[16] employed with the MIT bag model. The solution, $\zeta =$

---

[16] A bag parameter $B^{1/4}/(\hbar c)^3 = 150$ MeV characteristic of the lower limit leads to $\zeta = 1.039$ and densities $n_n = 0.16$ fm^{-3}, $n_q = 0.86$ fm^{-3}, $\rho_n = 2.69 \times 10^{17}$ kg/m^3, $\rho_q = 4.90 \times 10^{17}$ kg/m^3. The resulting neutron density is comparable to the saturation density of nuclear matter for which it is already evident that nucleons do not disrupt to form free quarks.

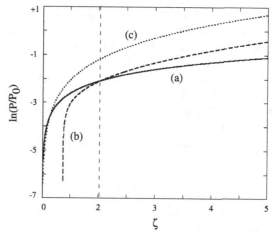

Fig. 8.10. Plot of the log of pressure (in units of $P_0 = m_n c^2 / \lambda_n^3 = 1.65 \times 10^{37}$ Pa) as a function of the neutron chemical potential parameter $\zeta = \mu_n / m_n c^2$ for **(a)** quark pressure P_q determined from the bag model (8C.36); **(b)** neutron pressure determined from the equation of state (8C.21); and **(c)** neutron pressure determined from the degenerate fermion gas model (8C.24) with degeneracy $g = 2$. The latter does not lead to a physically valid neutron–quark phase transition

2.049, results in particle densities $n_n = 1.43$ fm^{-3}, $n_q = 6.57$ fm^{-3} and mass densities ($\rho = \varepsilon/c^2$) of $\rho_n = 3.40 \times 10^{18}$ kg/m^3 and $\rho_q = 6.01 \times 10^{18}$ kg/m^3. It would seem, therefore, that at a neutron density roughly 8 or 9 times that of the saturation density of nuclear matter it may be possible for neutrons to 'melt' to form a denser phase of unbound quarks. However, whether a dense quark core surrounded by a neutron shell results in a stable star or not is at present an open question requiring more accurate equations of state for both nuclear and quark matter and solution of the TOV equation or perhaps a more appropriate gravitational relation for hydrostatic equilibrium. These considerations go beyond the intended scope of this appendix.

Appendix 8D   Energy Balance in the Creation of the Universe

Although a thorough treatment of almost any cosmological problem requires the use of general relativity, a simple model based on Newtonian gravity suffices to demonstrate that the positive mass–energy and negative gravitational energy in the accessible universe are close in magnitude. We must note at the outset, however, several important cosmological findings established primarily from measurements of the cosmological microwave background radiation (CMB) and the distance–brightness relation of Type Ia supernovae.

The CMB is the redshifted relic electromagnetic radiation from the formation of the universe over ten billion years (Gyr) ago. Satellite-based measurements[17] of the spectral distribution have established a radiation temperature T very close to 2.7 K. Measurements of the angular size of temperature fluctuations across the sky have established that the geometry of the universe is most likely flat (i.e., of zero curvature) in agreement with predictions based on cosmological models with inflation.

According to general relativity (Einstein–de Sitter model), the mass density of a universe with zero curvature is the critical density

$$\rho_c = \frac{3H_0^2}{8\pi G},$$

(8D.39)

in which the Hubble parameter H_0 is the fractional rate of increase of the size of the universe. More precisely,

$$H_0 = \frac{1}{R}\frac{dR}{dt},$$

where R is the scale factor giving the size of the universe at some time before the present at which $R(0) = 1$. Operationally, the Hubble parameter relates the speed of recession of galaxies to their distance from our own galaxy (Milky Way). In this regard, Type Ia supernovae serve as standard candles whose intrinsic brightness is known and whose distance (and therefore the distance of the galaxies in which they are embedded) can be determined from the observed brightness. The speed of recession of a galaxy is deducible from the redshift of known spectral lines. Although at one time highly contentious, the value of H_0 is now considered to be close to 73 km/s per megaparsec (1 parsec $\sim$ 3.3 light-years), or $H_0 \sim 2.3 \times 10^{-18}$ s^{-1} [283]. It then follows from (8D.39) that $\rho_c \sim 6$ amu/m^3. (1 amu is essentially the mass of a proton $m_p = 1.67 \times 10^{-27}$ kg.)

Let us model the matter in the universe as a homogeneous gas comprising particles of mass $m = 1$ amu filling a spherical volume of radius R with uniform number density n and mass density $\rho = mn$. Correspondingly, the radiation in the universe is taken to be a uniform photon gas of energy density $\varepsilon_\gamma = (\sigma/4c)T^4$ where $\sigma = 5.67 \times 10^{-8}$ W/m^{2}K^4 is the Stefan–Boltzmann constant and $T \sim 3$ K. The energy density of matter, which for the most part is nonrelativistic, is then $\varepsilon_m = \rho c^2$. Comparing energy densities of radiation and matter (taking the current value to be ρ_c) leads to $\varepsilon_\gamma/\varepsilon_m \sim 8 \times 10^{-5}$. For our present purposes we can disregard the energy contribution from radiation.

Using the result of Appendix 8A to calculate the gravitational energy density

$$\varepsilon_g = -\frac{3GM^2/5R}{4\pi R^3/3}$$

---

[17] The most extensive measurements of the CMB to date have been made by the Wilkinson Microwave Anisotropy Probe (WMAP) [282].

of a uniformly dense sphere of matter of radius R, our model for the total energy density of the universe leads to

$$\varepsilon_T = \rho c^2 \left(1 - \frac{4\pi}{5} \frac{G\rho R^2}{c^2} \right) . \tag{8D.40}$$

For ε_T to vanish, the mean density of matter in the universe must be

$$\rho = \frac{5c^2}{4\pi G R^2} = \frac{5}{4\pi G t^2} , \tag{8D.41}$$

in which $R = ct$, where t is the age of the universe.

Lower limits to the age of the universe can be estimated from the age of the oldest star clusters, the age of the oldest white dwarf stars, and from the age of various chemical elements. At the time of writing, the most recent and precise estimate of this age, independent of uncertainties associated with CMB fluctuations or models of stellar evolution, is $t \sim 14.5 \pm 2$ Gyr determined from the $^{238}U/^{232}Th$ abundance ratio in meteorites in conjunction with observations of low-metallicity stars in the halo of the Milky Way [284]. Substitution of this value into (8D.41) yields $\rho \sim 17$ amu/m^3, which differs from ρ_c by merely a factor of 2.8 – as opposed to orders of magnitude, which could have been the case.

To within experimental uncertainties of our knowledge of the mean mass density and the age of the universe, it appears that the total energy of the universe may well be close to zero.

Appendix 8E   Particle Resorption in a Schwarzschild Geometry

As in the process of cosmological particle creation, the cost of producing positive matter is a negative gravitational potential energy. The gravitational field, therefore, lowers the chemical potential of the particles. When the gravitational field is sufficiently strong that the chemical potential vanishes, then no net energy is expended to create or remove a particle from the system, and particle number is no longer a conserved quantity (provided no violation of electrical charge conservation occurs). The condition of chemical equilibrium is then

$$\mu_n + \mu_g = 0 , \tag{8E.42}$$

where the first term on the left hand side is the (positive) contribution to the chemical potential of the particles (assumed to be neutrons) from mass, motion, and compression, and the second term is the (negative) contribution from gravity.

For a description of dense nuclear matter, we again adopt relations (8C.19)–(8C.21) for the energy density ε_n, chemical potential μ_n, and degeneracy pressure P_n, respectively, which we express below in the dimensionless forms

$$\frac{\varepsilon_n}{m_n c^2 n_0} = \eta + \eta^\nu , \tag{8E.43}$$

$$\frac{\mu_n}{m_n c^2} = 1 + \nu \eta^{\nu-1} , \tag{8E.44}$$

$$\frac{P_n}{m_n c^2 n_0} = (\nu - 1)\eta^\nu , \tag{8E.45}$$

where

$$\eta \equiv \frac{n_n}{n_0} . \tag{8E.46}$$

It is to be recalled that this equation of state for dense nuclear matter is defined by two parameters, a characteristic particle density n_0, which we take to be 15 times the saturation density of nuclear matter ($n_{sat} \sim 0.15$ fm^{-3}), and an exponent ν, which we take to be 2.5.

From the perspective of Einstein's theory of general relativity, gravity is not a force between particles, but a consequence of the geometrical structure of space-time. The gravitational field of a nonrotating spherically symmetric distribution of matter is described uniquely by the Schwarzschild solution to Einstein's field equations. Based on this solution, the gravitational potential energy of a sphere of uniform mass density ρ and radius R in a static Schwarzschild geometry is defined to be [261]

$$U_g = \rho c^2 \int_0^R \left[1 - \left(1 - \frac{8\pi}{3c^2} G\rho r^2 \right)^{-1/2} \right] 4\pi r^2 dr . \tag{8E.47}$$

The gravitational energy per volume $\varepsilon_g = dU_g/dV$, where the differential volume element dV must be that of the Schwarzschild geometry (given in Sect. 8.3), can be written in the dimensionless form

$$\frac{\varepsilon_g}{m_n n_0 c^2} = -\left[1 - (1 - \beta\eta)^{1/2} \right] , \tag{8E.48}$$

where we have approximated the mass density by $\rho = m_n n_n$ and defined the dimensionless ratio $\beta = (R/\Lambda)^2$ of the radius to a characteristic length parameter

$$\Lambda = \left[\frac{8\pi}{3} \left(\frac{m_n}{m_P} \right)^2 n_0 \lambda_n \right]^{-1/2} . \tag{8E.49}$$

As encountered before, m_P is the Planck mass and λ_n is the neutron reduced Compton wavelength. The gravitational contribution to the chemical potential, calculated from (8E.48), is then

$$\frac{\mu_g}{m_n c^2} = \frac{1}{m_n c^2} \frac{d\varepsilon_g}{dn_n} = -\left[1 - \frac{1 - 3\beta\eta/2}{(1 - \beta\eta)^{1/2}} \right] . \tag{8E.50}$$

From the sum of (8E.44) and (8E.50) one arrives at the condition

$$\nu\eta^{\nu-1} + \frac{1 - 3\beta\eta/2}{(1 - \beta\eta)^{1/2}} = 0 \qquad (8E.51)$$

for the chemical equilibrium of a particle (neutron) in a gravitational field corresponding to the Schwarzschild solution of Einstein's field equations. Before continuing with the analysis, let us relate (8E.51), which is not a familiar result, to a more widely known application of nonrelativistic statistical physics. Expansion of $\mu = \mu_n + \mu_g$ in a Taylor series in $\beta\eta$ (which must be less than 1) and truncation at first order leads to

$$\mu = \mu_n + \mu_g \sim m_n c^2 + \mu_n' + m_n\phi(R) \,,$$

where μ_n' is the chemical potential of the particles exclusive of mass–energy and $\phi(R) = -GM(R)/R$ is the Newtonian gravitational potential. Applied to a particle of mass m at a height z above the surface of a spherical mass M of radius R, with the zero of potential defined at the surface, one obtains the 'textbook' condition for equilibrium in a weak gravitational field [252, p. 72]: $\mu_n' + mgz = $ constant with acceleration of gravity $g = GM/R^2$.

Rearrangement of (8E.51) leads to a quadratic equation for β with the solution

$$\beta(\eta) = \frac{2}{9\eta^2}\left(3\eta - \nu^2\eta^{2\nu-1} + \nu\eta^\nu\sqrt{3 + \nu^2\eta^{2\nu-2}}\right), \qquad (8E.52)$$

from which is determined the stellar radius R as a function of density n_n (or the inverse relation). A quadratic equation has two solutions, but the physically significant solution here is the one for which $\beta \to 0$ for $\eta \gg 1$. The stellar mass, and therefore the Schwarzschild radius, corresponding to a given density is calculable from (8.27). Figure 8.11 shows the variation of the stellar radius and the Schwarzschild radius as a function of particle density. In contrast to the example based on Newtonian gravity treated in the text, the surface of the collapsed star at equilibrium lies outside its event horizon.

There still remains the condition of hydrostatic equilibrium to be satisfied. In a rigorous calculation, one ordinarily proceeds by assigning a value to the central density and integrating the TOV equation outward to a radius where the pressure vanishes. This radius then marks the surface of the star, and a family of solutions is obtained for different values of central density. We will proceed differently in the spirit of the simple model expounded here by equating the neutron degeneracy pressure (8E.45) and the central pressure (8.36)

$$\frac{P_g}{m_n c^2 n_0} = \eta\left[\frac{1 - (1 - \beta\eta)^{1/2}}{3(1 - \beta\eta)^{1/2} - 1}\right], \qquad (8E.53)$$

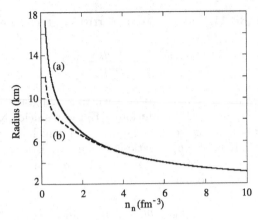

Fig. 8.11. Plot of **(a)** stellar radius and **(b)** Schwarzschild radius as a function of neutron density calculated from (8E.52) derived from a model of gravitationally-induced particle resorption in a Schwarzschild geometry. The parameters used in the nuclear equation of state are $n_0 = 2.25$ fm^{-3} and $\nu = 2.5$

Table 8.8. Stellar radius, mass, and Schwarzschild radius obtained for several values of the nuclear density parameter n_0 (with $n_{sat} = 0.15$ fm^{-3})

n_0/n_{sat}	n_n [fm^{-3}]	R [km]	M/M_{Sun}	R_S [km]
10	1.61	7.62	2.49	7.41
15	2.41	6.22	2.03	6.05
20	3.21	5.39	1.76	5.24

obtained by solution of the TOV equation for a uniformly dense star and re-expressed here in terms of the dimensionless variables pertinent to our model. The condition of hydrostatic equilibrium then becomes

$$(\nu - 1)\eta^{\nu-1} + \frac{1 - (1 - \beta\eta)^{1/2}}{3(1 - \beta\eta)^{1/2} - 1} = 0 , \qquad (8E.54)$$

where at equilibrium $\beta(\eta)$ is given by (8E.52).

If the expressions (8E.48) for energy and (8E.53) for pressure are not to become singular, then the product $\beta\eta$ is restricted to the range $1 > \beta\eta > 8/9$. Solving (8E.54) numerically or graphically yields $\eta = 1.07$ for $\nu = 2.5$ in the nuclear equation of state. The stellar radius, mass, and Schwarzschild radius obtained for several values of the nuclear density parameter n_0 (with $n_{sat} = 0.15$ fm^{-3}), are summarized in Table 8.8.

The stellar radius and mass decrease as the nuclear density parameter increases, i.e., as the equation of state stiffens (the degeneracy pressure is larger for a given particle density). Because the stellar surface lies marginally outside the event horizon, technically such a star would not be a black hole although, as remarked previously, it remains to be seen what becomes of the Schwarzschild horizon in some future quantum theory of gravity.

The significant outcome, however, is that the gravitationally-induced particle resorption leads to equilibrium states of finite size irrespective of the initial baryon particle number.

References

1. R.P. Crease: *The Most Beautiful Experiment*, Physics World, September 2002, http://physicsweb.org/article/world/15/9/2
2. R.P. Feynman: *The Character of Physical Law* (MIT Press, Cambridge, 1965) p. 129
3. A. Tonomura, J. Endo, T. Matsuda, and T. Kawasaki: Demonstration of single-electron build-up of an interference pattern, Am. J. Phys. **57**, 117–120 (1989)
4. M. Born and E. Wolf: *Principles of Optics*, 4th edn. (Pergamon, Oxford, 1970) pp. 267–268
5. See, for example, M. Jammer: *The Conceptual Development of Quantum Mechanics*, (McGraw-Hill, NY, 1966); M. Jammer: *The Philosophy of Quantum Mechanics* (Wiley, NY, 1974); G. Ghirardi: *Sneaking a Look at God's Cards* (Princeton University Press, NJ, 2004)
6. J.C. Maxwell: *A Treatise on Electricity & Magnetism*, 3rd edn., Vol. 2 (Dover, NY, 1954) pp. 232, 257
7. M.P. Silverman: *Waves and Grains: Reflections on Light and Learning* (Princeton Universiy Press, NJ, 1998)
8. E.J. Konopinski: What the electromagnetic vector potential describes, Am. J. Phys. **46**, 499–502 (1978)
9. Y. Aharonov and D. Bohm: Significance of electromagnetic potentials in the quantum theory, Phys. Rev. **115**, 485–491 (1959)
10. W. Ehrenberg and R.E. Siday: The refractive index in electron optics and the principles of dynamics, Proc. Phys. Soc. Lond. B **62**, 8–21 (1949)
11. P.A.M. Dirac: Quantised singularities in the electromagnetic field, Proc. Roy. Soc. A **133**, 60–72 (1931)
12. H. Goldstein: *Classical Mechanics*, 2nd edn. (Addison-Wesley, Reading, MA, 1980) pp. 346, 361
13. E.M. Purcell: *Electricity and Magnetism*, 2nd edn. (McGraw-Hill, NY, 1985) pp. 226–231
14. R.P. Feynman, R.B. Leighton, and M. Sands: *The Feynman Lectures on Physics*, Vol. 2 (Addison-Wesley, Reading MA, 1965), pp. 15.12–15.14
15. See, for example, G. Baym: *Lectures on Quantum Mechanics* (Benjamin, NY, 1969) pp. 77–79; J.J. Sakurai: *Advanced Quantum Mechanics* (Addison-Wesley, Reading, MA, 1967) pp. 16–18

16. P. Bocchieri and A. Loinger: Nonexistence of the Aharonov–Bohm effect, Il Nuovo Cimento **47** A, 475–482 (1978); Nonexistence of the Aharonov–Bohm effect II: Discussion of the experiments, Il Nuovo Cimento **51** A, 1–17 (1979)

17. S. Olariu and I. Popescu: The quantum effects of electromagnetic fluxes, Rev. Mod. Phys. **57**, 339–436 (1985)

18. R.L. Liboff: The correspondence principle revisited, Physics Today **37**, No. 2, 50–55 (1984)

19. T.T. Wu and C.N. Yang: Concept of nonintegrable phase factors and global formulation of gauge fields, Phys. Rev. D **12**, 3845–3857 (1975)

20. M. Peshkin and A. Tonomura: *The Aharonov–Bohm Effect*, Lecture Notes in Physics, Vol. 340 (Springer-Verlag, Berlin, 1989)

21. W. Bayh: Messung der kontinuierlichen Phasenschiebung von Elektronen-wellen im kraftfeldfreien Raum durch das magnetische Vektorpotential einer Wolfram-Wendel, [Measurement of the continuous phase shift of electron waves in force-free space by the magnetic vector potential of a tungsten solenoid], Zeitschrift f. Physik **169**, 492–510 (1962)

22. H. Buchholz: *Elektrische und Magnetische Potentialfelder* (Springer, Berlin, 1957) Chap. 6, pp. 112, 270

23. N. Osakabe, T. Matsuda, T. Kawasaki, J. Endo, A. Tonomura, S. Yano, and H. Yamada: Experimental confirmation of Aharonov–Bohm effect using a toroidal magnetic field confined by a superconductor, Phys. Rev. A **34**, 815–822 (1986)

24. D.R. Tilley and J. Tilley: *Superfluidity and Superconductivity*, 2nd edn. (Adam Hilger Ltd, Bristol, 1986) pp. 145–150

25. M.P. Silverman: Physics in Perspective **7**, 496–498 (December 2005)

26. H.B. Callen: *Thermodynamics* (Wiley, New York, 1965) 277

27. L. Hackermueller, K. Hornberger, B Brezger, A. Zeilinger, and M. Arndt: Decoherence of matter waves by thermal emission of radiation, Nature **427** (19 February 2004) 711–714

28. J.A. Wheeler: Law without law. In: *Quantum Theory and Measurement*, ed. by J.A. Wheeler and W.H. Zurek (Princeton University Press, Princeton NJ, 1983) pp. 190–194

29. P. Carruthers and M.M. Nieto: Phase and angle variables in quantum mechanics, Rev. Mod. Phys. **40**, 411 (1968)

30. V. Jacques, E. Wu, F. Grosshans, F. Treussart, P. Grangier, A. Aspect, J.-F. Roch: Experimental realization of Wheeler's delayed-choice gedanken experiment, Science **315**, 966 (2007)

31. A. Einstein, B. Podolsky, and N. Rosen: Can quantum mechanical description of physical reality be considered complete?, Phys. Rev. **47**, 777–780 (1935)

32. N. Bohr: Can quantum mechanical description of physical reality be considered complete?, Phys. Rev. **48**, 696–702 (1935)

33. M. Jammer: *The Philosophy of Quantum Mechanics* (Wiley, New York, 1974) p. 187

34. J.F. Clauser and A. Shimony: Bell's theorem: Experimental tests and implications, Rep. Prog. Phys. **41**, 1881–1927 (1978)

35. B. d'Espagnat: The quantum theory and reality, Scientific American **247**, 158–181 (November 1979)

36. T.A. Heppenheimer: Experimental quantum mechanics, Mosaic **17**, 19–27 (1986)

37. A. Shimony: The reality of the quantum world, Scientific American **256**, 46–53 (January, 1988)
38. G. Ghirardi: *Sneaking a Look at God's Cards: Unraveling the Mysteries of Quantum Mechanics* (Princeton University Press, Princeton NJ, 2005)
39. R. Hanbury Brown and R.Q. Twiss: Correlation between photons in two coherent beams of light, Nature **177**, 27–29 (1956)
40. R. Hanbury Brown: *The Intensity Interferometer* (Taylor and Francis, New York, 1974) p. 7
41. W. Heitler: *The Quantum Theory of Radiation*, 3rd edn. (Oxford University Press, London, 1954)
42. R.P. Feynman: *Quantum Electrodynamics* (W.A. Benjamin, New York, 1962)
43. P.A.M. Dirac: *The Principles of Quantum Mechanics*, 4th edn. (Oxford University Press, London, 1958). The first edition was in 1930
44. E.M. Purcell: The question of correlation between photons in coherent light rays, Nature **178**, 1449–1450
45. A. Einstein: Zum gegenwärtigen Stand des Strahlungsproblems [On the current state of the radiation problem], Phys. Zeit. **10**, 185–193 (1909)
46. R. Loudon: *The Quantum Theory of Light*, 2nd edn. (Oxford, New York, 1983) pp. 157–160
47. See, for example, the proceedings of the XXth Solvay Conference on Physics: *Quantum Optics*, ed. by P. Mandel, Physics Reports **219** (North Holland, Amsterdam, 1992)
48. M.P. Silverman: Distinctions between quantum ensembles, Annals of the New York Academy of Sciences **480**, 292 (1986)
49. G. Breit: Rev. Mod. Phys. **5**, 117 (1933)
50. J. Macek: Phys. Rev. Lett. **23**, 1 (1969)
51. A. Ruzmaikin: Sov. Astron. **19**, 702 (1976)
52. E.B. Aleksandrov, O.V. Konstantinov, V.N. Kulyasov, A.B. Mamyrin, and V.I. Perel: Sov Phys. JETP **34**, 1210 (1972)
53. M.P. Silverman: Nuovo Cimento **6** D, 283 (1985)
54. T. Hadeishi and W.A. Nierenberg: Phys. Rev. Lett. **14**, 891 (1965)
55. I.A. Sellin et al.: Phys. Rev. **184**, 56 (1969)
56. M.P. Silverman: *Probing The Atom: Interactions of Coupled States, Fast Beams, and Loose Electrons* (Princeton University Press, New York, 2000)
57. J.N. Dodd, W.J. Sandle, and D. Zissermann: Proc. Phys. Soc. London **92**, 497 (1967)
58. M.P. Silverman, S. Haroche, and M. Gross: Phys. Rev. A **18**, 1507, 1517 (1978)
59. R. Loudon: *The Quantum Theory of Light* (Oxford, New York, 1983)
60. V. Weisskopf and E. Wigner: Z. Phys. **63**, 54 (1930) and **65**, 18 (1930)
61. P.A.M. Dirac: *The Principles of Quantum Mechanics*, 4th edn. (Oxford University Press, London, 1958) p. 9
62. M.P. Silverman: Applications of photon correlation techniques to fermions, *OSA Proceedings on Photon correlation Techniques and Applications*, Vol. 1, ed. by J.B. Abbiss and A.E. Smart (OSA, Washington DC, 1988) pp. 26–34
63. M.P. Silverman: Second-order temporal and spatial coherence of thermal electrons, Nuovo Cimento B **99**, 227 (1987)
64. D. Gabor: Light and information. In: Progress in Optics **1**, 109 (1961)
65. R.C. Bourret: Nuovo Cimento **18**, 347 (1960)

66. Y. Kano and E. Wolf: Proc. Phys. Soc. **80**, 1273 (1962); C.L. Mehta and E. Wolf, Phys. Rev. A **134**, 1143, 1149 (1964)

67. M.P. Silverman: Second-order temporal and spatial coherence of thermal electrons, Nuovo Cimento B **99**, 227 (1987)

68. R. Hanbury Brown: *The Intensity Interferometer* (Taylor and Francis, London, 1974)

69. M.P. Silverman: Effects of potentials on fermion antibunching, Schrödinger Centenary Conference (Imperial College, London, 1987); Theoretical study of electron antibunching in a field-emission beam, Meeting of the American Physical Society, Arlington VA (1987); Fermion ensembles that manifest statistical bunching, Phys. Lett. A **124**, 27 (1987)

70. D.H. Boal, C.-K. Gelbke, and B.K. Jennings: Intensity interferometry in subatomic physics, Rev. Mod. Phys. **62**, 553 (1990)

71. P. Hawkes and E. Kasper: *Electron Optics*, Vol. 2 (Academic Press, New York, 1989), p. 271

72. M.P. Silverman: Distinctive quantum features of electron intensity correlation interferometery; Il Nuovo Cimento B **97**, 200 (1987)

73. R.H. Fowler and L.W. Nordheim, Proc. Roy. Soc. A **119**, 173 (1928)

74. L.W. Nordheim: Proc. Roy. Soc. A **121**, 626 (1928)

75. J.W. Gadzuk and E.W. Plummer: Rev. Mod. Phys. **45**, 487 (1973)

76. J.C.H. Spence, W. Qian, and M.P. Silverman: Electron source brightness and degeneracy from Fresnel fringes in field emission point projection microscopy, J. Vac. Sci. Technol. A **12**, 542 (1994)

77. J.T. Kiehl and K.E. Trenberth: Earth's annual global mean energy budget, Bulletin of the American Meteorological Society **78**, 197 (1997)

78. B. Lengyel: *Lasers* (Wiley, New York, 1971) p. 138

79. H.W. Fink: Point source for ions and electrons, Physica Scripta **38**, 260 (1988); P.A. Serena, L. Escapa, J.J. Saenz, N. Garcia, and H. Rohrer: Coherent electron emission from point sources, J. Microscopy **152**, 43 (1988)

80. M. Rezeq, J. Pitters, and R. Wolkow: Tungsten nanotip fabrication by spatially controlled field-assisted reaction with nitrogen, J. Chem. Phys. **124**, 204716-1 (2006)

81. M. Henny et. al.: The fermionic Hanbury Brown and Twiss experiment, Science **284**, 296 (1999); W. Oliver et al.: Hanbury Brown and Twiss-type experiment with electrons, Science **284**, 299 (1999)

82. M.P. Silverman: *A Universe of Atoms, an Atom in the Universe* (Springer, NY 2002)

83. H. Kiesel, A. Renz, and F. Hasselbach: Observation of Hanbury Brown–Twiss anticorrelations for free electrons, Nature **418**, 392 (2002)

84. M.P. Silverman et al.: Quantum test of the distribution of composite physical measurements, Europhysics Letters **57**, 572–578 (2004); M.P. Silverman et al.: The distribution of composite measurements: How to be certain of the uncertainties in what we measure, Am. J. Phys. **72**, 1068–1081 (2004)

85. M.P. Silverman: Two-solenoid Aharonov–Bohm experiment with correlated particles, Phys. Lett. A **148**, 154 (1990)

86. A. Zeilinger: General properties of lossless beam splitters in interferometry, Am. J. Phys. **49**, 882 (1981)

87. M.P. Silverman: More than one mystery: Quantum interference with correlated charged particles and magnetic fields, Am. J. Phys. **61**, 514 (1993)

88. M.P. Silverman: Quantum interference effects on fermion clustering in a fermion interferometer, Physica B **151**, 291 (1988)

89. R.J. Glauber: Optical coherence and photon statistics, in: *Quantum Optics and Electronics*, ed. by C. DeWitt et al. (Gordon and Breach, New York, 1965) pp. 65–185

90. M.P. Silverman: Fermion ensembles that show statistical bunching, Phys. Lett. A **124**, 27–31 (1987)

91. M.P. Silverman: On the feasibility of observing electron antibunching in a field-emission beam, Phys. Lett. A **120**, 442 (1987)

92. M.P. Silverman: Gravitationally induced quantum interference effects on fermion antibunching, Phys. Lett. A **122**, 226 (1987)

93. R. Colella, A.W. Overhauser, and S.A. Werner: Observation of gravitationally induced quantum interference, Phys. Rev. Lett. **34**, 1472 (1975)

94. B. Yurke: Input states for enhancement of fermion interferometer sensitivity, Phys. Rev. Lett. **56**, 1515 (1986)

95. E.B. Aleksandrov: Optical manifestations of the interference of nondegenerate atomic states, Sov. Phys. Uspekhi **15**, 436 (1973)

96. M.P. Silverman: On measurable distinctions between quantum ensembles, Ann. of the N.Y. Acad. Sci. **480**, 292 (1986)

97. G. Breit: Quantum theory of dispersion (continued). Parts VI and VII, Rev. Mod. Phys. **2**, 91–140 (1933)

98. P.A. Franken: Interference effects in the resonance fluorescence of 'crossed' excited atomic states, Phys. Rev. **121**, 508 (1961); E.B. Aleksandrov: Quantum beats of resonance luminescence under modulated light excitation, Opt. Spectrosc. (USSR) **14**, 233–234 (1963) [Engl. trans. of Opt. Spektr. **14**, 436–437 (1963); A. Corney and G.W. Series: Theory of resonance fluorescence excited by modulated or pulsed light, Proc. Phys. Soc. (London) **83**, 207 (1964)

99. J.N. Dodd, W.J. Sandle, and D. Zisserman: Proc. Phys. Soc. **92**, 497 (1964); E.B. Aleksandrov: Luminescence beats induced by pulse excitation of coherent states, Opt. Spectrosc. (USSR) **14**, 522–523 (1964) [Engl. trans. of Opt. Spektr. **14**, 957–958 (1964)

100. E. Hadeishi and W.A. Nierenberg: Direct observation of quantum beats due to coherent excitation of nondegenerate excited states by pulsed electron impact, Phys. Rev. Lett. **14**, 891 (1965)

101. I.A. Sellin, C.D. Moak, P.M. Griffin, and J.A. Biggerstaff: Periodic intensity fluctuations of Balmer lines from single-foil excited fast hydrogen atoms, Phys. Rev. **184**, 56 (1969); M.P. Silverman: Optical electric resonance investigation of a fast hydrogen beam (PhD Thesis, Harvard University, 1973)

102. M.P. Silverman, S. Haroche, and M. Gross: General theory of laser-induced quantum beats. I. Saturation effects of single laser excitation, Phys. Rev. A **18**, 1507 (1978)

103. M.P. Silverman, S. Haroche, and M. Gross: General theory of laser-induced quantum beats. II. Sequential laser excitation; effects of external static fields, Phys. Rev. A **18**, 1517 (1978)

104. T.F. Gallagher: *Rydberg Atoms* (Cambridge University Press, NY, 1994)

105. M.P. Silverman: Anomalous fine structure in sodium Rydberg states, Amer. J. Phys. **43**, 244 (1980)

106. J.P. Barrat and C. Cohen-Tannoudji: J. Phys. Rad. **22**, 329, 443 (1961)

107. See, for example, C. Cohen-Tannoudji: Optical pumping with lasers, in: *Atomic Physics 4*, ed. by G. Zu Putlitz, E.W. Weber, and A. Winnacker (Plenum, New York, 1975) pp. 589–614

108. M.D. Crisp and E.T. Jaynes: Radiative effects in semiclassical theory, Phys. Rev. **179**, 1253 (1969)

109. S. Haroche: Quantum beats and time-resolved fluorescence spectroscopy, in: *Topics in Applized Physics* Vol. 13: *High-Resolution Laser Spectroscopy*, ed. by K. Shimoda (Springer-Verlag, Heidelberg, 1976) p. 253

110. V. Weisskopf and E. Wigner: Berechnung der natürlichen Linienbreite auf Grund der Diracschen Lichttheorie [Calculation of the natural linewidth based on the Dirac theory of light], Z. Physik **63**, 54 (1930); Über die Natürliche Linienbreite in der Strahlung des harmonishcen Oszillators [Concerning the natural linewidth in the radiation of the harmonic oscillator], **65**, 18 (1930)

111. M.P. Silverman and F.M. Pipkin: Radiation damping of atomic states in the presence of an external time-dependent potential, J. Phys. B: Atom. Molec. Phys. **5**, 2236 (1972)

112. M. Ducloy: Nonlinear effects in optical pumping of atoms by a high-intensity multimode gas laser. General theory, Phys. Rev. A **8**, 1844 (1973)

113. M. Rosatzin, D. Suter, W. Lange, and J. Mlynek: Phase and amplitude variations of optically induced spin transients, J. Opt. Soc. Am. B **7**, 1231 (1990)

114. M.P. Silverman: *And Yet It Moves: Strange Systems and Subtle Questions in Physics* (Cambridge University Press, New York, 1993)

115. E. Luc-Koenig: Doublet inversion in alkali-metal spectra: Relativistic and correlation effects, Phys. Rev. A **13**, 2114 (1976)

116. S. Haroche, M. Gross, and M.P. Silverman: Observation of fine-structure quantum beats following stepwise excitation in sodium *D* states, Phys. Rev. Lett. **33**, 1063 (1974)

117. See, for example, H.E. White: *Introduction to Atomic Spectra* (McGraw-Hill, New York, 1934) Chap. 10

118. E. Schrödinger: Proc. Cambridge Philos. Soc. **31**, 555 (1935); **32**, 446 (1936)

119. J.S. Bell: On the Einstein–Podolsky–Rosen paradox, Physics **1**, 195 (1964)

120. Z.Y. Ou and L. Mandel: Violation of Bell's inequality and classical probability in a two-photon correlation experiment, Phys. Rev. Lett. **61**, 50 (1988)

121. M.P. Silverman: Quantum interference in the atomic fluorescence of entangled electron states, Phys. Lett. A **149**, 413 (1990)

122. P.B. Medawar: *The Art of the Soluble* (Methuen & Co, London, 1967) p. 7

123. I.I. Rabi: Space quantization in a gyrating magnetic field, Phys. Rev. **51**, 652 (1937)

124. J.S. Rigden: *Rabi, Scientist and Citizen* (Basic Books, New York, 1987) pp. 94–95

125. W.E. Lamb and R.C. Retherford: Fine structure of the hydrogen atom by a microwave method, Phys. Rev. **72**, 2412 (1947); Fine structure of the hydrogen atom, Part I, ibid. **79**, 549 (1950); Fine structure of the hydrogen atom, Part II, ibid. **18**, 222 (1951)

126. W.E. Lamb: Anomalous fine structure of H and He$^+$, Rept. Prog. Phys. **14**, 19 (1951)

127. W.E. Lamb and T.M. Sanders: Fine structure of short-lived states of hydrogen by a microwave–optical method. I, Phys. Rev. **119**, 1901 (1960); L.R. Wilcox and W.E. Lamb: Fine structure of short-lived states of hydrogen by a microwave–optical method. II, Phys. Rev. **119**, 1915 (1960)

128. M.P. Silverman: Optical electric resonance of a fast hydrogen beam, Ph.D. Thesis (Harvard University, Cambridge, 1973)

129. M.P. Silverman and F.M. Pipkin: Optical electric resonance investigation of a fast hydrogen beam I: Theory of the atom–RF field interaction, J. Phys. B: Atom. Molec. Phys. **7**, 704 (1974); II: Theory of the optical detection process, ibid. **7**, 730 (1974); III: Experimental procedure and analysis of H($n = 4$) states, ibid. **7**, 747 (1974)

130. C.W. Fabjan, F.M. Pipkin, and M.P. Silverman: Radiofrequency spectroscopy of hydrogen fine structure in $n = 3$, 4, 5, Phys. Rev. Lett. **26**, 347 (1971)

131. E.P. Wigner: Relativistic invariance and quantum phenomena, in: *Symmetries and Reflections* (Indiana University Press, Bloomington, 1967) p. 51

132. M.P. Silverman and F.M. Pipkin: Interaction of a decaying atom with a linearly polarized oscillating field, J. Phys. B: Atom. Molec. Phys. **5**, 1844 (1972)

133. M.P. Silverman: The curious problem of spinor rotation: Eur. J. Phys. **1**, 116 (1980) (and references contained therein)

134. F. Bloch and A. Siegert: Magnetic resonance for nonrotating fields, Phys. Rev. **57**, 522 (1940)

135. A.F. Stevenson: On the theory of the magnetic resonance method of determining nuclear moments, Phys. Rev. **58**, 1061 (1940)

136. W.E. Lamb: Fine structure of the hydrogen atom. III, Phys. Rev. **85**, 259 (1952)

137. D.A. Andrews and G. Newton: Observation of Bloch–Siegert shifts in the $2^2S_{1/2}$–$2^2P_{1/2}$ microwave resonance in atomic hydrogen, J. Phys. B: Atom. Molec. Phys. **8**, 1415 (1975)

138. P.A.M. Dirac: *The Principles of Quantum Mechanics*, 4th edn. (Oxford University Press, London, 1958) p. 148

139. A. Messiah: *Quantum Mechanics*, Vol. II, (Wiley, New York, 1961) p. 535

140. H. Rauch, A. Zeilinger, G. Badurek, A. Wilfing, W. Bauspiess, and U. Bonse: Verification of coherent spinor rotation of fermions, Phys. Lett. **54** A, 425 (1975); S. Werner, R. Colella, A. Overhauser, and C. Eagen: Observation of the phase shift of a neutron due to precession in a magnetic field, Phys. Rev. Lett. **35**, 1053 (1975)

141. J. Byrne: Young's double beam interference experiment with spinor and vector waves, Nature **275**, 188 (1979)

142. M.P. Silverman: The distinguishability of 0 and 2π rotations by means of quantum interference in atomic fluorescence, J. Phys. B: Atom. Molec. Phys. **13**, 2367 (1980)

143. E.D. Bolker: The spinor spanner, Amer. Math. Monthly 977 (November 1973)

144. See, for example, N.F. Ramsey: *Molecular Beams* (Oxford University Press, Oxford, 1956); N.F. Ramsey: Experiments with separated oscillatory fields and hydrogen masers, Rev. Mod. Phys. **62**, 541–552 (1990)

145. B.G. Levi: Atoms are the new wave in interferometers, Physics Today (July 1991) 17–20; M. Sigel and J. Mlynek: Atom optics, Physics World (February 1993) 36–42

146. D.W. Keith, C. Ekstrom, Q. Turchette, D.E. Pritchard: An interferometer for atoms, Phys. Rev. Lett. **66**, 2693 (1991)

147. P.L. Gould, G. Ruff, and D.E. Pritchard: Diffraction of atoms by light: The near-resonant Kapitza–Dirac effect, Phys. Rev. Lett. **56**, 827 (1986)

148. M. Kasevich and S. Chu: Atomic interferometry using stimulated Raman transitions, Phys. Rev. Lett. **67**, 181 (1991)

149. See, for example, C.J. Pethick and H. Smith: *Bose–Einstein Condensation in Dilute Gases* (Cambridge University Press, New York, 2002)

150. E. Fischbach and C. Talmadge: Six years of the fifth force, Nature **392**, 207 (1992); [114], Chap. 5

151. T.T. Wu and C.N. Yang: Concept of nonintegrable phase factors and global formulation of gauge fields, Phys. Rev. D **12**, 3845 (1975)

152. M.P. Silverman: Optical manifestations of the Aharonov–Bohm effect by ion interferometry, Phys. Lett. A **182**, 323 (1993)

153. M.P. Silverman: Aharonov–Bohm effect of the photon, Phys. Lett. A **156**, 131 (1991)

154. Ch.J. Bordé: Atomic interferometer with internal state labelling, Phys. Lett. A **140**, 131 (1991)

155. P.L. Kapitza and P.A.M. Dirac: The reflection of electrons from standing light waves, Proc. Cambridge Phil. Soc. **29**, 297 (1933); H. Schwarz: Zeit. f. Physik **204**, 276 (1967)

156. M.P. Silverman: Circular birefringence of an atom in uniform rotation: The classical perspective, Amer. J. Phys. **58**, 310 (1990)

157. G. Baym: *Lectures on Quantum Mechanics* (Benjamin, New York, 1969) p. 327

158. R. Gurtler and D. Hestenes: Consistency in the formulation of the Dirac, Pauli, and Schroedinger theories, J. Math. Phys. **16**, 573 (1975)

159. D. Hestenes: Spin and uncertainty in the interpretation of quantum mechanics, Am. J. Phys. **47**, 399 (1979)

160. Q. Wang and G.E. Stedman: Spin-assisted matter–field coupling and lanthanide transition intensities, J. Phys. B: At. Mol. Opt. Phys. **26**, 1415 (1993)

161. K.B. Wiberg: *Physical Organic Chemistry* (Wiley, New York, 1964) p. 9

162. M.P. Silverman: Broken symmetry of the charged planar rotator in electric and magnetic fields, Am. J. Phys. **49**, 871 (1981)

163. M.P. Silverman: Angular momentum and rotational properties of a charged particle orbiting a magnetic flux tube, *Fundamental Questions in Quantum Mechanics*, ed. by L. Roth and A. Inomata (Gordon & Breach, New York, 1986) pp. 177–190

164. M.P. Silverman: Exact spectrum of the two-dimensional rigid rotator in external fields I. Stark effect, Phys. Rev. A **24**, 339 (1981)

165. M.P. Silverman: Exact spectrum of the two-dimensional rigid rotator in external fields, Part II: Zeeman effect, Phys. Rev. A **24**, 342 (1981)

166. D.S. Carlstone: Factorization types and SU(1,1), Am. J. Phys. **40**, 1459–1468 (1972)

167. S. Flügge: *Practical Quantum Mechanics I* (Springer, New York, 1971) pp. 110–112

168. M.P. Silverman: Experimental consequences of proposed angular momentum spectra for a charged spinless particle in the presence of long-range magnetic flux, Phys. Rev. Lett. **51**, 1927 (1983)

169. M.P. Silverman: Quantum interference test of the fermionic rotation properties of a charged boson–magnetic-flux-tube composite, Phys. Rev. D **29**, 2404 (1984)

170. F. Wilczek: Magnetic flux, angular momentum, and statistics, Phys. Rev. Lett. **48**, 1144 (1982); Quantum mechanics of fractional-spin particles, Phys. Rev. Lett. **49**, 957 (1982)

171. R. Jackiw and A.N. Redlich: Two-dimensional angular momentum in the presence of long-range magnetic flux, Phys. Rev. Lett. **50**, 555 (1983)
172. See, for example, L.I. Schiff *Quantum Mechanics*, 3rd edn. (McGraw-Hill, New York, 1968) Chap. 8, pp. 244–255, 263–268
173. N.W. McLachlan: *Theory and Application of Mathieu Functions* (Dover, New York, 1964) p. 10
174. L. Brillouin: *Wave Propagation in Periodic Structures* (Dover, New York, 1953) Chap. 8
175. E.T. Whittaker and G.N. Watson: *A Course of Modern Analysis* (Cambridge, London, 1969) p. 417
176. See, for example, I.V. Schensted: *A Course on the Application of Group Theory to Quantum Mechanics* (Neo Press, Maine, 1976) p. 119
177. A. Messiah: *Quantum Mechanics II* (Wiley, New York, 1962), pp. 696 and 709
178. H.V. McIntosh: On accidental degeneracy in classical and quantum mechanics, Am. J. Phys. **27**, 620 (1959)
179. D. Bohm and B.J. Hiley: On the Aharonov–Bohm effect, Nuovo Cimento **52** A, 295 (1979)
180. K-H. Yang: Gauge transformations and quantum mechanics I. Gauge invariant interpretation of quantum mechanics, Ann. Phys. (NY) **101**, 62 (1976)
181. M.P. Silverman: Rotation of a spinless particle in the presence of an electromagnetic potential, Lett. Nuovo Cimento **41**, 509–512 (1984)
182. E. Merzbacher: Single valuedness of wave functions, Am. J. Phys. **30**, 237–247 (1962)
183. J.M. Blatt and V.F. Weisskopf: *Theoretical Nuclear Physics* (Wiley, New York, 1952) pp. 783, 787
184. M.P. Silverman: On the use of multiple-valued wave functions in the analysis of the Aharonov–Bohm effect, Lett. Nuovo Cimento **42**, 376 (1985)
185. J.D. Bjorken and S.D. Drell: *Relativistic Quantum Fields* (McGraw-Hill, New York, 1965) pp. 170–172
186. C. Itzykson and J.-B. Zuber: *Quantum Field Theory* (McGraw-Hill, New York, 1980) pp. 149–151
187. J.S. Lomont: *Applications of Finite Groups* (Academic, New York, 1959) p. 259
188. Y.-S. Wu: General theory for quantum statistics in two dimensions, Phys. Rev. Lett. **52**, 2103 (1984)
189. B. Halperin, J. March-Russell, and F. Wilczek: Consequences of time-reversal symmetry violation in models of high-Tc superconductors, Phys. Rev. B **40**, 8726 (1989)
190. R. Gilmore: *Lie Groups, Lie Algebras, and Some of Their Applications* (Wiley Interscience, New York, 1974) p. 129
191. C. Bernido and A. Inomata: Topological shifts in the Aharonov–Bohm effect, Phys. Lett. A **77**, 394 (1980)
192. See, for example, S. Washburn and R.A. Webb: Aharonov–Bohm effect in normal metal quantum coherence and transport, Advances in Physics **35**, 375 (1986); R.A. Webb and S. Washburn: Quantum interference fluctuations in disordered metals, Physics Today **41**, 46 (December 1988); S. Washburn: Conductance fluctuations in loops of gold, Am. J. Phys. **57**, 1069 (1989)
193. D.Yu. Sharvin and Yu.V. Sharvin: Magnetic flux quantization in a cylindrical film of a normal metal, JETP Lett. **34**, 272 (1981)

194. A.D. Stone and Y. Imry: Periodicity of the Aharonov–Bohm effect in normal-metal rings, Phys. Rev. Lett. **56**, 189 (1986)

195. D.C. Walker (Ed.): *Origins of Optical Activity in Nature* (Elsevier, Amsterdam, 1979)

196. *Chiral Symmetry Breaking in Physics, Chemistry, and Biology*, Biosystems **20**, No. 1 (1987)

197. M.P. Silverman: Reflection and refraction at the surface of a chiral medium: Comparison of gyrotropic constitutive relations invariant or noninvariant under a duality transformation, J. Opt. Soc. A **3**, 830 (1986)

198. M.P. Silverman et al.: Experimental configurations employing optical phase modulation to measure chiral asymmetries in light specularly reflected from a naturally gyrotropic medium, J. Opt. Soc. Am. A **5**, 1852 (1988)

199. L.D. Barron: *Molecular Light Scattering and Optical Activity* (Cambridge University Press, New York, 1982)

200. M.-A. Bouchiat and L. Pottier: Optical experiments and weak interactions, Science **234**, 1203 (1986)

201. E.A. Power and T. Thirunamachandran: Optical activity as a two-state process, J. Chem. Phys. **33**, 5322 (1971)

202. E.A. Power and S. Zienau: Coulomb gauge in non-relativistic quantum electrodynamics and the shape of spectral lines, Trans. Roy. Soc. (Lond). A **251**, 427 (1959)

203. See, for example, G. Baym: *Lectures on Quantum Mechanics* (W.A. Benjamin, New York, 1969) Chap. 13

204. M.P. Silverman, J. Badoz, and B. Briat: Chiral reflection from a naturally optically active medium, Optics Letters **17**, 886 (1992).

205. M.A. Bouchiat and C.C. Bouchiat: Weak neutral currents in atomic physics, Phys. Lett. **48** B, 111 (1974)

206. R.A. Hegstrom, J.P. Chamberlain, K. Seto, and R.G. Watson: Mapping the weak chirality in atoms, Am. J. Phys. **56**, 1086 (1988)

207. H. Goldstein: *Classical Mechanics*, 2nd edn. (Addison-Wesley, Reading MA, 1980) pp. 232–235

208. M.P. Silverman: Rotational degeneracy breaking of atomic substates: A composite quantum system in a noninertial reference frame, Gen. Rel. & Grav. **21**, 517 (1989); M.P. Silverman: Rotationally induced optical activity in atoms, Europhysics Letters **9**, 95 (1989)

209. M.P. Silverman: Optical activity induced by rotation of atomic spin, Nuovo Cimento **14** D, 857 (1992)

210. M. Born and E. Wolf: *Principles of Optics*, 4th edn. (Pergamon, Oxford, 1970) p. 87

211. S.C. Read, M. Lai, T. Cave, S.W. Morris, D. Shelton, A. Guest, and A.D. May: Intracavity polarimeter for measuring small optical anisotropies, J. Opt. Soc. Am. B **5**, 1832 (1988)

212. M.P. Silverman: Circular birefringence of an atom in uniform rotation: The classical perspective, Am. J. Phys. **58**, 310 (1990)

213. J.L. Staudenmann, S.A. Werner, R. Colella, and A.W. Overhauser: Gravity and inertia in quantum mechanics, Phys. Rev. A **21**, 1419 (1980)

214. M.P. Silverman: Effect of the Earth's rotation on the optical properties of atoms, Phys. Lett. A **146**, 175 (1990)

215. G.E. Stedman and H.R. Bilger: Could a ring laser reveal the QED anomaly via vacuum chirality?, Phys. Lett. A **122**, 289 (1987); G.E. Stedman: Ring laser tests of fundamental physics and geophysics, Rep. Prog. Phys. **60**, 615 (1997)

216. http://www.phys.canterbury.ac.nz/research/laser/ring_2000.shtml

217. P.S. Farago: Electron optic dichroism and electron optical activity, J. Phys. B **14**, L743 (1981)

218. D.M. Campbell and P.S. Farago: J. Phys. B **20**, 5133 (1987)

219. S. Mayer and J. Kessler: Experimental verification of electron optic dichroism, Phys. Rev. lett. **74**, 4803 (1995)

220. S. Mayer, C. Nolting, and J. Kessler: Electron scattering from chiral molecules, J. Phys. B **29**, 3497 (1996)

221. K. Ray, S.P. Ananthavel, D.H. Waldeck, and R. Naaman: Asymmetric scattering of polarized electrons by organized organic films of chiral molecules, Science **283**, 814 (1999)

222. M.P. Silverman: *Fundamental Studies of Chiral Media* (Plan of research submitted to the John Simon Guggenheim Memorial Foundation, New York, 1994)

223. M.P. Silverman: Specular light scattering from a chiral medium, Lett. Nuovo Cimento **43**, 378 (1985)

224. M.P. Silverman, N. Ritchie, G.M. Cushman, and B. Fisher: Experimental configurations using optical phase modulation to measure chiral asymmetries in light specularly reflected from a naturally gyrotropic medium, J. Opt. Soc. Am. A **5**, 1852 (1988)

225. E.N. Fortson and L.L. Lewis: Atomic parity nonconservation experiments, Phys. Rep. **113**, 289 (1984)

226. M.P. Silverman and J. Badoz: Light reflection from a naturally optically active birefringent medium, J. Opt. Soc. Am. A **7**, 1163 (1990)

227. M.P. Silverman and J. Badoz: Large enhancement of chiral asymmetry in light reflection near critical angle, Optics Commun. **74**, 129 (1989)

228. M.P. Silverman: Differential amplification of circularly polarised light by enhanced internal reflection from an active chiral medium, Optics Commun. **74**, 134 (1989)

229. M.P. Silverman and J. Badoz: Multiple reflection from isotropic chiral media and the enhancement of chiral asymmetry, J. Electromagnetic Waves and Applications **6**, 587 (1992)

230. A. Ghosh and P. Fischer: Chiral molecules split light: Reflection and refraction in a chiral liquid, Phys. Rev. Lett. **97**, 173002 (2006)

231. M.P. Silverman and R.B. Sohn: Effects of circular birefringence on light propagation and reflection, Am. J. Phys. **54**, 69 (1986)

232. H. Eyring, J. Walter, G. Kimball: *Quantum Chemistry* (Wiley, New York, 1944) p. 337

233. L. Landau and E.M. Lifshitz: *Electrodynamics of Continuous Media* (Pergamon, New York, 1960) pp. 334, 340

234. B. Whitney: How to make a massive star, Nature **437**, 37 (2005); M.R. Krumholz, C.F. McKee, and R.I. Klein: The formation of stars by gravitational collapse rather than competitive accretion, Nature **438**, 332 (2005)

235. C.E. Rolfs and W.S. Rodney: *Cauldrons in the Cosmos: Nuclear Astrophysics* (University of Chicago, Chicago, 1988) p. 117

236. S. Chandrasekhar: M.N.R.A.S. **95**, 91 (1931) and Ap. J. **74**, 81 (1931)
237. J.R. Oppenheimer and G.M. Volkoff: On massive neutron cores, Phys. Rev. **55**, 374 (1939)
238. C.W. Misner, K.S. Thorne, and J.A. Wheeler: *Gravitation* (Freeman, San Francisco, 1973), p. 823. Note that the authors employ so-called 'natural units' in which $G = c = \hbar = 1$. The Schwarzschild radius is then $2M$, where M is the stellar mass
239. A.N. Whitehead: *Alfred North Whitehead: An Anthology* (Macmillan, New York, 1953)
240. M.P. Silverman: Fermion condensation in a relativistic degenerate star, Int. J. Mod. Phys. D (2007). To be published
241. M.P. Silverman: Quantum stabilization of a relativistic degenerate star beyond the Chandrasekhar mass limit, Int. J. Mod. Phys. D **13**, 2281 (2004)
242. M.P. Silverman: Gravitationally-induced particle resorption into the vacuum: A general solution to the problem of black hole collapse, Int. J. Mod. Phys. D **14**, 2285 (2005)
243. S.W. Hawking: Black hole explosions?, Nature **248**, 30 (1974)
244. M. Bojowald et al.: Black hole mass threshold from nonsingular quantum gravitational collapse, Phys. Rev. Lett. **95**, 091302-1 (2005)
245. C.A. Regal et al.: Observation of resonance condensation of fermionic atom pairs, Phys. Rev. Lett. **92**, 040403 (2004); R. Grimm: Low temperature physics: A quantum revolution, Nature **435**, 1035 (2005); M.W. Zwierlein et al.: Vortices and superfluidity in a strongly interacting Fermi gas, Nature **435**, 1047 (2005); J. Kinast et al.: Evidence for superfluidity in a resonantly interacting Fermi gas, Phys. Rev. Lett. **92**, 150402-1 (2004)
246. B.G. Levi: Ultracold fermionic atoms team up as molecules: Can they form Cooper pairs as well?, Physics Today **56**, 18 (2003)
247. L. Pitaevskii and S. Stringari: The quest for superfluidity in Fermi gases, Science **298**, 2144 (2002); K.M. O'Hara et al.: Science **298**, 2179 (2002); J. Kinast et al.: Heat capacity of a strongly interacting Fermi gas, Science **307**, 1296 (2005)
248. D. Pines and M.A. Alpar: Superfluidity in neutron stars, Nature **316**, 27 (1985)
249. B. Link: Constraining hadronic superfluidity with neutron star precession: Phys. Rev. Lett. **91**, 101101 (2003)
250. A.G.W. Cameron: Canadian J. Phys. **35**, 1021 (1957); Ap. J. **129**, 676 (1959); Ap. J. **130**, 884 (1959); G. Ingrosso, D. Grasso, and R. Ruffini: Astron. Astrophys. **248**, 481 (1991); G.Q. Li, C-H. Lee, and G.E. Brown: Kaon production in heavy-ion collisions and maximum mass of neutron stars, Phys. Rev. Lett. **79**, 5214 (1997)
251. M.P. Silverman: Fermion condensation in a stellar black hole, *Gravity Research Foundation* 2006 (Honorable Mention Essay); M.P. Silverman: Condensates in the cosmos: Quantum stabilization of degenerate stars and dark matter, 2006 Biennial Meeting of the International Association for Relativistic Dynamics, University of Connecticut, Storrs CT, 12–14 June 2006
252. L. Landau and L. Lifshitz: *Statistical Physics* (Pergamon, London, 1958) p. 170
253. M.P. Silverman and R.L. Mallett: Dark matter as a cosmic Bose–Einstein condensate and possible superfluid, Gen. Rel. & Grav. **34**, 633 (2002)

254. M.W. Zwierlein, C.H. Schunck, A. Schirotzek, and W. Ketterle: Direct observation of the superfluid phase transition in ultracold Fermi gases, Nature **442**, 54 (2006)
255. E.P. Gross: Nuovo Cimento **20**, 454 (1961); J. Math. Phys. **4**, 195 (1963); L.P. Pitaevskii, Sov. Phys. JETP **13**, 451 (1961); L.P. Pitaevskii and S. Stringari: *Bose–Einstein Condensation* (Oxford, NY, 2003)
256. C.J. Foot: *Atomic Physics* (Oxford, NY, 2005) pp. 234–235
257. W. Greiner, L. Neise, and H. Stocker: *Thermodynamics and Statistical Mechanics* (Springer, NY, 1994) p. 320
258. C.J. Pethick and H. Smith: *Bose–Einstein Condensation in Dilute Gases* (Cambridge, New York, 2002) p. 21
259. A. Bulgac, J.E. Drut, and P. Magierski: Spin-1/2 fermions in the unitary regime: A superfluid of a new type, Phys. Rev. Lett. **96**, 090404 (2006)
260. W. Greiner et al.: *Thermodynamics and Statistical Mechanics* (Springer, New York, 1995) p. 358
261. C. Misner, K. Thorne, and J. Wheeler: *Gravitation* (Freeman, San Francisco, 1973) p. 604
262. H.B. Callen: *Thermodynamics* (Wiley, New York, 1960) p. 47
263. M. Schwarzschild: *Structure and Evolution of the Stars* (Dover, NY, 1958) p. 96
264. C. Schaab et al.: Astronomy & Astrophysics **335**, 596 (1998)
265. M.P. Silverman: Fermion condensation in a relativistic degenerate star, Int. J. Mod. Phys. D (2007). To be published
266. J.J. More, B.S. Garbow, and K.E. Hillstrom: *User's Guide to Minpack I*, Argonne National Laboratory publication ANL-80-74 (1980), cited in *Mathcad Plus 6 User's Guide* (MathSoft Inc, Cambridge MA, 1996) p. 601
267. H. Heiselberg: Fermi systems with long scattering lengths, Phys. Rev. A **63**, 043606 (2001)
268. K. Winkler, G. Thalhammer, F. Lang, R. Grimm, J. Denschlag, A. Daley, A. Kantian, H. Buechler, and P. Zoller: Repulsively bound atom pairs in an optical lattice, Nature **441**, 853 (2006)
269. J.C. Collins and M.J. Perry: Superdense matter: Neutrons or asymptotically free quarks?, Phys. Rev. lett. **34**, 1353 (1975); G. Baym and S.A. Chin: Can a neutron star be a giant MIT bag?, Phys. Lett. **62** B, 241 (1976); H. Heiselberg and M. Hjorth-Jensen: Phases of dense matter in neutron stars, Physics Reports **328**, 237 (2000); E.S. Fraga, R.D. Pisarski, and J. Schaffner-Bielich: Small, dense quark stars from perturbative QCD, Phys. Rev. D **63**, 121702 (2001)
270. N.K. Glendenning: First-order phase transitions with more than one conserved charge: Consequences for neutron stars, Phys. Rev. D **46**, 1274 (1992)
271. J. Drake et al.: Is RX J1856.5-3754 a quark star?, Ap. J. **572**, 996 (2002); R.X. Xu: A thermal featureless spectrum: Evidence for bare strange stars?, Ap. J. **570**, L65 (2002)
272. J. Cottam, F. Paerels, and M. Mendez: Gravitationally redshifted absorption lines in the X-ray burst spectra of a neutron star, Nature **420**, 51 (2002)
273. Top Story Goddard Space Flight Center: Exotic innards of a neutron star revealed in a series of explosions, http://www.gsfc.nasa.gov/topstory/20021003nsexplosion.html
274. See, for example, A.K. Harding and D. Lai: Physics of strongly magnetized neutron stars, Rep. Prog. Phys. **69**, 2631 (2006)

275. R.C. Duncan and C. Thompson: Formation of very strongly magnetized neutron stars: Implications for gamma-ray bursts, Astro. J. **392**, L9–L13 (1992)

276. S.W. Hawking and G.F.R. Ellis: *The Large Scale Structure of Space-Time* (Cambridge University Press, London, 1973) pp. 303–305

277. H.J. Lee, V. Canuto, H.-Y. Chiu, and C. Chiuderi: New state of ferromagnetism in degenerate electron gas and magnetic fields in collapsed bodies, Phys. Rev. Lett. **23**, 390 (1969)

278. E.W. Kolb and M.S. Turner: Grand unified theories and the origin of the baryon asymmetry, Ann. Rev. Nucl. Part. Sci. **33**, 645 (1983)

279. D.F. Figer: An upper limit to the masses of stars, Nature **434**, 192 (10 March 2005)

280. L. Chandrasekhar: Our Song, published in *Black Holes and Relativistic Stars*, ed. by R.M. Wald, (University of Chicago Press, Chicago, 1998) p. 274

281. R.C. Malone, M.B. Johnson, and H.A. Bethe: Neutron star models with realistic high-density equations of state, Astro. J. **199**, 741 (1975)

282. http://lambda.gsfc.nasa.gov/product/map/current/map_bibliography.cfm

283. A.G. Riess et al.: Cepheid calibrations from the Hubble Space Telescope of the luminosity of two recent type Ia supernovae and a redetermination of the Hubble constant, Astro. J. **627**, 527 (2005)

284. N. Dauphas: The U/Th production ratio and the age of the Milky Way from meteorites and galactic halo stars, Nature **435**, 1203 (2005)

Index

A

AB (Aharonov-Bohm) effect   13–16
  and winding number   248
  in a Mach-Zehnder interferometer
    125
  in mesoscopic rings   112
  of planar rotator   229
  with entangled electrons   112
  with ions   206
AB-EPR experiment   112, 117
AB-HBT experiment   122
angular momentum
  canonical   241, 242
  kinetic   241, 242
antibunching (of particles)   59, 87
anticommutation relations (of fermions)
  124
anyon   247
aperture function   34

B

Back-Goudsmit effect   167
baryon number   341
Bayh experiment   18, 19
BEC (Bose-Einstein condensate)   310,
  311
Big Bang (origin of Universe)   341
black hole   307, 333
Bloch-Siegert shift   182
brightness (axial)   89, 92
bunching (of particles)   59, 87

C

cell (of phase space)   89–91, 128
Chandrasekhar limit   327, 343
chaotic source   60, 127
charge transfer interaction   170,
  172–173
chemical potential   78, 317
  and Fermi energy   91
  in boson condensate   312
  in gravitational field   357
  of neutrons   250
chirality   257
circular dichroism   260
coherence (in density matrix)   144
coherence (second-order)   56, 69
coherence area   90, 101, 128
coherence function   81
coherence length   7
  of thermal electrons   87
coherence time   7, 56
  of thermal electrons   87
commutation relations
  annihilation-creation operators   71
  canonical angular momentum   241
  coordinate-momentum   46
  kinetic angular momentum   242
Compton wavelength   6
Cooper pairs   21
Coriolis force   277, 278
correlation function (first order)   78
  first order   78
  second order

of electrons   81, 102
of light   56, 69, 72
correspondence principle   17
cosmological concordance model   340
Coulomb gauge   14
COW (Colella-Overhauser-Werner)
  experiment   128
CP violation   341
critical density (cosmological)   354
cyclic transition   194
cyclotron frequency   236

D

de Broglie wavelength   6
decay operator   64
degeneracy
  accidental   230
  Stark   223
  Zeeman   238
degeneracy parameter   89, 90
  of neutron   315
delayed choice experiment   39
density matrix   141
density operator   48, 178, 209
detection operator   66, 144, 209
diagonalization (of matrix)   176
dissymmetry   257
Doppler broadening   139, 166
dynamical observable   14, 126

E

electron 2-slit experiment   3
electron microscope (field-emission)   3, 4
enantiomer   258
energy operator   241
entanglement   35, 36
entropy   62, 63
EPR paradox   45
equilibrium conditions (stellar)   320

F

Fermi momentum   107, 315, 342
fermicon stars   333
ferromagnetism (stellar)   339
fine structure constant   181
flux quantization   21

fluxon   16, 21, 113, 210
forward scattering
  of electrons   283
  of light   263
Fowler-Nordheim theory (of field
  emission)   95
free-induction decay   151–154, 217–218

G

g-factor (magnetic moment)   337
gauge transformation   12, 240–241
generalized   215
gravitational binding energy   346
group theory   230
groups
  Four [*Vierer*] V 230, 234
  homotopy   245, 249–250
  Special Orthogonal $SO(2)$ 230
  Special Unitary $SU(1, 1)$ 221
  Symmetric S_2 230

H

Hamiltonian
  as time-displacement operator   28
  effective (light shift)
    light shift   154
    Stark shift   155
  in interaction representation   13–14, 180
  in rotating reference frame   274
  magnetic   155, 254
  of 2D rotator   220
  of unstable states   183
  under gauge transformation   240
Hanle effect   151
Hawking radiation   341
HBT (Hanbury Brown-Twiss) experiment   54
helicity   263
Hubble parameter   354

I

intensity interferometer   54
ion interferometry   206

K

Kapitza-Dirac effect   211

L

Lamb-Retherford experiment   166
Landau levels   236
Land factor   158
Larmor frequency   236
lepton number   341
light shift   143
linear momentum
    canonical   13
    kinetic   14
long-distance beats   159
Lorentz force   11

M

Mach-Zehnder interferometer   122
Malus' law   23
Mathieu functions   226
Mathieu's equation   225, 234
Maxwell's equations   11
motional electric field   167

N

nanotip   98–99
neutron star   307–309
    crust of   326
    magnetic fields in   335, 336
    mass of   309
    origin of   336
    superfluidity in   311, 320
nonintegrable phase factor   17, 209

O

OER (optical electric resonance)   168
of photons   47
optical activity   257
optical pumping   137, 141–149
optical rotation   260
optical rotatory strength   264, 269
oscillating field theory   182–183
OV (Oppenheimer-Volkov) limit   308,
    310, 327, 332, 333, 335

P

parity violation (in atoms)   262,
    271–272

particle resorption   310, 340, 341
Paschen-Back effect   167
Pauli spin matrices   63, 178
perturbation theory   224–225
pi-electron ring   220
Planck length   308
Planck mass   309
Planck's constant   6
polarizability   224
polarization (of light)
    circular   259
    elliptical   260
pumping time   142

Q

quantum beats
    and energy indeterminacy   136
    and resonance   194
    in entangled states   160
    laser-induced   139
    multi-atom   164
    regeneration   149
    Zeeman   156
quantum boost   137
quantum vacuum   72, 183, 190, 341
quark deconfinement   349
quark star   333–334, 349

R

Rabi "flopping" formula   165, 182
raising/lowering operators   129
random walk   149, 250
resonance
    line shape   166, 171–173
    narrowing   204–205
response time (of detector)   100
retarded time   107, 164
ring-laser interferometer   279–281
rotating-wave approximation   182
Rydberg (unit of energy)   180
Rydberg state   139, 156, 207

S

saturation density (nuclear)   334, 336,
    349
saturation parameter   147
saturation regeneration (of beats)   149

Schwarzschild geometry   316–317
Schwarzschild radius   307, 313
separated oscillating fields   199–205
signal-to-noise ratio   104, 159
Silverman-Pipkin shift   187
spectral density   142
spin-orbit coupling   156
spin-statistics connection   247
spinor rotation   190–191
spinor spanner   198–199
Stark effect   167
Stokes' law   12
Stokes' parameters   63
Stokes' theorem   240
SU(2) operators   129, 130

T

thermal wavelength   315
thermionic emission   90
Thomas-Fermi approximation   312
three-level atom   145
time-reversal

of electromagnetic fields   262
of light in a gyrotropic medium   303
of paths in a ring   253
of scattering matrix   284
TOV (Tolman-Oppenheimer-Volkov)
    equation   318–319
two-level system   165, 174, 264

V

vector potential   11
visibility function   8

W

weak neutral current   270
white dwarf   307
winding number   248, 249, 253

Z

Z_0 boson   270
Zeeman effect   233, 273
zero-point energy   220

THE FRONTIERS COLLECTION

Series Editors:
A.C. Elitzur M.P. Silverman J. Tuszynski R. Vaas H.D. Zeh

Information and Its Role in Nature
By J. G. Roederer

Relativity and the Nature of Spacetime
By V. Petkov

Quo Vadis Quantum Mechanics?
Edited by A. C. Elitzur, S. Dolev,
N. Kolenda

Life – As a Matter of Fat
The Emerging Science of Lipidomics
By O. G. Mouritsen

Quantum–Classical Analogies
By D. Dragoman and M. Dragoman

Knowledge and the World
Challenges Beyond the Science Wars
Edited by M. Carrier, J. Roggenhofer,
G. Küppers, P. Blanchard

Quantum–Classical Correspondence
By A. O. Bolivar

Mind, Matter and Quantum Mechanics
By H. Stapp

Quantum Mechanics and Gravity
By M. Sachs

Extreme Events in Nature and Society
Edited by S. Albeverio, V. Jentsch,
H. Kantz

**The Thermodynamic
Machinery of Life**
By M. Kurzynski

**The Emerging Physics
of Consciousness**
Edited by J. A. Tuszynski

Weak Links
Stabilizers of Complex Systems
from Proteins to Social Networks
By P. Csermely

Mind, Matter and the Implicate Order
By P.T.I. Pylkkänen

Quantum Mechanics at the Crossroads
New Perspectives from History,
Philosophy and Physics
Edited by J. Evans, A.S. Thorndike

Particle Metaphysics
A Critical Account of Subatomic Reality
By B. Falkenburg

**The Physical Basis of the Direction
of Time**
By H.D. Zeh

**Asymmetry: The Foundation
of Information**
By S.J. Muller

Mindful Universe
Quantum Mechanics
and the Participating Observer
By H. Stapp

**Decoherence and the
Quantum-To-Classical Transition**
By M. Schlosshauer

Quantum Superposition
Counterintuitive Consequences of
Coherence, Entanglement, and Interference
By Mark P. Silverman

The Nonlinear Universe
Chaos, Emergence, Life
By A. Scott